T0317759

NANOMATERIAL CHARACTERIZATION

NANOMATERIAL CHARACTERIZATION

An Introduction

Edited by

RATNA TANTRA

Published by John Wiley & Sons, Inc., Hoboken, New Jersey
Published simultaneously in Canada

Library of Congress Cataloging-in-Publication data has been applied for

ISBN: 9781118753590

Typeset in 10/12pt TimesLTStd by SPi Global, Chennai, India

10 9 8 7 6 5 4 3 2 1

This book is gratefully dedicated to my father, I Wayan Tantra

CONTENTS

LIST OF CONTRIBUTORS

D. Bartczak, LGC Limited, Middlesex TW11 0LY, UK

N. A. Belsey, Analytical Science Division, National Physical Laboratory, Teddington TW11 0LW, UK

E. Bolea, Group of Analytical Spectroscopy and Sensors (GEAS), Institute of Environmental Sciences (IUCA), Universidad de Zaragoza, 50009 Zaragoza, Spain

H. Bouwmeester, Toxicology and Bioassays, RIKILT – Wageningen University & Research Center, 6708 WB Wageningen, The Netherlands

C.A. Charitidis, School of Chemical Engineering, Laboratory Unit of Advanced Composite, Nanomaterials and Nanotechnology, National Technical University of Athens, Athens 15780, Greece

C. A. David, Departament de Química and Agrotecnio, Universitat de Lleida, 25198 Lleida, Spain

J-M Dogné, Department of Pharmacy, University of Namur (UNamur), 5000 Namur, Belgium

D.A. Dragatogiannis, School of Chemical Engineering, Laboratory Unit of Advanced Composite, Nanomaterials and Nanotechnology, National Technical University of Athens, Athens 15780, Greece

D. Drobne, *Biotechnical Faculty*, Department of Biology, University of Ljubljana, 1000 Ljubljana, Slovenia

D. Fujita, Advanced Key Technologies Division, National Institute for Materials Science, Tsukuba 305-0047, Japan

H. Goenaga-Infante, LGC Limited, Middlesex TW11 0LY, UK

D. Gohil, Advanced Engineered Materials Group, Materials, National Physical Laboratory, Teddington TW11 0LW, UK

J. C. Jarman, Quantitative Surface Chemical Spectroscopy Group, Analytical Science, National Physical Laboratory, Teddington TW11 0LW, UK

A. Jemec, *Biotechnical Faculty*, Department of Biology, University of Ljubljana, 1000 Ljubljana, Slovenia

K. A. Jensen, Danish Centre for Nanosafety, National Research Centre for the Working Environment, Copenhagen, Denmark

E.P. Koumoulos, School of Chemical Engineering, Laboratory Unit of Advanced Composite, Nanomaterials and Nanotechnology, National Technical University of Athens, Athens 15780, Greece

F. Laborda, Group of Analytical Spectroscopy and Sensors (GEAS), Institute of Environmental Sciences (IUCA), Universidad de Zaragoza, 50009 Zaragoza, Spain

J. Laloy, Department of Pharmacy, University of Namur (UNamur), 5000 Namur, Belgium

M. Levin, Danish Centre for Nanosafety, National Research Centre for the Working Environment, Copenhagen, Denmark; Department of Micro- and Nanotechnology, Technological University of Denmark, Lyngby, Denmark

J. Li, School of Chemistry and Chemical Engineering, South China University of Technology, Guangzhou 510640, PR China

J. J. Liu, School of Chemistry and Chemical Engineering, South China University of Technology, Guangzhou 510640, PR China; Institute of Particle Science and Engineering, School of Chemical and Process Engineering, University of Leeds, Leeds LS2 9JT, UK

C. Y. Ma, Institute of Particle Science and Engineering, School of Chemical and Process Engineering, University of Leeds, Leeds LS2 9JT, UK

T. Mesarič, *Biotechnical Faculty*, Department of Biology, University of Ljubljana, 1000 Ljubljana, Slovenia

C. Minelli, Analytical Science Division, National Physical Laboratory, Teddington TW11 0LW, UK

W. Österle, Department of Materials Engineering, BAM Federal Institute for Materials Research and Testing, 12200 Berlin, Germany

C. Oksel, Institute of Particle Science and Engineering, School of Chemical and Process Engineering, University of Leeds, Leeds LS2 9JT, UK

G. Orts-Gil, European Office – Spanish Foundation for Science and Technology, Spanish Embassy, 10787 Berlin, Germany

D. Perivoliotis, School of Chemical Engineering, Laboratory Unit of Advanced Composite, Nanomaterials and Nanotechnology, National Technical University of Athens, Athens 15780, Greece

C. Rey-Castro, Departament de Química and Agrotecnio, Universitat de Lleida, 25198 Lleida, Spain

K. N. Robinson, Quantitative Surface Chemical Spectroscopy Group, Analytical Science, National Physical Laboratory, Teddington TW11 0LW, UK

T. Sainsbury, Materials Processing and Performance Group, Materials, National Physical Laboratory, Teddington TW11 0LW, UK

K. Sepčić, *Biotechnical Faculty*, Department of Biology, University of Ljubljana, 1000 Ljubljana, Slovenia

A.G. Shard, Analytical Science Division, National Physical Laboratory, Teddington TW11 0LW, UK

M. Sopotnik, *Biotechnical Faculty*, Department of Biology, University of Ljubljana, 1000 Ljubljana, Slovenia

R. Tantra, Quantitative Surface Chemical Spectroscopy Group, Analytical Science, National Physical Laboratory, Teddington TW11 0LW, UK

A. K. Undas, Toxicology and Bioassays, RIKILT – Wageningen University & Research Center, 6708 WB Wageningen, The Netherlands

M. van der Zande, Toxicology and Bioassays, RIKILT – Wageningen University & Research Center, 6708 WB Wageningen, The Netherlands

X. Z. Wang, School of Chemistry and Chemical Engineering, South China University of Technology, Guangzhou 510640, PR China; Institute of Particle Science and Engineering, School of Chemical and Process Engineering, University of Leeds, Leeds LS2 9JT, UK

O. Witschger, Laboratoire de Métrologie des Aérosols Département Métrologie des Polluants, Institute National de Recherche et de Sécurité, Nancy, France

EDITOR'S PREFACE

To measure is to know. If you cannot measure it, you cannot improve it

Lord Kelvin (1824–1907)

Since joining the National Physical Laboratory (NPL) (UK's national measurement institute) in 2004, I have been fortunate enough to have worked in numerous projects related to nanoscience and nanotechnology. During this time, the nature of my research activities varied widely across different disciplines, from the applications of nanomaterials in surface-enhanced Raman spectroscopy to understanding their potential toxicological implications. A critical part of the research throughout the years, however, has been the need to characterize physicochemical properties of the nanomaterials. This has not always been trivial.

The idea for this book came from my involvement in a European Commission Framework 7 research project entitled MARINA (Managing Risks of Nanomaterials). One of the goals of this project was to harmonize activities and to establish a common platform to ultimately support the scientific infrastructure for risk management of nanomaterials. Although the relevance of MARINA is for nanosafety, the idea of having a common approach can be extended to other application areas. This, coupled with my interest in measurement science, ultimately laid the foundation for this multi-authored book.

The book begins with a general introduction, which aims to give the reader a solid foundation to nanomaterial characterization. Chapters 2 and 3 focus on two principal topics: nanomaterial synthesis and reference nanomaterials, which serve as useful background for the rest of the book. Chapters 4–10 constitute the very heart of this book, dedicated to key physicochemical properties and their measurements. Undoubtedly, it is beyond the scope of the book to cover all properties and only several key properties, such as particle size distribution by number, solubility, surface area, surface chemistry, mechanical/tribological, and dustiness, are covered. Chapters 11–13 are devoted to state-of-the-art techniques, in which three very different sets of characterization tools are presented: (i) scanning tunneling microscopy operated under extreme conditions; (ii) novel strategy for biological characterization of nanomaterials; and (iii) methods to handle and visualize multidimensional nanomaterial characterization data.

Most of the chapters this book begin by giving an overview of the topic area before a case study is presented. The purpose of the case study is to demonstrate how the reader may make use of background information presented to them and show how this can be translated to solve a nano-specific application scenario. Thus, it will be useful for researchers to help them design experimental investigations.

The book is written in such a way that both students and experts in other fields of science will find the information useful. My intention is that it will appeal to a range of audience outside the research field, whether they are in academia, industry, or regulation and is particularly useful for readers whose analytical background may be limited. There is also an extensive list of references associated with every chapter, to encourage further reading.

Finally, it has taken me just less than 2 years to complete this book and so, I must say a few words of thanks. First, I am grateful to all of the authors for their chapter contributions. Second, I thank the many people who have encouraged me to publish this book: my Wiley editor, my husband Keith F. E. Pratt, family, and friends. Special thanks go to Sinta Tantra, for her generosity in donating her artwork, which has been used for the cover of this book. The cover is abstract art that depicts the image of a nanomaterial surface at atomic resolution!

Portsmouth, England
16 December, 2015

1

INTRODUCTION

R. TANTRA, J. C. JARMAN, AND K. N. ROBINSON

Quantitative Surface Chemical Spectroscopy Group, Analytical Science, National Physical Laboratory, Teddington TW11 0LW, UK

1.1 OVERVIEW

Over the course of the past few decades, the word "nanomaterial" started to shine in reporting and publishing; nanomaterial thus became the new buzzword, giving the impression of a new type of technology. In fact, nanomaterials are not new at all and can be found in everyday lives, with most people not being aware of their existence. Nanomaterials exist in nature, for example, in volcanic ashes, sea sprays and smoke [1]. In relation to manufactured nanomaterials, they have existed as early as the 4th century. The Lycurgus Cup, a glass cup made with tiny proportions of gold and silver nanoparticles is an example of Roman era nanotechnology. The use of nanoparticles for beautiful art continued ever since, and by 1600s it is not uncommon for alchemists to create gold nanoparticles for stained glass windows. These days, there are far more uses; nanomaterials thus represent a growing class of material already introduced into multiple business sectors. For example, in early 20th century, tire companies used carbon black in car tires, primarily for physical reinforcement (e.g., abrasion resistance, tensile strength) and thermal conductivity to help spread heat load. Although nanomaterials have been around for a long time, it was only the invention of the scanning tunneling microscope in 1981 and the discovery of fullerenes in 1986 that really marked the beginning of the current nanoscience revolution. This led nanoscientists to conduct research, to study their behavior, so as to control their properties and harness their power.

Nanomaterial Characterization: An Introduction, First Edition. Edited by Ratna Tantra.

Over the past few decades, research activity on nanomaterial has gained considerable press coverage. The use of nanomaterials has meant that consumer products can be made lighter, stronger, more aesthetically pleasing, and less expensive. The huge impact of nanomaterials to improve quality of life is clear, resulting in faster computers, cleaner energy production, target driven pharmaceuticals, and better construction materials [2, 3]. In particular, carbon nanotubes (CNTs) have been hailed as the wonder nanomaterial of the 21st century. CNTs are composed entirely of carbon and classed as high-aspect-ratio nanomaterial. They can be visualized as a single layer of carbon atoms in a hexagonal lattice called graphene and subsequently rolled to form a seamless cylinder/s. CNTs are classed as either single-walled carbon nanotubes (SWCNTs) or multi-walled carbon nanotubes (MWCNTs). As the name suggests, the former are in the form of a single tube, whereas the latter consist of multiple rolled layer or concentric tubes. CNTs typically have a diameter of 1–20 nm and a length that can be many millions of times longer. MWCNTs are normally thicker than SWCNTs, with a maximum diameter exceeding 100 nm.

According to the National Science Foundation's National Nanotechnology Initiative (NNI), the global nanotechnology market could be worth $1.2 trillion by 2020 [4]. There is huge demand for CNTs alone, with a worldwide commercial interest being reflected in its production capacity, estimated in 2011 to be 4.5 kt/year [5]. This represents a huge growth from the production of around 0.25 kt/year in 2006. Bulk, purified MWCNTs sell for approximately $1 per gram, between 1 and 10 times as expensive as carbon fibers. SWCNTs, in contrast, are currently several orders of magnitude more expensive than MWCNTs [5].

Most commercial applications of CNTs involved incorporating the powders to produce composite material with special properties, for example, electrically conductive plastics and lithium-ion batteries in laptops. A more recent exploitation of CNTs is when they are used as materials for sporting equipment. For example, CNT-based frame was used in a bicycle that won the Tour de France in 2005. The incorporation of CNTs into the material improved stiffness and fracture toughness without compromising other properties. The result is a bicycle that features minimal weight and maximal strength.

Although it is clear that nanomaterial holds great potential to form the basis of new products with novel or improved properties, concerns surrounding their potential harmful effects on health and the environment have been the topic of much debate. In over a decade, a scientific discipline called nanotoxicology [6] has emerged, which aims at understanding potential hazards posed by nanomaterials and subsequent risk implications, should, for example, they enter the human body through inhalation, ingestion, skin uptake, or injection. The field is thus interdisciplinary in nature and at the interface of biology, chemistry, and material science.

Undoubtedly, nanomaterial research spans across different disciplines, from material science to nanotoxicology. Common to all of these disciplines, however, is the need to measure physicochemical properties of nanomaterials. As mentioned in the Preface section, the goal of the book is to lay a common foundation, giving an introduction to nanomaterial characterization, thus allowing the reader to build background knowledge on this topic. This chapter gives an overview and focuses on generic topics/issues of relevance to nanomaterial characterization. It is sub-divided into four parts. The first part discusses why nanomaterials are unique in relation to their physicochemical properties. The second part presents the relevant terminology, such as the definition on what constitute a nanomaterial and what the different properties actually mean. Terminology is important as it avoids misunderstandings and ensures that the correct term is being used among stakeholders such as researchers, manufacturers, and regulators. The third part of this chapter focuses on good measurement practices; like any other research there is a need to generate reliable and robust data. In order to promote an integrated approach to quality assurance in the data being generated, topics such as method validation and standardization are covered. The last part of the chapter presents some of the common practices that are carried out in nanomaterial research, such as sub-sampling and dispersion. Although this chapter is intended to give a general overview for readers coming from different disciplines, many of the specific examples presented are of relevance to nanotoxicology.

1.2 PROPERTIES UNIQUE TO NANOMATERIALS

Undoubtedly, nanomaterials can exhibit unique physical and chemical properties not seen in their bulk counterparts. An important characteristic that distinguishes nanomaterial from bulk is associated with reduction of scale, which results in materials having unique properties arising from their nanoscale dimensions.

The most obvious effect associated with reduction of scale is the much larger specific surface area or surface area per unit mass [7]. An increase in surface area implies the existence of more surface atoms. As surface atoms have fewer neighbors than atoms in bulk, an increase in surface area will result in more atoms having lower coordination and unsatisfied bonds. Such surface atoms are overall less stable than bulk atoms, which means that the surface of nanomaterials is more reactive than their bulk counterparts [8].

Note that increase in specific surface area due to a reduction in size is an example of what is termed as scalable property. Scalable properties are those that can change continuously and smoothly with size, with no size limit associated with a sudden change in the properties. In addition to scalable properties, nanomaterials can also exhibit non-scalable properties; by this we refer to those properties that can change drastically when a certain size limit is reached. In this respect, nanomaterials cannot be simply thought of as another step in miniaturization. An example of non-scalable property is quantum confinement effects [9], which can be exemplified by some nanomaterials such as quantum dots. Quantum dots are semiconducting nanoparticles, for example, PbSe, CdSe, and CdS, with particle sizes usually smaller than ~10 nm [10]. Similar to all semiconductors, quantum dots possess a band gap; a band gap is an energy gap between valence and conduction bands in which electrons cannot occupy. In the corresponding bulk material and when at room temperature, electronic transitions across the band gap are the main mechanism by which semiconducting materials absorb or emit photons. These transitions are excited by photons of specific wavelengths, which correspond to the energy of the band gap and generate an excited electron in the conduction band and a hole in the valence band. Photons can be emitted by the recombination of these electron–hole pairs across the band gap, in which the wavelength and hence color of the emitted light will depend on the size of the gap. If not recombined, the electron–hole pairs exist in a bound state, forming quasiparticles called excitons. In quantum dots, the particle size is usually 2–10 nm, thus approaching Bohr exciton radius. The reduction in size thus results in the quantum confinement effect, in which the edges of the nanoparticle confine the excitons in three dimensions. This has the effect of increasing band-gap energy as the particle is made smaller, causing the previously continuous valence and conduction bands to split into a set of discrete energy levels, similar to those present in atomic orbitals. This is why quantum dots are sometimes called "artificial atoms." Hence, in quantum dots, band-gap energy can be tuned simply by changing the particle size. The color of the absorbed and emitted light can thus also be varied by altering the size of quantum dots. With such special properties, it is not surprising that quantum dots have applications in LEDs, solar cells, medical imaging and many other fields [11].

Another interesting nonscalable property that can be associated with nanoparticles is localized surface plasmon resonance (LSPR). This can be observed, for example, if we decrease the size of gold [12], small enough to result in a color change from gold color (as in bulk) to a variety of colors. In the bulk form, gold is shiny and reflects yellow light, whereas at 10 nm, gold absorbs green light and appears red. As the particle size increases, red light is absorbed and blue light transmitted, resulting in a pale blue or purple color. This phenomenon can be explained by the fact that the mechanism for generating color is quite different between bulk and nanoscale gold. In bulk, an electronic transition between atomic orbitals (5d and 6s) absorbs blue light, giving gold its yellow color, while the reflectivity is due to the presence of free electrons in the conduction band of the metal. If the size of the gold nanoparticles is reduced, it can restrict the motion of these free electrons, as they will be confined to a smaller region of space, that is, to the nanoparticle. If the particles are small enough, all of the free electrons can oscillate together. When resonance occurs, this leads to a strong absorption of certain frequencies of light that corresponds to the resonant frequency of the electron oscillation. This resonant frequency is highly dependent on the particle size, shape and the medium it is suspended in, for example, 50-nm spherical gold nanoparticles in water gives the suspension a cherry-red color due to the strong absorption of green–blue light. Overall, the LSPR is a phenomenon that occurs due to the collective oscillation of surface electrons with incident light at a specific wavelength. It is worth mentioning that the LSPR phenomenon is different from the quantum mechanical effect as observed in quantum dots, as the mechanism of producing color in metal nanoparticles is different from that in semiconducting ones.

1.3 TERMINOLOGY

1.3.1 Nanomaterials

The term "nano" has long been used as a prefix, as exemplified by nanoliter, nanomanufacturing, nanolithography, nanosystems, and so on. In science and engineering, "nano" refers to one billionth (10^{-9}) of a unit and thus a nanometer being defined as 1 billionth of a meter.

Historically, the word nanomaterial has been used to refer to products derived from nanotechnology. The term nanotechnology itself has been defined as far back as 1974 by Professor Norio Taniguchi, to mean a direct extension of silicon machining down into the regions of smaller than 1 μm [13]. In recent years, several definitions of the term nanomaterial have been proposed by various international organizations and committees (as summarized in Table 1.1), to include International Organization for Standardization (ISO), Comité Européen de Normalisation (CEN), that is, the European Committee for Standardisation, Organisation for Economic Cooperation and Development (OECD), EU Scientific Committee on Emerging and Newly Identified Health Risks (SCENIHR), EU Scientific Committee on Consumer Products (SCCP), and American Chemistry Council (ACC) and European Commission (EC).

In addition to those listed in Table 1.1, national authorities and organizations from other countries such as Australia have also provided their own definitions. Although our findings seem to indicate that there are variations in the definition of what constitute a nanomaterial, all definitions have indicated so far an upper dimension limit of 100 nm. However, this is not always the case. The Soil Association, for example, sets this upper limit to be 200 nm, whereas the limit is 300 nm with Friends of the Earth. Unless stated otherwise and to avoid confusion, the book will adopt the

TABLE 1.1 Nanomaterial as Defined by Different Organizations

Source	Definition [14]
ISO TS 80004-1 CEN ISO/TS 27687	"Material with any external dimension in the nanoscale or having internal structure or surface structure in the nanoscale". Nanoscale here has been defined as "size range from approximately 1 nm to 100 nm".
OECD	"Material which is either a nano-object or is nanostructured." Here, nanoobject is a "material confined in one, two, or three dimensions at the nanoscale. Nanostructured here is defined as having an internal or surface structure at the nanoscale; nanoscale is defined as size range typically between 1 nm and 100 nm".
EU SCENIHR	"Any form of a material that is composed of discrete functional parts, many of which have one or more dimensions of the order of 100 nm or less".
EU SCCP	"Material with one or more external dimensions, or an internal structure, on the nanoscale, which could exhibit novel characteristics compared to the same material without nanoscale features. Here nanoscale means having one or more dimensions of the order of 100 nm or less."

(continued)

TABLE 1.1 ***(Continued)***

Source	Definition [14]
EC: Cosmetic Products Regulation	"An insoluble or biopersistant and intentionally manufactured material with one or more external dimensions, or an internal structure, on the scale from 1 to 100 nm."
ACC	"An Engineered Nanomaterial is any intentionally produced material that has a size in 1, 2, or 3- dimensions of typically between 1 – 100 nanometres. It is noted that neither 1 nm nor 100 nm is a 'bright line' and data available for materials outside of this range may be valuable. Buckyballs are also included even though they have a size <1 nm." However, the following are excluded: 1. "Materials that do not have properties that are novel/unique/new compared to the non-nanoscale form of a material of the same composition 2. Materials that is soluble in water or in biologically relevant solvents. Solubility occurs when the material is surrounded by solvent at the molecular level. The rate of dissolution is sufficiently fast that size is not a factor in determining a toxicological endpoint. 3. For those particles that have a particle distribution such that exceeds the 1 – 100 nm range (e.g. 50 – 500 nm) if less than 10% of the distribution falls between 1 – 100 nm it may be considered as non Engineered Nanomaterial. The 10% level may be on a mass or surface area basis, whoever is more inclusive. 4. Micelles and single polymer molecules."
EC	"A natural, incidental or manufactured material containing particles, in an unbound state or as an aggregate or as an agglomerate and where, for 50 % or more of the particles in the number size distribution, one or more external dimensions is in the size range 1 nm – 100 nm. In specific cases and where warranted by concerns for the environment, health, safety or competitiveness the number size distribution threshold of 50 % may be replaced by a threshold between 1 and 50 %. By derogation from the above, fullerenes, graphene flakes and single wall carbon nanotubes with one or more external dimensions below 1 nm should be considered as nanomaterials."
EC : for novel foods (amending Regulation (EC) No 258/97), under discussion	"Any intentionally produced material that has one or more dimensions of the order of 100 nm or less or is composed of discrete functional parts, either internally or at the surface, many of which have one or more dimensions of the order of 100 nm or less, including structures, agglomerates or aggregates, which may have a size above the order of 100 nm but retain properties that are characteristic to the nanoscale."

ISO definition as in Table 1.1. ISO has been especially chosen as it operates on an international level and most recognized globally.

In addition to the definition of nanomaterial, there is also a need to differentiate some other similar terms. In particular, nanomaterials and nanoparticles are often used interchangeably, but they are clear differences. According to the ISO definition, nanoparticle is a "nano-object with all three external dimensions in the nanoscale"; nano-object here is a "material with one, two or three external dimensions in the nanoscale." Nanomaterial, however, is a *material with any external dimension in the nanoscale or having internal structure or surface structure in the nanoscale*. In both cases, the nanoscale is referred to as a *size range from approximately 1–100 nm* [15]. In this book, the terms nanomaterial and nanoparticle will be differentiated accordingly, in accordance to ISO definitions.

1.3.2 Physicochemical Properties

An important part of nanomaterial research is to identify what the relevant physicochemical properties that one should measure and define the corresponding measurands, that is, the quantity intended to be measured. However, this depends on the scientific field and nanospecific application. In some cases, these have already been defined by the relevant scientific community and are published in standard documents. Let's consider the field of nanotoxicology. In this community, physicochemical properties of relevance have already been defined, in accordance to published ISO standard document and OECD guidelines [16, 17]. Having two separate guidelines can cause some confusion, and it is wise to read both and make comparison. There are several things worth highlighting when comparing the two:

a) OECD refers to "endpoints," as opposed to ISO's "properties."
b) OECD also has a much longer list of endpoints, that is, 16, compared to ISO's 8.
c) Some of the OECD's endpoints can be categorized under the same umbrella as an ISO property. For example, the OECD *particle size distribution – dry and in relevant media* and *representative TEM images*, is similar to ISO *particle size and particle size distribution*. In addition, the ISO *surface chemistry* can potentially encompasses quite a number of OECD endpoints: *surface chemistry (where appropriate)*, *redox potential*, *radical formation potential*, *photocatalytic activity*, *octanol-water partition coefficient*.
d) Some OECD endpoints have not been taken into account within the ISO document. *Dustiness* and *pour density*, for example, cannot be categorized under any of the ISO properties, even though they are highly relevant in nanotoxicology. In nanotoxicology, the property of *dustiness* is important as it relates to the properties of airborne nanomaterial and thus of relevance in workplace hazard/risk scenarios.
e) The OECD endpoint *representative TEM images* is unusual as this is specific to an analytical technique rather than a physicochemical property. This endpoint can be incorporated under various ISO properties, such as *particle size/size distribution* and *shape*.

Table 1.2 aims to summarize and integrate the information from ISO and OECD guidelines. A limitation of the OECD guideline is that the measurand is less well defined. As a result, the measurands (apart from dustiness and pour density) in Table 1.2 are those that have been defined by ISO [16].

1.4 MEASUREMENT OF GOOD PRACTICE

There is a network of organizations in Europe called Eurachem, whose main mission is to promote best practice in analytical measurement. According to Eurachem, "analytical measurements should be made to satisfy an agreed requirement, that is, to a defined objective and should be made using methods and equipment which have been tested to ensure that they are fit for purpose" [18]. To achieve this, there is a need to understand several key terms such as method validation and standard documents.

1.4.1 Method Validation

The term "*fit for purpose*" implies that the method must be sufficiently reliable and robust [19, 20]. To ensure that a method is fit for purpose, a validation process must take place.

The process of validation may not be straightforward as it is hard to tell when method development ends and validation begins. The two processes can be considered as an interactive process and will not be differentiated here. The first step in method validation is to be clear on the stated objectives for carrying out the analysis and subsequently to establish what the analytical requirements are. The analytical requirements are often related to factors such as specificity, selectivity, accuracy, repeatability/reproducibility, robustness (e.g., not sensitive to operator and day-to-day variability), and analysis time. Other practical issues may also be taken into account such as speed of analysis, costs, technical skill requirements, availability, and laboratory safety. A method can then be developed by choosing the best analytical technique in which parameters such as sample type (matrix) and size, data requirements, for example, qualitative or quantitative, expected level of analytes, and likely interferences, will be taken into account.

As part of the method development step, it is necessary to conduct a literature research to check if suitable methods already exist as existing methods can potentially be used and modified, if necessary. Once a method is developed, it must be refined to demonstrate that it is fit for purpose. Hence, as part of the validation process, an assessment has to be made in order to verify whether the method fulfils the analytical requirements being set, in which round robin studies [21–24] are often carried out. Method validation is not trivial, and sometimes it may be necessary to conduct a prevalidation step to identify any necessary refinements that can be made to the method. Prevalidation study can be conducted among a few established/competent laboratories, preferably with registered/recognized validation authority (RVA), for example, European Centre for the Validation of Alternative Methods (ECVAM). The

TABLE 1.2 Physicochemical Properties of Relevance to Nanotoxicology Community, as Defined by ISO and OECD Guidelines

ISO Properties	Corresponding OECD "end-points"	Measurand; from ISO, Unless Stated Otherwise
Particle size and particle size distribution	Particle size distribution – dry and in relevant media Representative TEM images	"The physical dimensions of a particle and, for collections of particles, the distribution of the sizes of the particles determined by specified measurement conditions" "Equivalent spherical diameter, for particles displaying a regular geometry (unit m); the length of one or several specific aspects of the particle geometry, (unit m) the particle size distribution, the number of peaks and their width are a set of values, often displayed as a histogram, which for each of a number of defined size classes which shows the quantity of particles, being either the number of particles, or the cumulative length, area, or volume of these particles or the signal intensity they produce"
Aggregation/ Agglomeration state in relevant media	Agglomeration/ Aggregation Representative TEM images	Aggregate is "strongly bonded or fused particles where the resulting external specific surface area might be significantly smaller than the sum of known specific surface areas of primary particles". Agglomerate is "collection of weakly or loosely bound particles or aggregates or mixtures of the two in which the resulting external specific surface area is similar to the sum of the specific surface areas of the individual components" "Particle size (unit, m); number of aggregate (or agglomerate) particles in comparison to the total number of primary particles, unit (number/number); number of primary particles in the aggregate (or agglomerate), unit (number/number); distribution of number of primary particles per aggregate (or agglomerate)."

(continued)

TABLE 1.2 (*Continued*)

ISO Properties	Corresponding OECD "end-points"	Measurand; from ISO, Unless Stated Otherwise
Shape	Representative TEM images	"A description of the contour or outline of the surface of the nano-objects or collection of nano-objects, aggregates, agglomerates, that make up the material under investigation" "Size-independent descriptors of shape (examples are ratios of extensions in a different direction such as aspect ratio, unit (m/m) or fractal dimension); distribution of values of the size-independent shape descriptors"
Surface area/ mass-specific surface area/ volume-specific surface area	Specific surface area Porosity	This is the "quantity of accessible surface of a sample when exposed to either gaseous or liquid adsorbate phase. Surface area is conventionally expressed as a mass specific surface area or as volume specific area where the total quantity of area has been normalised either to the sample's mass or volume" "Specific surface area is defined as surface area of a substance divided by its mass, unit [m^2/g]; or surface area of a substance divided by its volume, unit [m^2/cm^3]. The research should also consider reporting results in both m^2/g and m^2/cm^3."
Composition	Crystallite size. Crystalline phase.	"Chemical information and crystal structure of the entire sample of nano-objects including: a) composition b) crystalline structure including lattice parameters and space group, and c) impurities, if any" "The number and identity of elements alone or in molecules (can be expressed as a chemical formula with a specific stoichiometry; crystalline state; crystallographic structure; chemical state of atoms/elements; molecular structure-conformation including dextrorotatory and levorotatory (handedness); spatial distribution of the above items."

(*continued*)

TABLE 1.2 ***(Continued)***

ISO Properties	Corresponding OECD "end-points"	Measurand; from ISO, Unless Stated Otherwise
Surface chemistry	Surface chemistry, where appropriate Redox potential Radical formation potential Photocatalytic activity Octonal–water partition coefficient, that is, to what degree colloidal suspended particle in the aqueous phase can also be suspended in a nonaqueous phase (such as octanol)	"Chemical nature, including composition, of the outermost layers of the nano-objects and their aggregates and agglomerates greater than 100 nm" "Elemental and molecular abundance unit [mole/mole], including thickness for fixed layers or [number of molecules/surface area] or [number of molecules bound /theoretical number of molecules bound with perfect reaction or perfect packing] for chemically reacted species that do not form a distinct phase; reactivity: standard chemical reaction rate concepts [mole/ (dm^3/s)] preferably of a species of toxicological interest or its surrogate. Measurement of reactivity is very specific to the measurement of the species to which it is reactive (such as reactive to water) and typically involves measuring products or by-products of that reaction."
Surface charge	Zeta-potential (surface charge)	"Electrical charge on a surface in contact with a continuous phase" "Net number of positive and negative charges per unit particle surface area, unit [Coloumb/m^2]; zeta potential, unit [V]"
Solubility/ dispersibility	Water solubility/ Dispersibility	"Solubility is the degree to which a material (the solute) can be dissolved in another material (the solvent) so that a single homogeneous phase results. Dispersibility is the degree to which a particulate material (the dispersed phase) can be uniformly distributed in another material (the dispersing medium or continuous phase) and resulting dispersion remains stable (for example one hour or one minute)"

(continued)

TABLE 1.2 ***(Continued)***

ISO Properties	Corresponding OECD "end-points"	Measurand; from ISO, Unless Stated Otherwise
		For solubility this is the "maximum mass or concentration of the solute that can be dissolved in a unit mass or volume of the solvent at specified (or standard) temperature and pressure, unit [kg/kg] or [kg/m^3] or mole/mole]". For dispersibility, this is "the maximum mass or concentration of the dispersed phase present in a unit mass of the dispersing medium (solvent) or in a unit volume of the dispersion (solvent plus dispersed phase) at specified (or standard) temperature and pressure, units [kg/kg], [kg/m^3], or [mole/mole].
N/A	Dustiness	OECD definition: This is defined as the "propensity of a material to generate airborne dust during its handling, and provides a basis for estimating the potential health risk due to inhalation exposure". [17] "The measurand of interest is the degree to which a given nanomaterial can remain in the air column before settling. This would require investigation and characterisation of interactions of nanomaterials with other common airborne particulate matter." [17]
N/A	Pour density	OECD definition: This is the "apparent density of a bed of material formed in a container of standard dimensions when a specified amount of the material is introduced without settling". "Determination of bulk density." [17]

purpose of the prevalidation is to assess protocol performance and carry out any subsequent actions needed to refine the protocol. After prevalidation, a formal validation trial with other RVAs or other appropriate sponsors can be carried out.

In nanomaterial research, every effort should be made towards method validation, as only when the conditions of method validation are met, only then a higher metrological standard of measurement, that is, making traceable measurements, can be considered. According to Eurachem/CITAC [21], traceability is *property of the result of a measurement or the value of a standard whereby it can be related to stated references, usually national or international standards, through an unbroken chain of comparisons, all have stated uncertainties*. The traceability framework thus focuses on two main activities: calibration and development of an uncertainty budget. Calibration is defined as the *comparison of an instrument against a reference or standard, to find any errors in the values indicated by the instrument* [25], whereas uncertainty of measurement *is the quantified doubt about the result of a measurement*, which can be established by evaluating the uncertainty budget. This chapter will not delve into the details on how to perform uncertainty budget analysis as this can be found elsewhere [18, 26]. In brief, in order to establish an uncertainty budget, major components contributing to the measurement uncertainty has to be identified and quantified as standard deviations (uncertainties). The contribution of each major component is then statistically combined and the combined uncertainty computed.

In metrology, the ability to make traceable measurements, ideally to the SI units of measurements, is always desirable. However, in some instances, it has to be appreciated that making traceable measurement is difficult and unachievable. In nanotoxicology research, for example, an incomplete traceability chain is likely as calibration is often being carried out under conditions too different from the application.

1.4.2 Standard Documents

A standard document provides "requirements, specifications, guidelines or characteristics that can be used consistently to ensure that materials, products, processes and services are fit for their purpose"[27]. According to BSI 0:2011, standards can aid in "a) facilitating trade, particularly in reducing technical barriers and artificial obstacles to international trade b) providing framework for achieving economies, efficiencies and interoperability c) enhancing consumer protection and confidence and; d) supporting public policy objectives and, where appropriate, offering effective alternatives to regulation"[28]. As such it is not surprising that standard documents on measurement and test methods, specifications, terminology, management, and management systems [29] exist.

So, what can be classified as standard documents?

Standard documents generally fall into one of the following two categories: formal and informal standards. Formal standards are made by official standard organizations, proceeding through government recognized National Standard Bodies (NSBs) at a national, regional or international level. NSBs include British Standards Institute (BSI, founded in 1901), Deutsches Institut fur Normung (DIN, 1917), Schweizerische Normen-Vereinigung (SNV in 1919), Standardiseringen I Sverige (SIS in 1922), Norges Standariseringsforbund (NSF in 1923), Den Danske Standardiserings Kommission (DS in 1926), L'Association francaise de normalisation (AFNOR in 1926), and so on. By the end of the 20th century, the work on regional and international standards became more prominent. In some cases, this had meant that standardization work previously carried out at a national level was transferred to regional, for example, European Committee for Standardization (CEN) or international working groups (WGs) in, for example, ISO [29].

Unlike formal standards, informal standards are published by Standards Development Organizations (SDOs), with some being well known and highly respected, for example, ASTM International (previously the American society for Testing

Materials), IEEE (previously the Institute of Electrical and Electronic Engineers), SAE (Society of Automotive Engineers), SEMI (Semiconductor Equipment and Materials International), TAPPI (formerly the Technical Association of the Pulp and Paper Industry), and OECD. The process to develop an informal standard is the same as those used for formal standards, the only difference being that development and approval is undertaken by members of the SDO rather than through a network of NSBs [30]. Although not technically categorized as "standard" in the real sense of the word, there also exists "private standards," which are developed for internal use, for example, used in companies. Obviously, such standards have less impact and global recognition and often not considered as a viable route.

In addition to the classification of formal versus informal standards, a further sub-classification can be made, on the basis whether the document is considered to be a normative or informative document. Normative documents are "those documents that contain requirements which must be met in order for claims of compliance with the standard to be certified." Informative documents on the other hand, are those that "do not contain any requirements and it therefore not possible for compliance claims to be certified" [30]. Normative means that it's an official formal part of the specification, whereas informative means that it's there to be helpful, for example, aid understanding but cannot be used in formal circumstance, such as appeal to it in a court of law or as part of an audit process.

So, what process is involved in developing a standard document?

In general, the development and publication of a formal document standard is often a long process. The first step involves identification of new work and begins with a proposer, which might be a corporate, public organization, individual, or a consortium [28]. The proposer must then decide if the standard should exist at a national, for example, British, regional, for example, European or at an international level, for example, ISO. The proposer then must demonstrate the need for the standard, that it will be widely/actively supported, that there are enough resources to complete the project in a reasonable time and there no conflict that exists with existing standards [28]. The work of drafting a standard can then be undertaken under by, for example, a suitable Working Group (WG). The members in the WG will then draft document and build a consensus before releasing the draft for comments. The final draft will then be put to a voting period, thus rely on consensus, that is, an agreement between people and organizations that will be affected. After successfully going through a voting period, a final document can then be published as a standard document. As a ball park figure, a national standard can take between 1 and 3 years to produce, whereas international standards usually require the consensus of a more number of participants and therefore usually take longer to publish [28]. Figure 1.1 schematically shows the different stages of standard development/publication and the corresponding relative level of impact associated with the different stages.

Before embarking on any standard development work, it is important that the proposer undertakes a review of existing standards applicable and to identify relevant technical committees and working groups. In relation to nanomaterial characterization, ISO is active in producing standards under Technical Committee on Nanotechnologies, TC229. This committee consists of four working groups and is actively

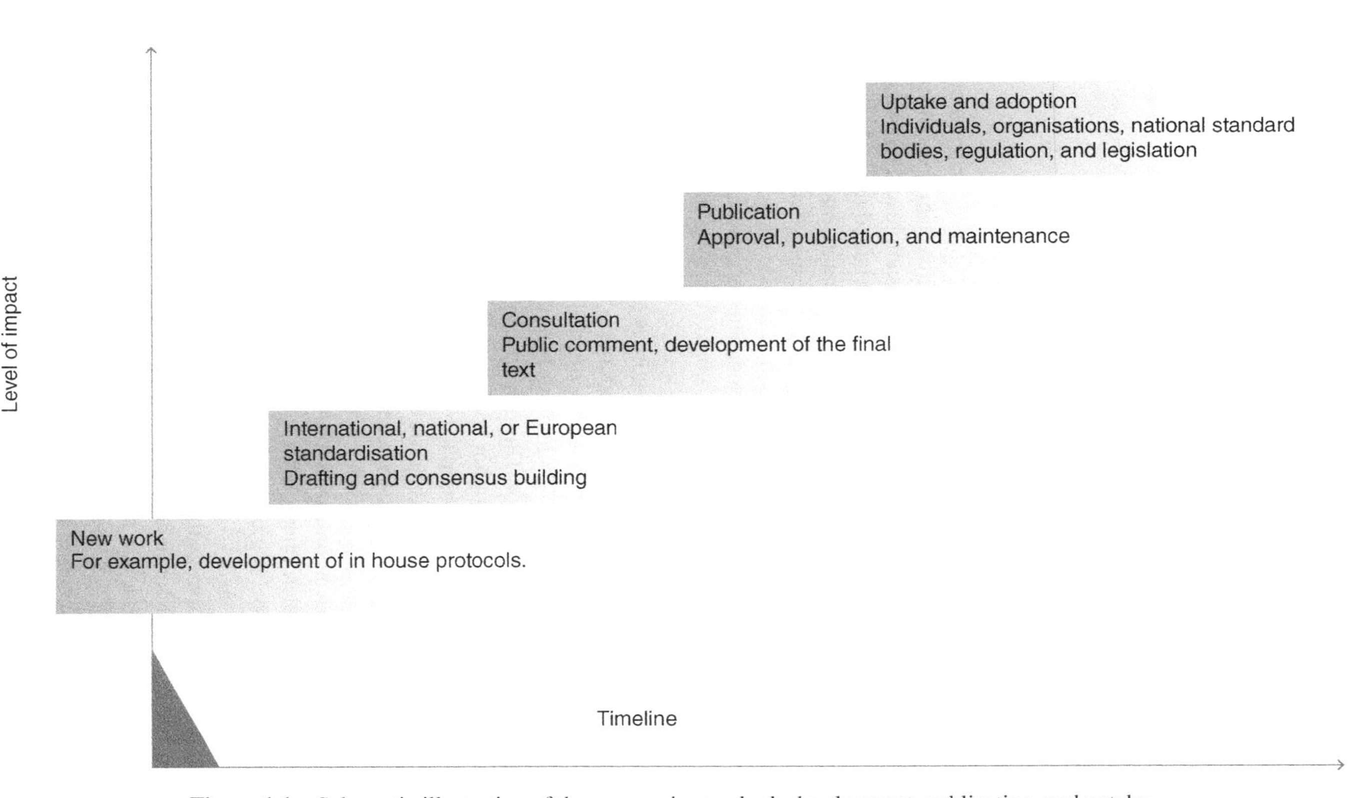

Figure 1.1 Schematic illustration of the process in standards development, publication, and uptake.

publishing on (i) terminology and nomenclature, (ii) measurement and characterization, (iii) health, safety, and the environmental, and (iv) material specifications. To date, the total number of published ISO standards related to this TC is 42. Further details of ISO standards can be found elsewhere [31].

1.5 TYPICAL METHODS

The following sections list two of the most common methods that can be associated with nanomaterial characterization, namely, sampling and dispersion.

1.5.1 Sampling

Sampling is defined as "a procedure whereby a part of a substance, material or product is taken to provide for testing or calibration a representative sample of the whole" [32]. The main purpose of sampling is to collect the entire sample and reduce its size for subsequent analysis. If a representative sample is not obtained through sampling, then this step in an experimental investigation can be a major source of data variation. Errors incurred through the sampling stage can be minimized if proper sampling methods are carried out. There are several ways to minimize sampling errors. First, there is a need for suitable mixing to ensure homogeneity prior to removal of sample aliquots. Second, there is a need to increase the sample size, by taking measurements from large number of sample increments.

When nanomaterial is in the form of dispersion, achieving reliable sampling is potentially straightforward. For example, the sampling step may involve sufficient agitation of the dispersion before an aliquot is extracted by using a pipette. In relation to powder material, the process of sampling is more challenging, as sampling will be more prone to segregation error. This can arise when particles are exposed to gravitational, rotational, vibratory/aeration operations, or other types of mechanical motion, resulting in fine particles to migrate to the bottom and larger particles being concentrated at the top [32]. Segregation error is more problematic with free or easily flowing powders and those having a significant range of particle size distribution.

In relation to powder sampling, there is a need to carefully consider the different options of sampling. The five common sampling methods are scoop sampler, cone and quartering, table sampler, chute riffler, and spinning riffler (also called rotary sample divider) [33]. Table 1.3 gives an overview of the sampling methods, along with their inherent limitations.

Allen and Khan [32] have evaluated the different sampling methods shown in Table 1.3, in which the performance of each method was assessed by sampling a known particle size distribution of sand mixtures. Their findings show that out of all the methods listed in Table 1.3, the spinning riffler is the most reliable method for sampling, as this method incorporates little operator bias. In addition to reliability, the spinning riffler is also most practical, especially when dealing with kilograms of materials. However, there is one prerequisite that must be fulfilled when using the riffler, in that powder must be free-flowing in nature.

TABLE 1.3 A Comparison of Powder Sample Reduction Methods

Sampling Device	Description	Limitations
Scoop sampler	This method is the simplest for sample division and involves an operator, using a scoop, to extract laboratory samples from some portion of the bulk sample. This method is reliable for samples that are homogeneous and exhibit poor flow characteristics	All of the bulk material do not go through sampling process. Operator dependent, for example, operator decides where to scoop and what quantity to extract. Sampling more likely to be atypical due to segregation of the material. Not suitable for heterogeneous sample
Cone and quartering	This technique involves placing the sample on a flat surface in the form of a conical heap. The heap is then spread out and flattened into a circular cake, which is then divided into approximately equal quarters. One pair of opposite quarters is removed, combined and formed into a new cone for the process to be repeated (with the other two quarters discarded). The process is repeated as many times as is necessary to obtain a sample of the required size. Good for powders with poor flow characteristics and minimal segregation	Operator dependent; errors can occur due to differences in the manner the heap is formed and sub-divided

(continued)

TABLE 1.3 ***(Continued)***

Sampling Device	Description	Limitations
Table sampler	This utilizes tilted surface (in which there is a series of holes and splitting prisms) over which a powder sample is allowed to slide. The prism breaks the stream into fractions and some of the powder will fall through the holes and then discarded. Ultimately, at the bottom of the plane, a decreased quantity of sample is collected. The method has the ability to separate large quantity of material	It is necessary that the incoming powder is uniform and consistent. Hence, the method is dependent on the initial feed being uniformly distributed, with a complete mixing after each separation; this is a condition not general achieved
Chute riffler	This utilizes chutes, that is, funneling or channeling device, to divide the powder. Unlike the spinning riffler, the chute riffler has no moving parts. It has the ability to separate large quantity of material and to reduce powder samples in half after one pass	The technique is subject to error and operator bias if segregation is allowed to occur in loading the bulk-sampling through
Spinning riffler (also called rotary sample divider)	This utilizes a series of smaller receivers (or collection tubes), mounted in such a way so as to collect a flowing powder stream over a very short time period. The powder flows from a "hopper" to a "vibratory chute" and then to a "receiver" that holds the containers, which are rotating in a circular motion at a constant speed. The method is able to separate large quantity of material. Spinning rifflers are available in different sizes, in which common commercial systems can provide samples ranging from about 0.5–300 g [34]. It has the ability to do large quantity of powder efficiently	Limited to free-flowing powder only

If a spinning riffler is to be used for nanomaterials, then it must be validated for different nanomaterials in accordance to ISO 14488 guideline [35]. This ISO standard document makes recommendations on how to choose and use a riffler. It also highlights the importance of validating the instrument for each new material to be riffled. According to the document, the simplest way to validate is by mass validation, which consists of several steps. First, there is a need to measure the mass of the gross sample together with the masses of all the increments; this is to be repeated three times. Then, there is a need to calculate the mean loss of the material. If the mean loss of material is larger than 1%, then the riffler is either not working properly or that the riffler is not a suitable method.

1.5.2 Dispersion

Dispersion of powder nanomaterial into a liquid matrix is another common practice in nanomaterial research [36] and basically involves three main stages:

a) Wetting of the nanomaterial powder. The purpose of this step is to substitute solid–air interface with solid–liquid interface, such that the particles are sufficiently "wetted." The efficiency of wetting will depend on the comparative surface tension properties of the nanomaterial and the liquid media, as well as the viscosity of the resultant mix. The wetting step can easily be achieved by mixing the powder with several drops of liquid media, to form a thick paste. In the case where a powerful ultrasonic probe is used, wetting may occur simultaneously during sonication step.
b) De-agglomeration of the nanomaterial, using a de-agglomeration tool. Here, sufficient shear energy is needed to break up loosely bound agglomerates in the powdered nanomaterial. There are various de-agglomeration tools that exist on the market to include mills (ball, stirred media, centrifugal, and jet mills), stirring (magnetic or overhead stirring), high-speed homogenizer, high pressure homogenizer, ultrasound sonicating bath, and ultrasound probe sonication or ultrasonic disruptor [37].
c) Stabilization of the dispersion. Stability can be impacted by the choice of the liquid dispersant and things added to the dispersant, for example, surfactant that can lead to marked changes in its interfacial properties [38].

To date, no standard guidelines exist that details how to disperse powdered nanomaterials into a liquid matrix. In some science areas such as nanotoxicology, the need to produce stable and reproducible dispersions is important for the purpose of data reliability and comparability. Recent works by Tantra et al. [39] have shown how huge data variability can arise from using different dispersion protocols. Their findings highlighted the importance of controlling the dispersion step, as factors such as particle concentration, sonicating time, can all influence dispersion quality.

1.6 POTENTIAL ERRORS DUE TO CHOSEN METHODS

Characterizing the property of nanomaterial (with current instrumentations) may not be straightforward. With some nanomaterial samples, getting reliable data is not easy to achieve and can lead to a situation in which experimental data can get reported without proper understanding of the associated errors and propagation of such errors. Sources of experimental errors may arise from a number of factors, to include polydispersity of the nanomaterial and the difficulty to measure such highly polydisperse samples.

Baalousha and Lead [40] have highlighted that most of the nanomaterials tested in nanotoxicology are far too polydisperse and that materials close to monodispersity are needed. The main issue with having a highly polydisperse sample is the lack of analytical techniques that can measure the corresponding properties accurately. Anderson et al. [41] show most routine methods can characterize particle size distributions of monomodal distribution. However, if the particle distribution is away from the ideal, then errors can be incurred during the measurements. A typical example to highlight this point is light-scattering-based methods such as Particle Tracking Analysis and Dynamic Light Scattering. Both techniques have been shown to mainly reliably detect a single population of particles corresponding either the largest or smallest particles in a multimodal sample. Clearly, the inadequacy of the instrumental methods to characterize highly polydisperse nanomaterials can pose a barrier to reliable measurement [42].

In addition to issues associated with polydispersity, nanomaterials dispersed in complex, for example, biological, environmental matrix, which contains other interferents, can also pose problems where measurement is concerned. Hence, an instrument with high selectivity and sensitivity may be needed. Apart from the presence of interferents, nanomaterial–media interactions can be dynamic in nature and in the example of nanotoxicology research, the physicochemical properties measured at a given time may not be directly associated with observed biological effects. Nanomaterials dispersed in complex medium may also be unstable, potentially resulting in agglomeration and sedimentation, which may pose further difficulties for the instrument to measure the sample under such conditions. Due to the analytical challenges posed in nanotoxicology, it is difficult to reliably assess the potential transformation of nanomaterials in an environmental or mammalian system [43].

1.7 SUMMARY

The potential benefits of nanomaterials to society and economy are clear and, as such, much research on nanomaterial is currently being conducted, covering a wide range of disciplines. This introductory chapter is a good starting point for readers, to get to grips with some of the key topics, to include terminology, measurement of good practice, issues/challenges, and so on. An important point to highlight is for researchers to choose the right methods and the need to validate such methods for their nanospecific applications. Researchers are encouraged to give careful thought in identifying potential sources of error associated with their measurements, which will lead to improved experimental design and methods employed during an investigation.

ACKNOWLEDGMENTS

Funding was provided by FP 7 MARINA.

REFERENCES

1. Kent JA. *Handbook of Industrial Chemistry and Biotechnology*. 12 ed. Vol. 1. New York: Springer; 2012.
2. European Commission. *Nanosciences and Nanotechnologies: An action plan for Europe 2005–2009*. Office for Official Publications of the European Communities, 2009; 2005.
3. The Project on Emerging Nanotechnologies. 2014. Consumer products inventory. Available at http://www.nanotechproject.org/cpi/. Accessed 2015 Oct 14.
4. Roco MC, Mirkin CA, Hersam MC. Nanotechnology research directions for societal needs in 2020: Summary of international study. J Nanoparticle Res 2011;13(3):897–919.
5. De Volder MFL, Tawfick SH, Baughman RH, Hart AJ. Carbon nanotubes: Present and future commercial applications. Science 2013;339(6119):535–539.
6. Oberdorster G, Oberdorster E, Oberdorster J. Nanotoxicology: An emerging discipline evolving from studies of ultrafine particles. Environ Health Perspect 2005;113(7):823–839.
7. Ju-Nam Y, Lead JR. Manufactured nanoparticles: An overview of their chemistry, interactions and potential environmental implications. Sci Total Environ 2008;400(1–3):396–414.
8. Li YM, Somorjai GA. Nanoscale advances in catalysis and energy applications. Nano Lett 2010;10(7):2289–2295.
9. Yu Y, Zhang P, Guo L, Chen Z, Wu Q, Ding Y, Zheng W, Cao Y. The design of TiO_2 nanostructures (nanoparticle, nanotube, and nanosheet) and their photocatalytic activity. J Phys Chem C 2014;118(24):12727–12733.
10. Petryayeva E, Algar WR, Medintz IL. Quantum dots in bioanalysis: A review of applications across various platforms for fluorescence spectroscopy and imaging. Appl Spectrosc 2013;67(3):215–252.
11. Chuang C-HM, Brown PR, Bulović V, Bawendi MG. Improved performance and stability in quantum dot solar cells through band alignment engineering. Nat Mater 2014;13(8):796–801.
12. Messersmith RE, Nusz GJ, Reed SM. Using the localized surface Plasmon resonance of gold nanoparticles to monitor lipid membrane assembly and protein binding. J Phys Chem C 2013;117(50):26725–26733.
13. Nimesh Surindra CR. *Theory, Techniques and Applications of Nanotechnology in Gene Silencing*. River Publishers; 2011.
14. Lövestam G, Rauscher H, Roebben G, Klüttgen BS, Gibson N, Putaud J-P, Stamm H. *Considerations on a Definition of Nanomaterial for Regulatory Purposes*. European Commission Joint Research Centre; 2010.

15. ISO. *ISO/TS 27687:2008 Nanotechnologies – Terminology and Definitions for Nano-Objects – Nanoparticle, Nanofibre and Nanoplate*. ISO; 2008.
16. ISO. *ISO/TR 13014:2012: Nanotechnologies – Guidance on Physico-Chemical Characterization of Engineered Nanoscale Materials for Toxicologic Assessment*. ISO; 2012.
17. OECD, Guidance on sample preparation and dosimetry for the safety testing of manufactured nanomaterials, in Working Party on Manufactured Nanomaterials. OECD; 2012.
18. Guide E-C. *Quantifying Uncertainty in Analytical Measurement*. EURACHEM; 2000.
19. Büttner J, Borth R, Boutwell JH, Broughton PM, Bowyer RC. Provisional recommendation on quality of control in clinical chemistry part 1. General principles and terminology. Clin Chim Acta 1975;63(1):F25–F38.
20. Horwitz W, Cohen S, Hankin L, Krett J, Perrin CH, Thornburg W. Quality assurance practices for health laboratories. In: Inhorn S, editor. *Analytical Food Chemistry*. Washington, DC: American Public Health Association; 1978. p 545–646.
21. Eurachem. *The Fitness for Purpose of Analytical Methods: A Laboratory Guide to Method Validation and Related Topics*. Laboratory of the Government Chemist; 1998.
22. Papadakis I, Wegscheider W. CITAC position paper: Traceability in chemical measurement. Accredit Qual Assur 2000;5(9):388–389.
23. Curren RD, Southee JA, Spielmann H, Liebsch M, Fentem JH, Balls M. The role of prevalidation in the development, validation and acceptance of alternative methods. Atla 1995;23:211–217.
24. Green JM. Peer reviewed: A practical guide to analytical method validation. Anal Chem 1996;68(9):305A–309A.
25. Bell, S.A., A beginner's guide to uncertainty of measurement, in Measurement Good Practice Guide. National Physical Laboratory; 2001.
26. Wüthrich J, Weber M. *ISO/IEC 17025. Double Accreditation Brings a New Class of CRMs*. Switzerland: Sigma-Aldrich Marketing Communications Europe; 2008. p 4.
27. ISO. Available at http://www.iso.org/iso/home/standards.htm. Accessed 2015 Nov 19.
28. British Standards Institute. *BS 0: 2011 A Standard for Standards – Principles of Standardization*. British Standards Institute; 2011.
29. Murphy CN, Yates J. *The International Organization for Standardization (ISO): Global Governance through Voluntary Consensus (Global Institutions)*. Routledge; 2008.
30. Hatto P. *Standards and Standardisation: A Practical Guide for Researchers*. European Commission; 2010. https://ec.europa.eu/research/industrial_technologies/pdf/practical-standardisation-guide-for-researchers_en.pdf.
31. ISO. http://www.iso.org/iso/home/store.htm. Accessed 2015 Nov 19.
32. Allen T, Khan AA. *Critical Evaluation of Powder Sampling Procedures*. The Chemical Engineer; 1970. p 108–112.
33. Allen T. *Particle Size Measurement: Volume 1: Powder Sampling and Particle Size Measurement*. Springer; 1997.
34. Jillavenkatesa A, Dapkunas SJ, Lum L-SH. *Particle Size Characterization*. Gaithersburg, MD: NIST Recommended Practice Guide; 2001.
35. ISO. *ISO 14488:2007: Particulate materials Sampling and sample splitting for the determination of particulate properties*. ISO; 2007.
36. Fischer EK. *Colloidal Dispersions*. Fischer Press; 2008.

37. Van Son N, Rouxel D, Vincent B. Dispersion of nanoparticles: From organic solvents to polymer solutions. Ultrason Sonochem 2014;21(1):149–153.
38. Vaisman L, Wagner HD, Marom G. The role of surfactants in dispersion of carbon nanotubes. Adv Colloid Interface Sci 2006;128:37–46.
39. Tantra R, Sikora A, Hartmann NB, Sintes JR, Robinson KN. Comparison of different protocols on the particle size distribution of TiO_2 dispersions. Particuology 2015;19:35–44.
40. Baalousha M, Lead JR. Nanoparticle dispersity in toxicology. Nat Nanotechnol 2013;8(5):308–309.
41. Anderson W, Kozaka D, Colemanb VA, Jämtingb ÅK, Traua M. A comparative study of submicron particle sizing platforms: Accuracy, precision and resolution analysis of polydisperse particle size distributions. J Colloid Interface Sci 2013;405:322–330.
42. Tantra R, Shard A. We need answers. Nat Nanotechnol 2013;8(2):71–71.
43. Fadeel B, Savolainen K. Broaden the Discussion. Nat Nanotechnol 2013;8(2):71–71.

2

NANOMATERIAL SYNTHESES

R. Tantra, K.N. Robinson, and J.C. Jarman

Quantitative Surface Chemical Spectroscopy Group, Analytical Science, National Physical Laboratory, Teddington TW11 0LW, UK

T. Sainsbury

Materials Processing and Performance Group, Materials, National Physical Laboratory, Teddington TW11 0LW, UK

2.1 INTRODUCTION

Nanomaterials can be synthesized by a wide variety of methods. These methods can be grouped into two general strategies, one in which a bottom–up approach is used and the other being a top–down approach [1]. A top–down approach involves reducing the size of the bulk material to form nanomaterials, whereas a bottom–up approach generates nanomaterials by assembling atoms or molecules via synthetic chemistry. Top–down approaches involve the attrition (wear) of the source material, whereas bottom–up approaches start with either a solution or a vapor of atoms, molecules, or a precursor that reacts to form the nanomaterial population.

The purpose of this chapter is twofold: first to give an overview of common methods for nanomaterial synthesis found in the two approaches and second to present a case study on the synthesis of gold nanoparticles by using various methods. As part of the overview, some theoretical considerations on particle nucleation and growth will be given. In addition, how carbon nanotubes (CNTs) are synthesized commercially will be presented. CNTs have been chosen as the specific example for further discussion, due to their use in a wide range of commercial applications, as discussed in Chapter 1. In relation to the case study, the effects of different bottom–up methods on

Nanomaterial Characterization: An Introduction, First Edition. Edited by Ratna Tantra.

the size characteristics of gold nanoparticles will be assessed. In particular, a method on how gold nanoparticles are traditionally synthesized, that is, "bulk" approach, will be compared to methods that employ microreactor-based technology.

2.2 BOTTOM–UP APPROACH

2.2.1 Arc-Discharge

In the arc-discharge synthesis of nanomaterials, a plasma is generated by an electrical discharge between two electrodes, which is then used to vaporize the electrode material [1]. The choice of material is dependent on what type of nanomaterial is required; for example, graphite electrodes can be used to synthezise CNTs. The vapor of atoms that is created will then condense to produce the nanomaterial. The arc discharge method proceeds either in gas (as used for the generation of CNTs) or in liquid-phase systems (as for silver nanoparticles [2]). Indeed, this method had been used to synthesize the first CNTs and since then has been applied to the synthesis of numerous other nanomaterials, for example, silver, silica-coated iron nanoparticles [3], and gold nanoparticles [4]. A schematic of the arc-discharge apparatus is shown in Figure 2.1.

2.2.2 Inert-Gas Condensation

In inert-gas condensation, metal atoms are evaporated into an inert carrier gas at high temperature [6]. Cooling the gas then creates a supersaturated vapor, and particles

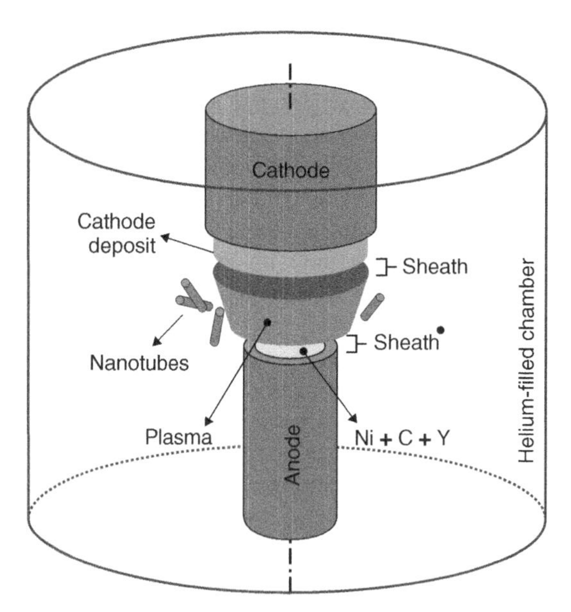

Figure 2.1 Schematic of common apparatus used in the arc-discharge synthesis of carbon nanotubes. Kundrapu et al. [5]. Figure 2.1. Reproduced with permission of IOP Publishing.

nucleate homogeneously from this vapor in the gas stream. Clusters of particles form and sinter, before both nucleation and sintering are stopped by cooling the gas further, either with another stream of gas, that is, the quench gas, or a chilled surface, for example, a liquid-nitrogen filled "cold finger." Particles often form agglomerates at this stage, which under the given conditions will be "loose" enough to be separated in a post-processing step. This method of synthesis has been used to generate carbon blacks, silicon dioxide, and titanium dioxide [6], for example.

2.2.3 Flame Synthesis

Flame synthesis is the most widely used method for the synthesis of commercial quantities of nanomaterials [7]. In flame synthesis, the process begins with the evaporation of a precursor, which is taken into a stream of inert gas [8]. This gas mixture is then mixed with fuel and an oxidizing agent before being injected into a flame. Nanomaterials are produced within the flame, in which the particle characteristics are controlled by the configuration and composition of the flame. Precursors can also be dissolved in flammable organic solvents and sprayed into the flame in a process called flame spray pyrolysis. Flame synthesis has been used to produce various nanopowders to include mixed oxides, nonoxides, fullerenes, and nanotubes [7, 8].

2.2.4 Vapor-Phase Deposition

There are two types of vapor-phase deposition used in nanoparticle synthesis: chemical vapor deposition (CVD) and physical vapor deposition (PVD) [9]. Both methods involve depositing material from the vapor phase, which then coalesces to form the desired nanomaterial. In CVD, a precursor gas is passed over a substrate in a furnace, where it reacts chemically with the surface, depositing the desired chemical species and subsequently resulting in the formation of the nanomaterial. While CVD involves the precursor undergoing a chemical reaction with the substrate, PVD deposits the material through a purely physical process, with no chemical reactions taking place. Both CVD and PVD have been used to fabricate nanomaterials, as well as to deposit thin surface coatings on various substrates. CVD, for example, is the primary process used for growing sapphire crystals. It also has been used to commercially produce CNTs [10].

The morphology of the substrate is often important in determining how the resulting nanomaterial grows. In CNT synthesis, for example, the substrate itself may consist of nanoparticles in order to control the growth of nanotubes. In a type of CVD method called vapor–liquid–solid (VLS), deposition uses catalyst nanoparticles on the surface of a substrate to allow the vapor to condense, forming a liquid, before solidifying. This has been used to synthesize different types of nanowires such as Si and III–V semiconductor nanowires [9].

2.2.5 Colloidal Synthesis

A common method of synthesizing colloidal dispersions of metal nanomaterials is to reduce metal complexes in dilute solutions [11]. The reduction reaction will form a supersaturated solution of metal atoms, which nucleates to form nanoparticles. Agglomeration of the nanoparticles is likely, but there are several steps to prevent this, such as ensuring that the concentration of nanomaterials is low enough or coating of nanomaterials with another chemical species that prevents agglomeration, such as a capping agent.

The colloidal synthesis method is commonly used to produce gold nanoparticles, which is achieved by adding a slight excess of sodium citrate to chloroauric acid ($HAuCl_4$) [12]. The sodium citrate reduces the Au^{3+} ions (present in the acid) to Au atoms. The synthesis here does not necessarily require the addition of capping agents because the citrate ions already present in the solution can also act as the stabilizing agent. This reaction results in the formation of a saturated solution of (nearly) spherical Au nanoparticles. Copper nanoparticles can be synthesized in a similar manner, this time by reducing cupric chloride with hydrazine. However, in this case a separate capping agent (the surfactant cetrimonium bromide (CTAB)) is usually added to prevent particle agglomeration [13].

In order to produce a homogeneous colloidal suspension, experimental conditions should favor diffusion-controlled growth. This can be done by controlling experimental variables such as localized concentration gradients, mixing time and temperature gradients. Microfluidic reactors are also useful in this respect, in that reactions taking place inside microchannels will enable the experimental conditions to be controlled more precisely than compared to reactions taking place in corresponding traditional bulk systems [14, 15].

2.2.6 Biologically synthesized nanomaterials

Bacteria, fungi, actinomycete, and viruses have all shown potential in synthesizing nanomaterials. Synthesis can happen inside (intracellular) or outside (extracellular) biological cells. Examples include the biological assembly of gold nanoribbons from a solution of nanoparticles with *Bacillus subtilis* [16] and synthesis of magnetite nanoparticles in *Magnetospirillum magnetotacticum* bacteria [17]. Magnetite is synthesized not only in this bacteria but also present in a wide range of organisms, including chitons, honeybees and homing pigeons [18, 19]. The synthesis of magnetite in some animals is thought to help its orientation in the Earth's magnetic field.

2.2.7 Microemulsion Synthesis

Micelles (or inverse micelles) can be used to synthesize nanomaterials [20]. For example, it may be possible to create micelles containing the two precursors required for growth of nanomaterial, then allowing them to collide with each other, as schematically represented in Figure 2.2. The precursors, such as a metal salt and a reducing agent, then can react to form the particle inside the micelle; the growth of the nanomaterial in microemulsion synthesis is influenced by the constraints imposed by the local medium. This method of synthesis has been shown to be well suited to the production of nanomaterials with a "core-shell" structure; by this we mean those nanomaterials that have a core made of one material surrounded by a shell made of a different material [21].

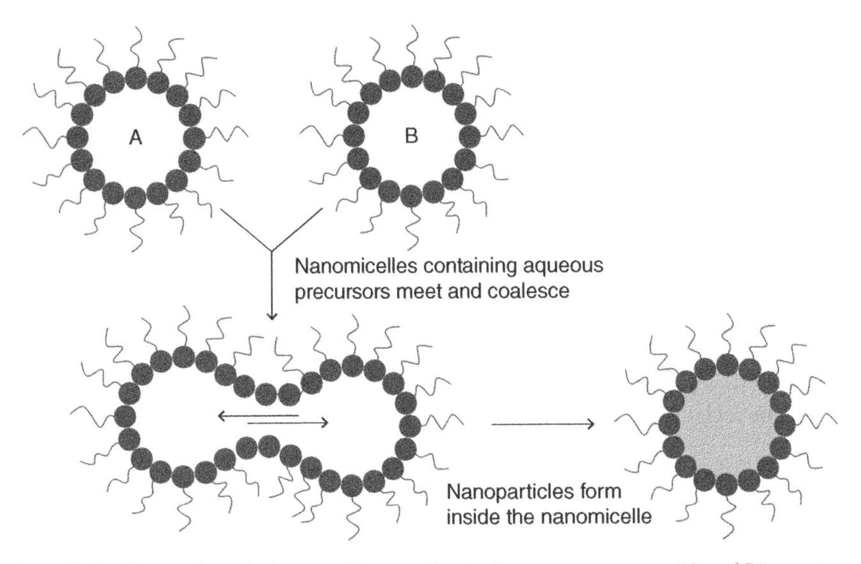

Figure 2.2 Illustration of microemulsion synthesis. Aqueous precursors (A and B) are encapsulated in nanomicelles, which collide and coalesce, mixing the precursors and leading to the formation of nanoparticles.

2.2.8 Sol–Gel Method

As the name implies, this method involves the formation of colloidal suspension, that is, the sol, followed by the gelation of the sol to form an inorganic network in a continuous liquid phase, that is, the gel. It is a wet-chemical technique, in which precursors, typically metal alkoxides (to form metal oxide clusters) first undergo hydrolysis. This is then followed by a condensation reaction, for example, polycondensation or polyesterification [22], which results in a dramatic increase in the viscosity of the solution. The size and morphology of the nanoparticles formed can be controlled by adjusting the reaction conditions. Incomplete reaction can lead to the inclusion of organic groups in the nanomaterial. The sol–gel method has the advantage of a low processing temperature and can be used to fabricate nanoparticles that are thermodynamically unfavorable. Most commonly, the method has been used for the synthesis of colloidal dispersions of metal oxide nanoparticles [23, 24].

2.3 SYNTHESIS: TOP–DOWN APPROACH

2.3.1 Mechanical Milling

This process involves grinding up a bulk material into smaller particles, reducing the size of the particles by attrition [24]. Mechanical milling requires a lot of energy and hours (or days) to complete. Among all of the different types of mechanical milling, ball milling has been widely used for the synthesis of various nanomaterials [25]. A ball mill consists of a rotating chamber, which is partially filled with balls, that is, the grinding media and the substance that is to be powdered. The process generates heat, and the combination with mechanical grinding can cause certain chemical reactions to occur, which is often desirable. For example, when manufacturing silicon nanomaterials, SiO_2 can be reduced by carbon in the mill, giving a pure silicon nanomaterial and CO_2. Another example is the production of Fe_2O_3 nanomaterials, produced by oxidizing iron in the presence of water [26].

The size distribution of particles resulting from ball milling is very broad, and the morphology of the formed particles tends to be quite varied. Hence, nanomaterial synthesized through this route is most suited for certain application areas, for example, in nanocomposites, where a broad size distribution is not problematic. It is, however, unsuited to produce nanomaterials for optical applications, for example, since properties of monomodal size distribution and uniform shape are highly desirable in such applications.

2.3.2 Laser Ablation

Laser ablation is a process of using high-energy nanosecond pulses of laser light to remove and vaporize material from a solid surface [27] to produce nanomaterials. The ionized particles are ejected into a plume before combining together to form the desired nanomaterial [28]. The resulting particles can potentially be deposited from the plume onto a substrate, as done in a PVD process. Although the process can be performed in the gas phase or under vacuum, it can also be done in liquid, that is, laser ablation of chemical precursors in liquid solvents. A wide variety of nanomaterials, for example, metals, metal oxides, semiconductors, and organic materials, have been synthesized this way [29].

2.4 BOTTOM–UP AND TOP–DOWN: LITHOGRAPHY

In strict terms, lithography could be classified as a being both bottom–up and top–down. It is a process used to produce nanoscale structures on the surface of materials, in which the structures are fabricated by selectively depositing a coating or by removing layers of existing material [9]. There are different forms of lithography [4]. Often, a resist (a layer selectively deposited by various means on top of a substrate material) is used to protect regions of the material from an etchant, which is used to remove material by a chemical or physical means. The process results in the removal of the uncoated material, thereby generating the nanostructure. Although the process is sometimes laborious and expensive, lithography can be used to generate intricate nanoscale structures [30], as exemplified in Figure 2.3.

2.5 BOTTOM–UP OR TOP–DOWN? CASE EXAMPLE: CARBON NANOTUBES (CNTs)

As discussed in Chapter 1, CNTs are molecular-scale cylindrical tubes of graphitic carbon six-member rings. Typically, CNTs can be grown to lengths up to several

Figure 2.3 An example of a complex structure with nanoscale features generated by lithography using a focused ion beam (FIB) and assembled using nanomanipulation, showing the possibilities of this technique. Jeon et al. [30]. Figure 2.3. Reproduced with permission of American Vaccuum Society.

tens of microns, with diameters in the nanometer range. Structurally, CNTs can be divided into two general types: single-walled (SW) and multi-walled (MW) CNTs. SWCNTs consist of graphene sheets rolled up into singly cylindrical tubes, with a typical diameter of ∼ 1 nm. MWCNTs, on the other hand, are stacks of SWCNTs nested inside one another to make concentric cylinders, with diameters in the range of 2–100 nm with a layer spacing of 0.3–0.4 nm and many microns in length [31]. CNTs can be produced by very rudimentary methods. For example, the soot from a candle flame is known to contain a range of carbon-based nanostructures, including nanotubes and fullerenes [32]. However, the need to better control the synthetic process has led to the use of the "bottom–up" arc-discharge method (as illustrated in Figure 2.1) as well as "top–down" laser ablation methods. In fact, these methods were the first to be used in the commercial synthesis of CNTs. Both SWCNTs and MWCNTs can be produced using either method [33]. However, these techniques require the use of extremely high temperatures (>3000 °C) to evaporate the carbon atoms, often producing tangled CNTs that are hard to purify [34].

Nowadays, commercial MCWNTs are often synthesized using "bottom–up" CVD methods. While arc-discharge and laser ablation are well-established methods, producing nearly perfect nanotube structures, CVD is a comparatively simple and more importantly relatively economical, as it does not require excessively high temperatures or pressures. As a result, it allows for the continuous and economic production of CNTs. In the production of CNTs using a CVD method, a hydrocarbon precursor, for example, carbon monoxide or ethylene, is vaporized and passed over a nanosize catalyst (at a temperature high enough to allow the precursor to decompose). The catalyst is chosen to have reasonable solubility and high diffusivity for carbon. Hence, metals such as Fe, Co, or Ni are often chosen, even though a range of other metals and nonmetallic catalysts (such as nano-diamond) are currently being investigated [6]. Although the exact nanoparticle growth mechanism is still debatable, it is thought that the mechanism involves hydrocarbon decomposition on the surface of the catalyst. Once the catalyst is saturated with carbon, nanotube formation then begins [33]. The diameter of the resulting CNTs is generally related to the size of the catalyst particles. In order to encourage the production of SWCNTs, the size of the catalyst particles can be reduced to a few nanometers in diameter and the reaction temperature elevated. While the catalyst is often mounted on a substrate made of silicon, glass or alumina, past studies have shown that the catalyst can also be suspended in a gas flow [33].

2.6 PARTICLE GROWTH: THEORETICAL CONSIDERATIONS

This section provides an introduction to theory of particle nucleation and growth, which is of most relevance to bottom–up nanomaterial synthesis. While the equations and theories presented in the following section are derived for spherical nanoparticles, similar considerations can be applied to all types of nanomaterials. However, it must be noted that such theories, which are based on modeling of thermodynamic properties, may not cover all eventualities, and other factors are likely to affect the process of particle growth. For example, if a particular crystal plane has a lower surface energy than another, then certain faces of a nanomaterial might grow preferentially to others that lead to facets.

2.6.1 Nucleation

The formation of nanoparticles from liquid or vaporized precursors is favorable because their formation lowers the Gibbs' free energy of that system (denoted by ΔG_V, which is expressed as per unit volume of a newly formed particle) [9]. However, the formation of a particle also involves the creation of a surface between the newly formed particle and the surrounding medium. This surface will have energy, γ associated with it (expressed as per unit area of the material). These two energies act in opposite directions, with ΔG_V resulting in a *reduction* in energy, while γ resulting in an *increase* in energy. Combining each contribution allows the overall energy change (ΔG) to be calculated:

$$\Delta G = -\frac{4}{3}\pi r^3 \Delta G_V + 4\pi r^2 \gamma \tag{2.1}$$

where the reduction in Gibbs' free energy is multiplied by the volume of the particle and the surface energy by the surface area, when considering a spherical particle. Hence, ΔG here is the sum of the free energy due to the formation of a new volume and the free energy due to the new surface created for spherical particles. Plotting this

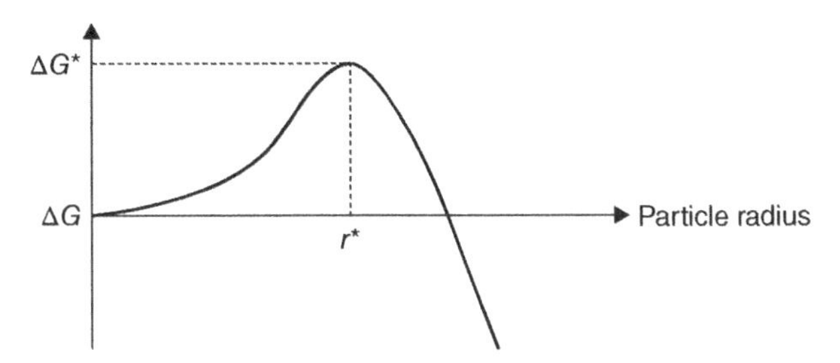

Figure 2.4 Change in free energy ΔG plotted against particle radius. The maximum value of ΔG corresponds to the critical radius of a cluster of atoms, r^*.

ΔG against the radius of the particle, r, indicates the presence of an energy barrier, which must be overcome before a particle can freely grow (Fig. 2.4). This process is called *nucleation* and governs the first step in particle growth. The height of the energy barrier is labeled ΔG^*. Any cluster of atoms that can overcome this barrier and reach the critical radius (r^*) will grow. A key challenge in synthesizing nanoscale particles lies in maximizing the rate of nucleation, while ensuring that the particles grow only to the desired size.

The rate of nucleation can be increased by

a) increasing ΔG_V, which will lower the energy barrier. This can be achieved by increasing the concentration of the precursor solution or vapor. When ΔG_V is zero, the solution becomes *saturated*. If the concentration is increased beyond this point, then this will create a *supersaturated solution*, which subsequently lowers the energy barrier to nucleation. Concentrations below the saturation concentration will never lead to nucleation because there is no energy reduction.
b) increasing the temperature of the system. This will increase the average energy of the atoms and molecules in the solution, resulting in a greater number of particles being able to overcome the energy barrier.

2.6.2 Particle Growth and Growth Kinetics

The growth of particles means that the concentration of precursors in the solution will be lowered to a level around the saturation concentration, where nucleation stops. While nucleation cannot occur under conditions of insufficient supersaturation, existing particles can continue to grow. The phenomena of *diffusion-limited growth* and *Ostwald ripening* can be used to further explain particle growth and growth kinetics, as further described in the following sections.

2.6.2.1 Diffusion-Limited Growth The rate of particle growth can be limited by the rate at which precursors diffuse to reach the growing particle [35]. In this scenario, particle growth is said to be diffusion limited, being essentially controlled by the diffusion of monomers to the surface of the particles. Achieving diffusion-limited growth is often desired as it causes particle size distributions to become more uniform over time. In diffusion-limited growth, larger particles will grow slower and so the sizes of smaller particles can catch up with those of larger ones, which focus the size distribution. In order to promote diffusion-limited growth, several approaches can be taken. These include increasing the viscosity of the reacting solution, lowering the temperature of the reaction, controlling the supply of growth species (via variations in the concentration precursors in solution) [9] and using a capping agent (such as a surfactant) that coats the surface of the nanoparticles, to present a barrier to diffusion.

2.6.2.2 ***Ostwald Ripening*** Ostwald ripening is the process that causes old ice-cream to take on a grainy texture, with ice crystals growing over time and coarsening the texture. In simple terms, Ostwald ripening refers to the growth of larger particles at the expense of smaller ones, thus conserving volume while reducing surface area [24]. The phenomenon is also of relevance to nanoparticle synthesis as it can prevent a desired particle size distribution to be maintained. As previously discussed, surfaces of nanoparticles have an associated energy, γ. Ostwald ripening is often favorable since the formation of a larger nanoparticle reduces the surface area, thus lowering total surface energy. The process is governed by the diffusion of smaller particles to larger ones. Although the reduced total surface energy means that Ostwald ripening is thermodynamically favorable, a chemical potential gradient still needs to exist in order for diffusion to occur. In this sense, the curvature of a surface is an important consideration as it is proportional to the chemical potential of atoms or molecules of that surface. Diffusion is thus driven by the chemical potential gradient that exists between the smaller particle (with high surface curvature) and the larger particle (with low surface curvature). A simple way to prevent Ostwald ripening is to use a capping agent as this will create a barrier to diffusion.

2.7 CASE STUDY: MICROREACTOR FOR THE SYNTHESIS OF GOLD NANOPARTICLES

2.7.1 Introduction

The ability to produce homogeneous (monodisperse) nanoparticles is of importance in many different applications. Several examples will now be given to show this. One example is in the area of drug delivery, in which a relatively homogeneous nanoparticle formulation is desired to achieve the correct pharmacodynamics and dissolution rates. It is also of analytical importance, especially during pharmaceutical development as it minimizes the need to have tools that can reliably measure highly polydisperse and inhomogeneous nanoparticles [36]. Another example is in the field of nanotoxicology. In the past, several researchers have highlighted the need to access test materials that are close to monodisperse for better reliability when studying behavior, dose, structure–activity relationships and the mechanisms of toxicity [37]. A final example is in relation to quantum dots, in which it is important to control the size of the particles synthesized as it is the particle size that will affect the optical band-gap variability and thus affecting their optical and electronic characteristics [38, 39].

Although the control of properties such as particle morphology and size distribution of the whole sample is one of the most important issues in the synthesis of well-defined nanoparticles, the ability to produce them is far from trivial. In the past, scientists have often used the bottom–up approach to produce nanoparticles of low polydispersity, in which the population of particles have more or less uniform size and shape [38]. Such a bottom–up approach should result in the creation of near-identical structures with precision limited only by the Boltzmann energy distribution of the system, but in reality this is not the case. Particles can exhibit some excess width in relation to the particle size distribution. This is attributed to the fact that chemical reactions are often carried out in large vessels and the reactants will thus be subjected to variations in temperature and concentration gradients, leading to disparate microchemical environments. In order to improve particle size homogeneity, other ways to synthesize particles have been considered. It is worthy to note that the term polydispersity (or monodispersity) is used here rather loosely, as requirements of monodispersity have yet to be quantified in different scientific fields.

One approach for minimizing particle size and shape variations that can arise from the traditional bulk processes is to better control the experimental environment. This can be achieved if the size of the reaction vessel is significantly reduced, for example, in the case of a microfluidic device. A microfluidic device is a miniaturized platform that allows flow of liquid to take place in channels with characteristic dimensions close to or smaller than the mean free path of reactant species. Microreactors are microfluidic reaction devices, designed for rapid mixing and reaction or two or more liquid reagent streams. As the flow within the channels is almost always laminar, this will lead to the creation of parallel flowstreams without turbulent mixing. Mixing in microfluidic devices is dominated by diffusion and thus better controlled. In addition, there is greater control over temperature and concentration gradients within the mixing chamber in microfluidic devices. The ability of microreactors to offer improved control over reaction environments can potentially be harnessed to enhance the controllability and reproducibility of nanoparticle synthesis. In fact, several past workers have reported that microfluidic devices have indicated improvement in the homogeneity of polymeric nanoparticles [40–42] . Microreactors have also been used to fine-tune physicochemical properties of nanoparticles including size, surface chemistry, surface charge [41, 43, 44]. Adoption of such microfluidic formats have been shown to significantly reduce the size distribution of nanoparticles, compared to the corresponding traditional batch process [43, 45–47].

Microreactor technology (involving liquids) appeared in the early 1990s [48]. Since that time, a wide variety of nanoparticles have been synthesized including quantum dots [39, 49–56], silver [46, 57–59], gold [58, 60], CNTs [61], platinum [62], silica [45, 63, 64], zinc oxide [65], titania [66], rubrene [67], polymer [68], and magnetic nanoparticles [69–71].

Undoubtedly, the microreactor platform is a promising technology for nanoparticle synthesis [15, 72], but there are several issues that must be overcome for successful uptake of such a technology and microfluidic devices in general [73, 74]. Becker et al. [75] have highlighted that in general, for successful commercial uptake, microfluidic devices must demonstrate significant improvements compared with their macroscale counterparts. In the case of gold synthesis, the macroscale traditional counterpart is the bulk or batch synthesis, in which nanoparticles are produced in a flask. Thus, there is a need to understand to what extent the microreactor technology has the competitive edge (e.g., homogeneity, reproducibility, and tunability) over batch synthesis.

In the case study presented here, gold nanoparticles will be synthesized using various methods encapsulating both microfluidic and batch formats. Gold nanoparticles will be synthesized by the reduction of chloroauric acid ($H[AuCl_4]$) using sodium borohydride ($NaBH_4$), that is, reduction of Au^{3+} ions to neutral gold atoms, precipitating as nanoparticles. The microreactor used will be simply constructed from PTFE-tubing connected using T-junctions. There are several advantages associated with the use of PTFE tubings rather than planar chip devices that are usually made up of glass, silicon or quartz. Firstly, the hydrophobic surface of the PTFE tubing will mean less wetting of such surfaces and thus will help to suppress the deposition of elemental gold within the reactor channels. This will help to prevent the nanoparticles sticking to the surface of the inner tubes during operation. The deposition of solid material within the microchannels is not desired due to potential blockages. Secondly, the tubing can be cut to any length, to govern the flow and mixing of fluid. Thirdly, such tubing is relatively inexpensive (compared to chip-based systems) and thus can be easily replaced.

For synthesis using microreactor platform, two types of flow will be employed: continuous and segmented flow. In the continuous flow approach, the mixing strategy involves introducing two types of reagent streams ($H[AuCl_4]$,$NaBH_4$) in a continuous manner, where they are mixed and allowed to react. The product then leaves the reactor as a continuous stream, with the flow rate and operation time determining the synthesis scale. Unlike continuous flow systems, segmented flow (or droplet-based) systems involve creating discrete volumes of reagents with the use of an immiscible phase. Often, a flow of perfluorinated oil is used to create a flow perpendicular to an aqueous liquid flow; the final result will be in the formation of microdroplets (of liquid suspensions) in oil phase. These droplets are typically created via Rayleigh–Plateau instability, which controls the formation of droplets that are of a characteristically uniform size. Synthesis of particles therefore occurs within the droplets that move at a constant linear velocity. The constant advection within the droplet, triggered by Newtonian flow and the concomitant no-slip layer at the droplet boundary, means that the droplet itself represents a single, uniform chemical microenvironment. Upon collection of the mixture, the two phases can be collected before separating them to extract the aqueous phase, which contains the nanoparticles. In the study, the gold particles produced by the three different methods, that is, batch, microfluidics continuous flow and microfluidics segmented flow, will be evaluated using dynamic light scattering (DLS) for particle size measurements.

2.7.2 Method

2.7.2.1 Materials Reagents of analytical grade or better were purchased from Sigma-Aldrich (UK) and used as received. Deionized (DI) water, with a resistivity value of 18.2 Mohm cm, was filtered (using a 0.2-μm-pore filter) prior to use.

The following stock solutions were prepared, as part of the sample preparation procedure:

a) Gold precursor stock solution of $HAuCl_4 \cdot 3H_2O$ (10 mM) in DI water; this was kept in an amber glass bottle at 4 °C. This stock solution was appropriately diluted with DI water to make 1 and 2 mM concentrations, to be used for batch and microfluidic synthesis, respectively. These solutions, which were made freshly before each experiment, will be referred to as gold "working" solutions in the remainder of the chapter.
b) $NaBH_4$ solution in DI water of 50 mM concentration for batch synthesis and 10 mM for microreactor synthesis; these solutions were made up and used within 2 h.
c) Aqua regia. This was used for cleaning purposes and made by adding 65% nitric acid to 37% hydrochloric acid in a volume ratio of 1:3. All reaction implementations that are used for the preparation of gold nanoparticle formation, for example, magnetic stirrer, beaker and glassware were cleaned with aqua regia and then rinsed with copious amounts of DI water. Aqua regia was made fresh before use. Due to the corrosive nature of aqua regia, appropriate safety precautions were taken.

As detailed below, the stock solutions prepared by methods a) and b) were used in the respective synthesis methods. In all three methods, the $HAuCl_4$ and $NaBH_4$ reagents were made up to ensure that their concentrations in the resulting reaction solution were 1 and 5 mmol/l, respectively.

2.7.2.2 Protocol: Nanoparticles Batch Synthesis Gold nanoparticles were prepared using the commonly employed $NaBH_4$ reduction method in which the reduction reaction takes place at room temperature. This involved adding 9 ml of the corresponding gold working solution (for batch synthesis) to a clean glass vial and stirring vigorously using a magnetic stir bar before 1 ml of $NaBH_4$ solution (50 mM) was added to the vial. The mixture was then stirred for a further 30 min before the magnetic stirrer bar was removed. The remaining colloidal suspension of gold nanoparticles was then stored at 4 °C until ready for analysis.

2.7.2.3 Protocol: Nanoparticle Synthesis via Continuous Flow Microfluidics
PTFE tubing (I.D of 250 μm) was used to connect a PEEK tee mixer (500 μm through bore, Upchurch Scientific, UK) to two 5 ml reagent loops (Asia Reagent Injector, Syrris, UK), filled with 2 mmol/l $HAuCl_4$ and 10 mmol/l $NaBH_4$ solutions, respectively, each being driven by syringe pump units (Asia Pump, Syrris, UK). The two precursor solutions were pumped through a tee mixer at 50 μl/min, equating to the total flow rate of 100 μl/min. The tee mixer was connected to PTFE tubing (500 μm I.D); this provided an outlet for collection of the mixture into a vial containing 5 ml of DI water and a magnetic stirrer bar (to provide constant stirring). The mixture was left to stir in the mixing chamber for 30 min before the magnetic stirrer bar was removed and the solution stored at 4 °C, until required for analysis. A schematic of the set-up is shown in Figure 2.5a.

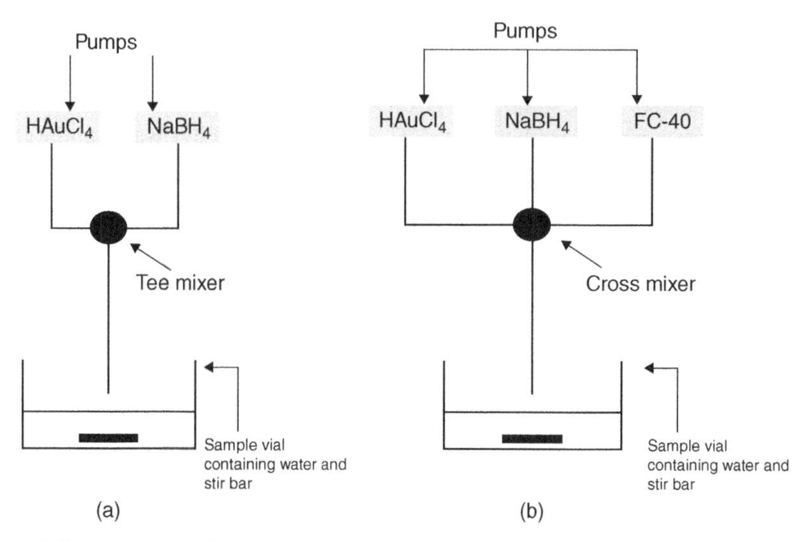

Figure 2.5 Schematic microreactor set-up under different synthesis strategies: (a) continuous flow, (b) microdroplet.

2.7.2.4 Protocol: Nanoparticles Synthesis via Droplet-Based Microfluidics For droplet-based experiments, a slightly modified version of the continuous flow set up was used. In this system a PEEK cross mixer (500 μm through bore diameter, Upchurch Scientific, UK) was used to combine the reactants, with the extra channel being used to deliver an immiscible carrier fluid (Fluorinert FC 40, 3M, USA) to enable reactants to be isolated into discrete droplets. The carrier fluid was delivered using a separate syringe pump (KDS Scientific, USA), fitted with a 20 ml disposable syringe. As before, the reactant solutions were pumped at 50 μl/min each, equating to a combined flow rate of 100 μl/min, while the carrier fluid was pumped at a rate of 300 μl/min. The output from the cross mixer was connected via PTFE tubing (500 μm I.D) into a sample collection vial containing 5 ml of DI water (under constant stirring conditions). As before, the mixture was left to stir for 30 min before the mixture was allowed to settle at room temperature. This resulted in a complete separation of two phases. The aqueous phase that contained the gold nanoparticles was collected and stored at 4 °C, until required for analysis. A schematic of the set-up is shown in Figure 2.5b.

2.7.2.5 Protocol: Dynamic Light Scattering Particle size characterization was carried out using a Zetasizer Nano ZS (Malvern Instruments, UK) equipped with a 633-nm red laser. Malvern Instruments Dispersion Technology software (Version 4.0) was used to analyze all data. A brief overview of the technique will now be given.

DLS, sometimes referred to as photon correlation spectroscopy (PCS), is used measure time-dependent correlation function of the scattered light arising from a collection of the particles undergoing random Brownian motion. Cumulant analysis is then used for analyzing the autocorrelation function generated by the DLS experiments, in which the diffusion coefficient of the particles can be deduced and subsequently used to give the particle diameter. The calculations related to cumulant analysis, which describes the suspended size distributions, have been defined in ISO [76]. In practice, the cumulant analysis yields only two values that are of importance: a mean value for the size (Z-average) and a width (or breadth) parameter known as the polydispersity index. The Z-average is intensity-based calculated value, as defined by the ISO standards. The PDI index is a number calculated from a simple two parameter fit to the correlation data. The PDI is dimensionless and scaled such that values range from 0 to 1, with values smaller than 0.05 being rarely seen other than with highly monodisperse reference standards. Values greater than 0.7 indicate that the sample has a very broad particle size distribution and is probably not suitable for the DLS technique [76, 77].

To acquire the DLS data, 100 μl of sample was pipetted into a disposable microcuvette (ZEN0118, Malvern Instruments, USA). The cuvette was then capped to prevent any further contamination. The analysis was carried out at 25 °C. The data collected were analysed to extract the mean and standard deviation of replicate measurements. Results were plotted as mean values, with error bars of one standard deviation.

2.7.3 Results, Interpretation, and Conclusion

Figure 2.6 shows the dynamic light scattering Z-average diameter for particles synthesized using microreactor (under both continuous and segmented flow) versus batch. The plot shows the average value of six different samples taken, with three replicate measurements conducted per sample. Results show that all particles synthesized appeared to have similar (Z-average) sizes, ranging from 7–12 nm, with microdroplet process yielding the largest average size (all being ~12 nm). Figure 2.6 also shows a representative particle size distribution acquired for the three synthesis modes. Results indicated that aggregates of particles >100 nm are present in batch and continuous flow but not when particles are synthesized using the droplet-based method.

Figure 2.7 shows the effect when adopting the three methods of gold nanoparticle synthesis on the PDI value. Results show that this value ranges between 0.14 and 0.32, with the droplet-based method giving the smallest PDI values (0.18 or less). However, inspection of the particle size distributions in Figure 2.6 indicates that the higher PDI values associated with batch and continuous flow can be explained by the existence of a bimodal size distribution and the presence of large aggregates size >100 nm.

The repeatability of dispersion quality was assessed by calculating RSD % from the corresponding particle size and PDI datapoints (of the six subsamples) associated with Figures 2.6 and 2.7. Figure 2.8 shows a bar graph of the RSD% associated with the Z-average and PDI data. Results clearly indicate that the droplet-based method

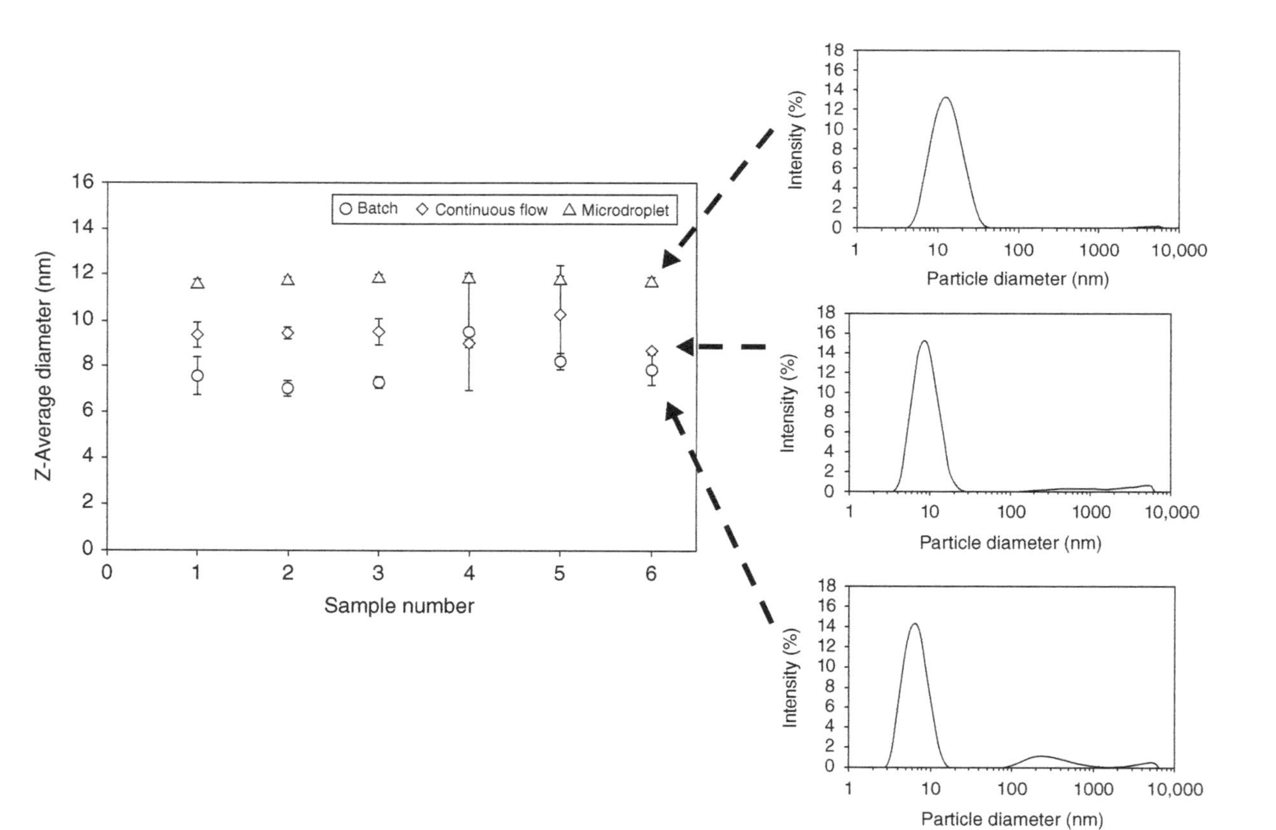

Figure 2.6 Typical DLS particle size distributions and the corresponding *Z*-average particle diameter of gold nanoparticles of six samples, as prepared by: traditional batch, microfluidic continuous flow, and microfluidic microdroplet flow. Each data point in the plot is the average of triplicate measurements; error bars represent the SD of those measurements.

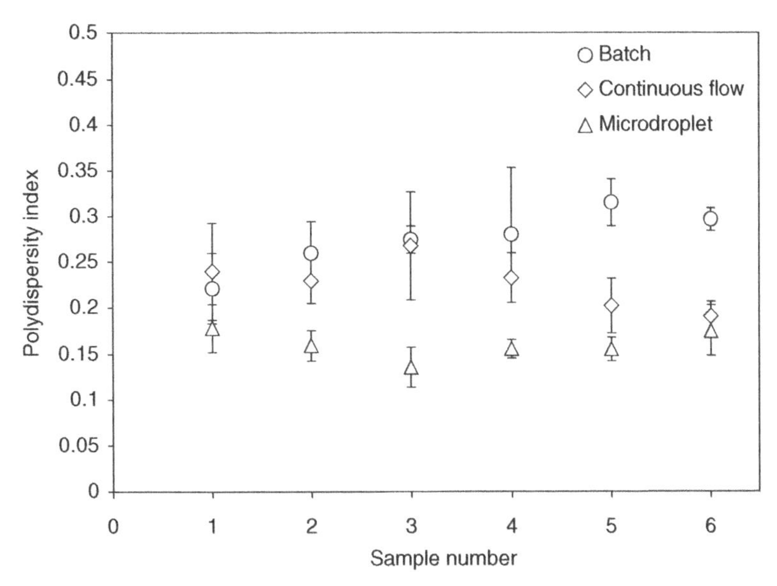

Figure 2.7 Polydispersity index from DLS measurements of six samples. Results compare the different types of nanoparticle synthesis methods of traditional batch, microfluidic continuous flow, and microfluidic microdroplet flow synthesis. Each data point in the plot is the average of triplicate measurements; error bars represent the SD of those measurements.

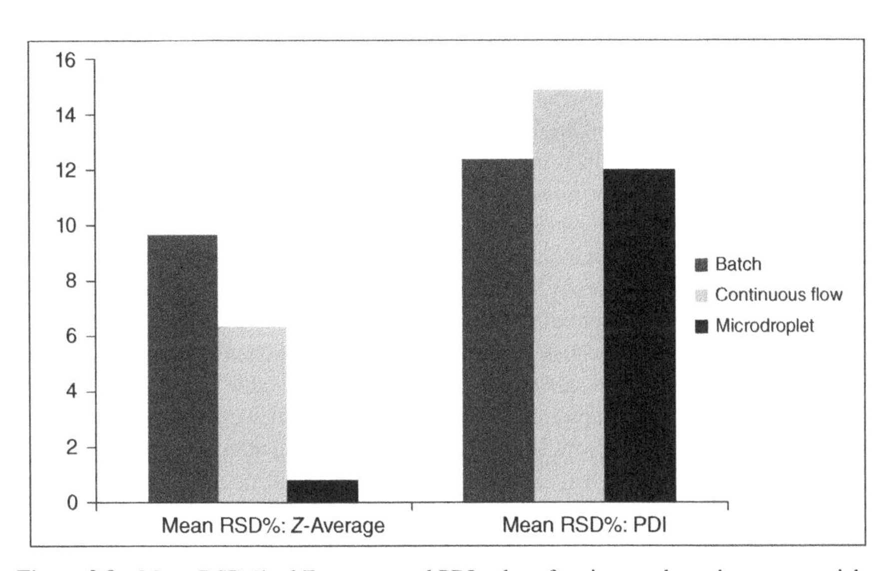

Figure 2.8 Mean RSD % of Z-average and PDI values for six samples, when nanoparticles are synthesized with (a) batch, (b) microreactor continuous flow, and (c) microreactor microdroplet.

gave the smallest RSD% values (hence less variation) for all values reported, compared to batch or continuous flow. The much reduced degree of variability was more evident for the Z-average values (rather than PDI); for the microdroplet this was 0.8%, whereas batch was 9.7%.

Overall, the results indicate that improvements in the quality of gold nanoparticles can be achieved using the droplet-based microfluidic approach compared to batch. Improvements in the quality of dispersion were attributed to much lower PDI values and improved repeatability in particular for the reported Z-average values. It is difficult to know for certain as to the potential causes for such improvements, as observed in the microdroplet method but not in either the batch or microfluidics continuous method. It is likely that the improvement is due to that fact that the microdroplet method was able to completely isolate the chemical reactions in compartments through the generation and manipulation of droplets inside the microchannels. The ability to "compartmentalise" allows the creation of the particles inside the droplets. Reactions within the droplets will mean that interactions of chemical precursors and synthesized particles with the surfaces of reaction vessels will be minimized. The ability for the microdroplet method to do this has meant that the reduction of gold ions to elemental gold will never take place at the walls of the reacting vessels, in which the surface walls itself can serve as nucleation starter.

2.8 SUMMARY

In this chapter, we introduced a general but nonexhaustive overview of nanomaterial synthesis to show how there are multiple ways of synthesizing nanomaterials. Undoubtedly, the final choice of what method to use will be governed by several factors such as type of nanomaterial to be produced, taking into account the specific requirements for the intended applications such as particle type, shape, size distributions, and surface functionalization. In addition to structural and functional aspects, the choice method can also be governed by factors such as economies of scale and the cost of production; this is the particular case when scaling up the process to commercial viability. Although a much desired effect in nanomaterial synthesis is to achieve homogeneity and monodispersity, this is not easy and potentially costly. Thus, it is not surprising that for the commercial synthesis of CNTs, methods are chosen to take into account that they are scalable to produce large commercial quantities, while ensuring that processes chosen can be carried out at the lowest achievable cost.

The second part of the chapter presented a case study, which compares the particle size properties of gold nanoparticles when synthesized using three different methods: batch, droplet microreactor and continuous flow microreactor. Our findings illustrate the potential of the microdroplet microreactor to produce gold nanoparticles with lower PDIs and improved repeatability in relation to Z-average values, compared to the other two methods. Such improvements in particle dispersion quality were attributed to the influence of confinement and compartmentalization of chemical reactions, which has the effect of minimizing the interaction of precursors and nanoparticles with the surfaces of the reacting vessels.

ACKNOWLEDGMENTS

The work leading to these results has received support from National Measurement Office, from the P13 project. The authors gratefully acknowledge Professor Andrew de Mello and Dr Robert Wootton from ETH Zurich for reviewing the manuscript and providing useful comments.

REFERENCES

1. Zhong W-H. *Nanoscience and Nanomaterials: Synthesis, Manufacturing and Industry Impacts*. DEStech Publications, Inc.; 2012.
2. Tien D, Liao CY, Huang JC, Tseng KH, Lung JK, Tsung TT, Kao WS, Tsai TH, Cheng TW, Yu BS. Novel technique for preparing a nano-silver water suspension by the arc-discharge method. Rev Adv Mater Sci 2008;18:750–756.
3. Fernández-Pacheco R, Arruebo M, Marquina C, Ibarra R, Arbiol J, Santamaría J. Highly magnetic silica-coated iron nanoparticles prepared by the arc-discharge method. Nanotechnology 2006;17(5):1188.
4. Feldman M. *Nanolithography: The Art of Fabricating Nanoelectronic and Nanophotonic Devices and Systems*. Woodhead Publishing Ltd; 2014.
5. Kundrapu M. A model of carbon nanotube synthesis in arc discharge plasmas. J Phys D Appl Phys 2012;45(31):315305.
6. Gogotsi Y. *Nanomaterials Handbook*. Taylor & Francis; 2006.
7. Bourdon J. *Growth and Properties of Metal Clusters. Applications to Catalysis and the Photographic Process*. New York: Elsevier; 1980. p xviii+549.
8. Lackner M. *Combustion Synthesis: Novel Routes to Novel Materials*. Bentham Science Publishers; 2010.
9. Cao G. *Nanostructures and Nanomaterials Synthesis, Properties and Applications*. World Scientific; 2004.
10. Tripathi N, Mishra P, Joshi B, Islam SS. Precise control over physical characteristics of carbon nanotubes by differential variation of Argon flow rate during chemical vapor deposition processing: A systematic study on growth kinetics. Mater Sci Semicond Process 2015;35:207–215.
11. Turkevich J, Stevenson PC, Hillier J. A study of the nucleation and growth processes in the synthesis of colloidal gold. Discuss Faraday Soc 1951;11(0):55–75.
12. Schmid G. *Clusters and Colloids: From Theory to Applications*. John Wiley & Sons; 2008.
13. Wu S-H, Chen D-H. Synthesis of high-concentration Cu nanoparticles in aqueous CTAB solutions. J Colloid Interface Sci 2004;273(1):165–169.
14. Starov V, Procházka K. Trends in colloid and interface science XXIV. Surf Colloid Sci 2011;138:39–43.
15. Zhao C-X, He L, Qiao SZ, Middelberg APJ. Nanoparticle synthesis in microreactors. Chem Eng Sci 2011;66(7):1463–1479.
16. He Y, Yuan J, Su F, Xing X, Shi G. Bacillus subtilis assisted assembly of gold nanoparticles into long conductive nodous ribbons. J Phys Chem B 2006;110(36):17813–17818.
17. Matsunaga T, Takeyama H. Biomagnetic nanoparticle formation and application. Supramol Sci 1998;5(3–4):391–394.

18. Qian X. Magnetite in the head of the homing pigeon. Wuli 2014;43(5):330–6.
19. Kirschvink, J.L., Biogenic magnetite (Fe $3O_4$): a ferrimagnetic mineral in bacteria, animals, and man. Ferrites. Proceedings of the ICF 3. Third International Conference on Ferrites, H. Watanabe, S. Iida, and M. Sugimoto, editors. 1982. p 135–138.
20. Rao CNR, Thomas PJ, Kulkarni G. *Nanocrystals: Synthesis, Properties and Applications*. Vol. 95. Springer; 2007.
21. Kumar CS. *Mixed Metal Nanomaterials*. Nanomaterials for the Life Sciences. Wiley VCH; 2009.
22. Corriu RNTA. *Molecular Chemistry of Sol–gel Derived Nanomaterials*. Wiley; 2009.
23. Sakka S. *Handbook of sol–gel science and technology. 1. Sol–gel processing*. Vol. 1. Springer; 2005.
24. Li YM, Somorjai GA. Nanoscale advances in catalysis and energy applications. Nano Lett 2010;10(7):2289–2295.
25. Sapkota B, Mishra S. A simple ball milling method for the preparation of p-CuO/n-ZnO nanocomposite photocatalysts with high photocatalytic activity. J Nanosci Nanotechnol 2013;13(10):6588–6596.
26. Janot R, Guérard D. One-step synthesis of maghemite nanometric powders by ball-milling. J Alloys Compd 2002;333(1):302–307.
27. Lu K. *Nanoparticulate Materials: Synthesis, Characterization, and Processing*. John Wiley & Sons; 2012.
28. Hartanto A, Ning X, Nakata Y, Okada T. Growth mechanism of ZnO nanorods from nanoparticles formed in a laser ablation plume. Appl Phys A 2004;78(3):299–301.
29. Tsuji T. Preparation of silver nanoparticles by laser ablation in polyvinylpyrrolidone solutions. Appl Surf Sci 2008;254(16):5224–5230.
30. Jeon J, Floresca HC, Kim M. Fabrication of complex three-dimensional nanostructures using focused ion beam and nanomanipulation. J Vac Sci Technol B 2010;28(3):549–553.
31. Tantra R, Cumpson P. The detection of airborne carbon nanotubes in relation to toxicology and workplace safety. Nanotoxicology 2007;1(4):251–265.
32. Su Z, Zhou W, Zhang Y. New insight into the soot nanoparticles in a candle flame. Chem Commun 2011;47(16):4700–4702.
33. Khare R, Bose S. Carbon nanotube based composites: A review. J Miner Mater Charact Eng 2005;4:31.
34. Echlin P. *Handbook of Sample Preparation for Scanning Electron Microscopy and X-ray Microanalysis*. Springer; 2009.
35. Rao CNR, Muller A, Cheetham AK. *Nanomaterials Chemistry: Recent Developments and New Directions*. Wiley-VCH; 2007.
36. Zhang J. Assessing the heterogeneity level in lipid nanoparticles for siRNA delivery: Size-based separation, compositional heterogeneity, and impact on bioperformance. Mol Pharm 2013;10(1):397–405.
37. Baalousha M, Lead JR. Nanoparticle dispersity in toxicology. Nat Nanotechnol 2013;8(5):308–309.
38. Zhitomirsky D, Kramer IJ, Labelle AJ, Fischer A, Debnath R, Pan J, Bakr OM, Sargent EH. Colloidal quantum dot photovoltaics: The effect of polydispersity. Nano Lett 2012;12(2):1007–1012.
39. Yang H. Synthesis of quantum dots via microreaction: structure optimization for microreactor system. J Nanopart Res 2011;13(8):3335–3344.

40. Johnson BK, Prud'homme RK. Mechanism for rapid self-assembly of block copolymer nanoparticles. Phys Rev Lett 2003;91(11):118302/1–4.
41. Karnik R, Gu F, Basto P, Cannizzaro C, Dean L, Kyei-Manu W, Langer R, Farokhzad OC. Microfluidic platform for controlled synthesis of polymeric nanoparticles. Nano Lett 2008;8(9):2906–2912.
42. Rhee M, Valencia PM, Rodriguez MI, Langer R, Farokhzad OC, Karnik R. Synthesis of size-tunable polymeric nanoparticles enabled by 3D hydrodynamic flow focusing in single-layer microchannels. Adv Mater 2011;23(12):H79–H83.
43. Valencia PM, Basto PA, Zhang L, Rhee M, Langer R, Farokhzad OC, Karnik R. Single-step assembly of homogenous lipid – polymeric and lipid – quantum dot nanoparticles enabled by microfluidic rapid mixing. ACS Nano 2010;4(3):1671–1679.
44. Lim J-M, Bertrand N, Valencia PM, Rhee M, Langer R, Jon S, Farokhzad OC, Karnik R. Parallel microfluidic synthesis of size-tunable polymeric nanoparticles using 3D flow focusing towards in vivo study. Nanomedicine 2014;10(2):401–409.
45. Gutierrez L, Navascues LG, Irusta S, Arruebo M, Santamaria J. Comparative study of the synthesis of silica nanoparticles in micromixer-microreactor and batch reactor systems. Chem Eng J 2011;171(2):674–683.
46. Patil GA, Bari ML, Bhanvase BA, Ganvir V, Mishra S, Sonawane SH. Continuous synthesis of functional silver nanoparticles using microreactor: Effect of surfactant and process parameters. Chem Eng Process 2012;62:69–77.
47. Jahn A, Reiner JE, Vreeland WN, DeVoe DL, Locascio LE, Gaitan M. Preparation of nanoparticles by continuous-flow microfluidics. J Nanopart Res 2008;10(6):925–934.
48. Wiles C, Watts P. Recent advances in micro reaction technology. Chem Commun 2011;47(23):6512–6535.
49. Byoung-Hwa K. Synthesis of ZnSe quantum dots using a continuous-flow microreactor and their white emission through energy transfer. ECS Solid State Lett 2013;2(8):R27–R30.
50. Gomez-de Pedro S, Martínez-Cisneros CS, Puyol M, Alonso-Chamarro J. Microreactor with integrated temperature control for the synthesis of CdSe nanocrystals. Lab Chip 2012;12(11):1979–1986.
51. Gómez-de PS, Puyol M, Izquierdo D, Salinas I, de la Fuente JM, Alonso-Chamarro J. A ceramic microreactor for the synthesis of water soluble CdS and CdS/ZnS nanocrystals with on-line optical characterization. Nanoscale 2012;4(4):1328–1335.
52. Baek J, Allen PM, Bawendi MG, Jensen KF. Investigation of indium phosphide nanocrystal synthesis using a high-temperature and high-pressure continuous flow microreactor. Angew Chem Int Ed 2011;50(3):627–630.
53. Peterson, D.A., P. Chandran, and B.K. Paul, A reverse oscillatory flow microreactor system for the synthesis of uniformly-size CdS nanoparticles. IEEE 11th International Conference on Nanotechnology; 2011. 666–770.
54. Kikkeri R, Laurino P, Odedra A, Seeberger PH. Synthesis of carbohydrate-functionalized quantum dots in microreactors. Angew Chem Int Ed 2010;49(11):2054–2057.
55. Watanabe K, Orimoto Y, Nagano K, Yamashita K, Uehara M, Nakamura H, Furuya T, Maeda H. Microreactor combinatorial system for nanoparticle synthesis with multiple parameters. Chem Eng Sci 2012;75:292–297.
56. Watanabe, K., Orimoto, Y., Yamashita, K., Uehara, M., Nakamura, H., Furuya, T., Maeda H. Development of automatic combinatorial system for synthesis of nanoparticles

using microreactors. 3rd International Congress on Ceramics (Icc3): Hybrid and Nano-Structured Materials, 2011. 18.

57. Ravi Kumar DV, Prasad BLV, Kulkarni AA. Segmented flow synthesis of Ag nanoparticles in spiral microreactor: Role of continuous and dispersed phase. Chem Eng J 2012;192:357–368.
58. Singh A, Shirolkar M, Lalla NP, Malek CK, Kulkarn SK. Room temperature, water-based, microreactor synthesis of gold and silver nanoparticles. Int J Nanotechnol 2009;6(5–6):541–551.
59. Xue Zhang L, Terepka AD, Hong Y. Synthesis of silver nanoparticles in a continuous flow tubular microreactor. Nano Lett 2004;4(11):2227–2232.
60. Wagner J, Kohler JM. Continuous synthesis of gold nanoparticles in a microreactor. Nano Lett 2005;5(4):685–691.
61. Wei-Chih L, Yao-Joe Y, Hsieh G-W, Ching-Hsiang T, Chien-Chen C, Chao-Chiun L. Selective local synthesis of nanowires on a microreactor chip. Sens Actuators A 2006;130–131:625–632.
62. Luty-Blocho M, Wojnicki M, Pacławski K, Fitzner K. The synthesis of platinum nanoparticles and their deposition on the active carbon fibers in one microreactor cycle. Chem Eng J 2013;226:46–51.
63. Chung, C.K., T.R. Shih, and B.H. Wu, Design of a novel microreactor for microfluidic synthesis of silica nanoparticles. 15th International Conference on Solid-State Sensors, Actuators and Microsystems. Transducers 2009. 1341–1344.
64. Chung CK, Shih TR, Chang CK, Lai CW, Wu BH. Design and experiments of a short-mixing-length baffled microreactor and its application to microfluidic synthesis of nanoparticles. Chem Eng J 2011;168(2):790–798.
65. Sue K, Kimura K, Arai K. Hydrothermal synthesis of ZnO nanocrystals using microreactor. Mater Lett 2004;58(25):3229–3231.
66. Park KY, Ullmann M, Suh YJ, Friedlander SK. Nanoparticle microreactor: Application to synthesis of titania by thermal decomposition of titanium tetraisopropoxide. J Nanopart Res 2001;3(4):309–319.
67. Génot V, Desportes S, Croushore C, Lefèvre J-P, Pansu RB, Delaire JA, von Rohr PR. Synthesis of organic nanoparticles in a 3D flow focusing microreactor. Chem Eng J 2010;161(1–2):234–239.
68. Kumar Yadav A, Barandiaran MJ, de la Cal JC. Synthesis of water-borne polymer nanoparticles in a continuous microreactor. Chem Eng J 2012;198–199:191–200.
69. Sue K, Hattori H, Sato T, Komoriya T, Kawai-Nakamura A, Tanaka S. Super-rapid hydrothermal synthesis of highly crystalline and water-soluble magnetite nanoparticles using a microreactor. Chem Lett 2009;38(8):792–793.
70. Abou-Hassan A, Neveu S, Dupuis V, Cabuil V. Synthesis of cobalt ferrite nanoparticles in continuous-flow microreactors. RSC Adv 2012;2(30):11263–11266.
71. Abou-Hassan A, Dufrêche J-F, Sandre O, Mériguet G, Bernard O, Cabuil V. Fluorescence confocal laser scanning microscopy for pH mapping in a coaxial flow microreactor: Application in the synthesis of superparamagnetic nanoparticles. J Phys Chem C 2009;113(42):18097–18105.
72. Singh A, Khan Malek C, Kulkarni SK. Development in microreactor technology for nanoparticle synthesis. Int J Nanosci 2010;9(1–2):93–112.

73. Tantra R, Robinson K, Sikora A. Variability of microchip capillary electrophoresis with conductivity detection. Electrophoresis 2014;35(2–3):263–270.
74. Tantra R, van Heeren H. Product qualification: A barrier to point of care microfluidics based diagnostics? Lab Chip 2013;13:2199–2201.
75. Becker H. Hype, hope and hubris: the quest for the killer application in microfluidics. Lab Chip 2009;9(15):2119–2122.
76. BS 3406–8:1997, ISO 13321:1996. *Methods for Determination of Particle Size Distribution. Photon Correlation Spectroscopy*. British Standard Institute and International Organization for Standardization; 1997.
77. BS ISO 22412:2008. *Particle Size Analysis. Dynamic Light Scattering (DLS)*. British Standard Institute; 2008.

3

REFERENCE NANOMATERIALS

G. ORTS-GIL

European Office-Spanish Foundation for Science and Technology, Spanish Embassy, Berlin 10787, Germany

W. ÖSTERLE

Department of Materials Engineering, BAM Federal Institute for Materials Research and Testing, Berlin 12200, Germany

3.1 DEFINITION, DEVELOPMENT, AND APPLICATION FIELDS

It is without a doubt that reference materials (RMs) are invaluable tools for an analyst. For example, reference materials have always been important for validating and controlling analytical methods. According to the ISO Guide 30 [1], a reference material (RM) is a "material, sufficiently homogeneous and stable with respect to one or more specified properties, which has been established to be fit for its intended use in a measurement process." In a higher level of standardization, ISO defines certified reference material (CRM) [2]: "material which is accompanied by a certificate, one or more of whose property values are certified by a procedure which establishes traceability to an accurate realization of the unit in which the property values are expressed, and for which each certified value is accompanied by an uncertainty at a stated level of confidence." Thus, the different concepts of an RM are related to different levels of traceability [3]. In the case of nanomaterial, such reference materials are also important, which shall be referred to as reference nanomaterial (RNM) for the remainder of the chapter. The term nanomaterial in this chapter is defined here in accordance to the 2011 European Commission definition, that is, "nanomaterial means a natural, incidental or manufactured material containing particles, in an

Nanomaterial Characterization: An Introduction, First Edition. Edited by Ratna Tantra.

unbound state or as an aggregate or as an agglomerate and where, for 50% or more of the particles in the number size distribution, one or more external dimensions is in the size range 1 nm–100 nm [4]."

The need of RNM has been evidenced mainly in two parallel areas: material science and nanotoxicology. In relation to material science, for example, there is an increasing demand for using nanomaterial as fillers in nanocomposites. As a result, techniques that are applicable to the measurement of single nano-objects have been developed. These methods have to be calibrated with the aid of nano-objects with well-defined properties, and hence the availability of RNMs in this sense becomes important. Furthermore, RNM is needed to determine the impact of nanoconstituents of a microstructure on the properties of a material by appropriate modeling. In the context of nanotoxicology research and hence risk assessment of nanomaterial, there is a need for better understanding of data. Traditionally, there is an assumption that the use of nanomaterial has been accepted as safe, unless scientific evidence proved their harm. However, in the past few years, this conception has been gradually changing and the "precautionary principle" (*the absence of knowledge about the dangers is taken as not safe*) may play a major role in how nanomaterials are regulated [5]. According to Tantra et al., a gap in nanotoxicology research is the lack of suitable RNM to assure reliability in the testing of possible risks for health and environment [6, 7]. Ideally, RNM should be representative of those nanomaterials present in emerging commercial products. However, challenges are associated with the ability to develop such RNMs.

In general, the development of RM has been traditionally performed through official governmental institutions such as the National Institute of Standards and Technology (NIST) in the United States and, more recently, the Joint Research Center (JRC) in Europe. The interlaboratory comparison is the most common way to develop RM and CRM [8]. Alternatively, some institutions such as the Federal Institute for Materials Research and Testing in Germany (BAM), the NIST, and companies such as Sigma-Aldrich [9] have developed CRMs, according to the ISO 17025 rules. A list of available reference materials is constantly growing and can be found in the repositories of JRC [10], NIST [11], and BAM [12]. Table 3.1 shows a list of worldwide available RNMs; this list has been compiled and published recently [7].

In this chapter, an overview is presented to give an insight into relevant aspects that need to be considered when developing RNM. In particular, silica NM will be considered in two separate case studies, that is, potentially for use in materials science and nanotoxicology research.

3.2 CASE STUDIES

3.2.1 Silica Nanomaterial as Potential Reference Material to Establish Possible Size Effects on Mechanical Properties

3.2.1.1 Introduction In materials science, information about the mechanical properties of nanoconstituents of a material is needed for modeling nanostructures

TABLE 3.1 List of Existent and in Progress Reference nanomaterials

Material[a]	Number	Reference or certified value[b]	Name	RM/CRM/SRM[c]		Status[d]
Silica	1	20 nm size	ERM-FD100	CRM	IRMM	For sale
	2	40 nm size	ERM-FD304	CRM	IRMM	For sale
Gold	3	10 nm size	RM 8011	RM	NIST	For sale
	4	30 nm size	RM 8012	RM	NIST	For sale
	5	60 nm size	RM 8013	RM	NIST	For sale
Polystyrene	6	60 nm size	SRM 1964	SRM	NIST	For sale
	7	100 nm size	SRM 1963[a]	SRM	NIST	For sale
TiO_2	8	55.55 m^2/g specific surface area	SRM 1898	SRM	NIST	For sale
SWCNTs	9	Elemental composition	SRM 2483	SRM	NIST	For sale
SWCNTs "bucky paper"	10	Elemental composition	RM 8282	RM	NIST	Production in progress
SWCNT	11	200, 400, and 800 nm length	RM 8281	RM	NIST	Fall 2012
Silver	12	10 nm size	RM8016	RM	NIST	In production
	13	35 nm, d_{90}, volume-weighted	BAM-N001	CRM	BAM	For sale
	14	75 nm size	RM8017	RM	NIST	In production

Source: Orts et al. [7], http://pubs.rsc.org/en/content/articlelanding/2013/ra/c3ra42112k#!divAbstract. Used under CC-BY 3.0 http://creativecommons.org/licenses/by/3.0/.

[a]Only nanomaterials with nominal sizes below 100 nm are considered.

[b]Values shown here are approximate since nominal size depends on experimental technique.

[c]RM = reference material; CRM = certified reference material; SRM = standard reference material certified by NIST.

[d]Until July 2012. This is a selection and not a complete list of ENPs.

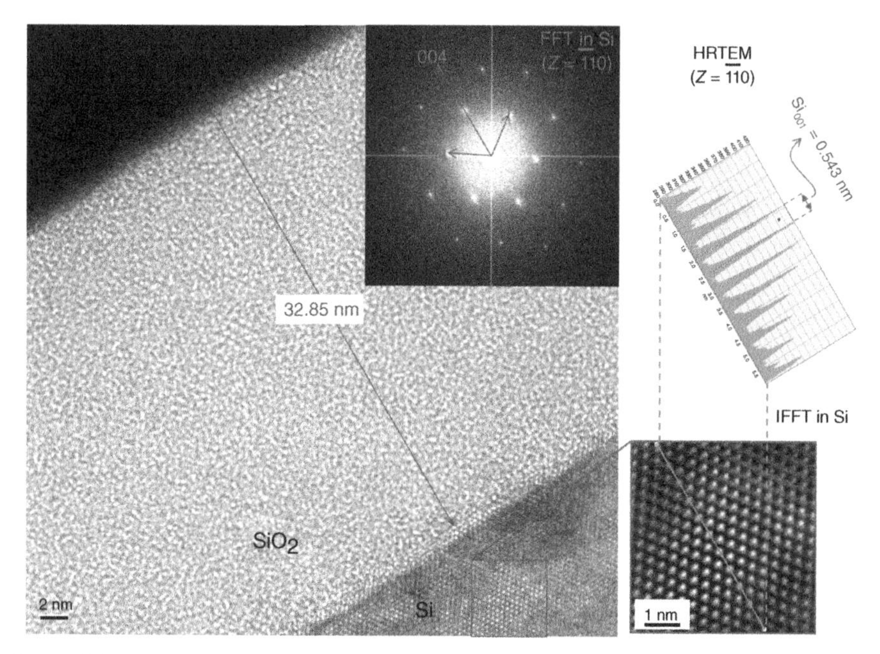

Figure 3.1 Determination of film thickness for SiO_2 on Si-wafer referring to (100) lattice spacing of Si (Unpublished work, see acknowledgements).

and nanocomposites. Therefore, nanoscale measuring methods have to be developed, and subsequently, RNMs are needed for the calibration of these methods. Of particular interest to researchers is understanding the relationship between particle size and its effects on mechanical properties. In relation to the measurement of particle size and the thickness of thin films, traceable length measurements can be performed by using lattice spacing of well-known crystal structures as internal standards in high-resolution transmission electron microscopic images, as shown in Figure 3.1. Although there are innumerable papers describing the size (as well as shape and surface functionalities) of silica nanoparticles, only very few information on the mechanical properties of such nanosized objects is available. In the past, Yan et al. have shown that, in principle, it is possible to determine the elastic modulus of soft and hard nanoparticles embedded in a polymeric matrix by nanoindentation in combination with finite element modeling [2]. Basu et al. have shown that not only elastic properties but also stress–strain curves can be derived from nanoindentation tests [13]. Gallego-Gómez et al. investigated the layers of colloidal submicron silica particles by nanoindentation and observed an impact of water bridges on mechanical properties [14]. Finally, Sun et al. calculated the contact forces between silica nanoparticles by applying molecular dynamics [15].

Despite such reported findings, one important issue remains unanswered: what is the effect of particle size on the mechanical properties of NMs? In order to scrutinize this, appropriate RNMs are needed. In the past, size effects have been observed with nanopillars prepared from crystalline materials [16–19], but these may be mainly attributed to differences of grain orientation in individual pillars. Li et al. also reported size effect for equally oriented silicon pillars, but they could show that bulk values were obtained after taking account of compressibility, which was different for the two considered pillar sizes [20].

In relation to amorphous silica nanomaterial, this may be the ideal candidate for the assessment of size effects of mechanical properties. At least orientation effects can be excluded while considering amorphous silica. This material also offers the opportunity of preparing nano-objects of different shape, because it is available not only as spherical particles but also as thin films and even free-standing membranes within a wide size range. In addition, silica nanoparticles and thin films are interesting for many technical applications, such as drug delivery [21, 22], wastewater treatment [23], polymer matrix composites [24, 25], and dielectric layers in the semiconductor industry [26], respectively. Hence, in this respect, silica NM is worthy of note for further consideration.

The evaluation of silica as potential future RNM is currently being evaluated in an on-going research project. Mechanical properties of nano-objects (MechProNO) is a 3-year collaborative project between European National Metrology Institutes (NMIs) to develop metrological traceability for the measurement of the mechanical properties (adhesion, stiffness, and elasticity) of nano-objects using AFM and other methods. The work plan of MechProNO is depicted in Figure 3.2 [27] and the following sections describe how the main objectives of MechProNO can be reached if amorphous silica is to be considered as the ideal RNM candidate.

Material Thermally grown amorphous silica on silicon substrate is available within the MechProNO project as thick layer (2 μm) as well as thin film (nominal 40 nm). The latter specimen, a TEM-grid of type Pelco® available from Plano GmbH [28], is a film supported by a silicon substrate, which was partly etched away thus providing nine windows $100 \times 100\,\mu m^2$ with free-standing silica membranes. Instrumented indentation testing (IIT) can be used for the measurement of both thick and thin films (supported by silicon substrate). The procedure of the IIT test is based on an ISO-standard that is currently under revision (ISO 14577 part 4), which means that reliable values of Young's modulus can be obtained from thin films supported by a substrate such as silicon wafer, in this case. Furthermore, it is possible to cut beams from the free-standing membranes bridging the windows. These beams make it possible for AFM to perform bending experiments; such beams are real nano-objects because at least one of the dimensions is smaller than 100 nm. In addition, micron- and submicron-sized pillars (fabricated from thick film by focused ion beam (FIB) machining) will be investigated.

3.2.1.2 Findings So Far The ultimate challenge of the MechProNO activity is to show traceability and perform reliable measurement of mechanical properties of spherical silica nanoparticles. It has been shown that such particles can be easily synthesized and homogeneously dispersed in liquid epoxy before curing, thus being able to produce epoxy-based nanocomposites. This procedure was extensively applied by Zhang et al., thus improving the tribological properties of conventional composites

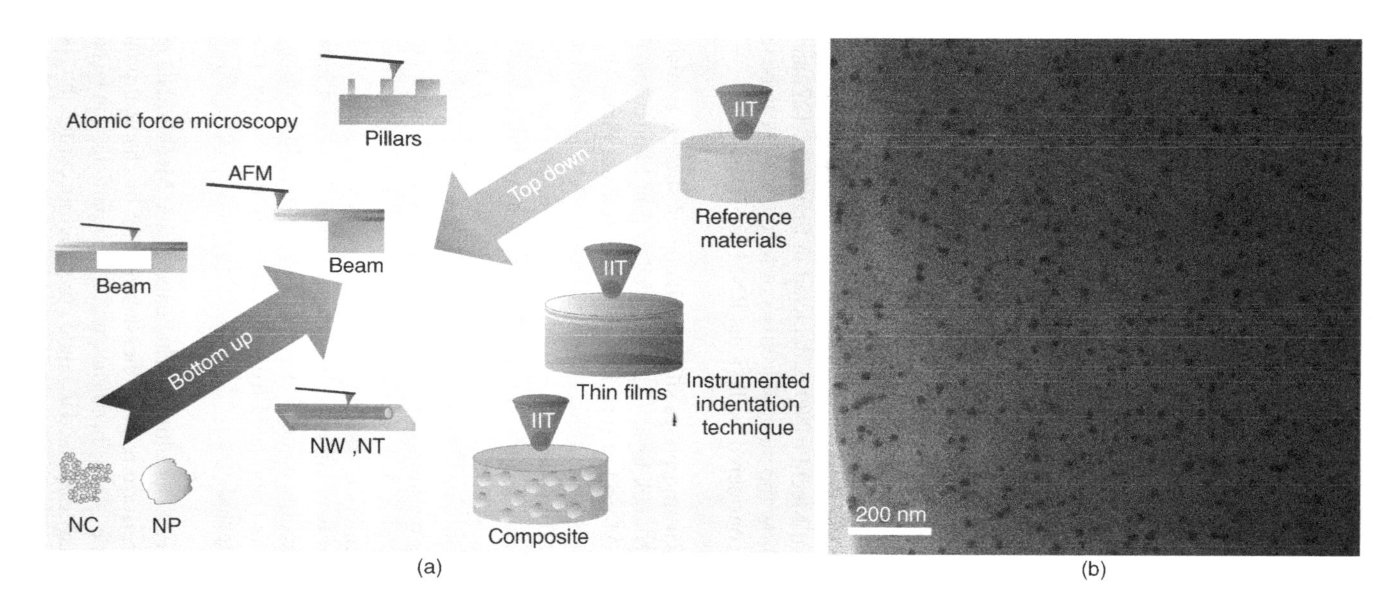

Figure 3.2 (a) Specimens and methods used in MechProNO [27] for the determination of mechanical properties of nanosized features. European Metrology Research Programme [27]. Reproduced with permission of European Metrology Research Programme. (b) Transmission electron micrograph showing silica nanoparticles embedded homogeneously in epoxy matrix. Thin foil prepared by ultramicrotomy of EP + 5% SiO_2 nanocomposite.

by introducing silica nanoparticles as additional filler [24, 29]. A model nanocomposite sample with 5 vol% of such particles embedded homogeneously in an epoxy resin has been prepared recently. Figure 3.2(b) shows a transmission electron micrograph of this model nanostructure with spherical silica nanoparticles with diameters 20–30 nm embedded in epoxy resin. Currently, work is under progress to bond single silica nanoparticles to a silicon substrate. Such a fixing is necessary because otherwise the particles will be pushed away by the AFM tip during measurement. The mechanical properties of the nanocomposite can be obtained easily by IIT or alternatively by tensile tests of bulk specimens. In order to be able to model the properties of the composite, it is necessary to know the properties of the individual nanoconstituents. By assuming bulk properties, it was not possible to fit experimental data in a satisfying way. A possible reason for this behavior could be that properties of nanoparticles must be different from bulk properties.

In addition to spherical silica particles, nanorods and nanowires are also under consideration under the MechProNO project. It would be most desirable for rounding up the program on silica nano-objects to incorporate features such as nanowires into the potential set of reference material. Although carpets of silicon wires on silicon wafer are well established already for photovoltaic applications [30], reports on silica nanowires are rare. Nevertheless, it is possible to prepare silica nanowires by thermal oxidation of silicon nanowires [31]. First results of IIT measurements with silica coatings on silicon substrate within the MechProNO project did not reveal a size effect for Young's modulus. Almost the same modulus of approximately 72 GPa (the same as for bulk silica) was measured for 2 μm thick and 30 nm thick coatings. Furthermore, measurements made on nanobeams within MechProNO suggested even higher modulus values, which coincide with the findings in the literature [31, 32]. In future, the project plans to study the effects of different specimen shapes of the same material (amorphous silica) with different methods. This will be carried out in order to assess the reliability of mechanical property determination on the nanometer scale. Hence, in addition to utilizing IIT and AFM, work is in progress to determine mechanical properties of pillars, beams, and wires by *in situ* force measurements within an FIB/SEM instrument [33]. The advantage of this method is that the experiment can be performed immediately after preparation of the nanosized object by FIB, and that it can be monitored by high-resolution SEM imaging.

Finally, the ability to employ different measurement tools on nano-objects of different shape and size of the same material will enable the proposal of one or more RNM candidates and will potentially lead to certified (or at least well-characterized) mechanical properties. It is likely that amorphous silica will be the first material that can fulfill such requirements.

3.2.2 Silica Nanomaterial as Potential Reference Material in Nanotoxicology

In relation to nanotoxicology, there are some specific requirements when developing RNMs. Ideally, RNMs should (i) be representative of existent materials, which has in the past led to the formulation of priority lists [34], (ii) present some toxicity, which can be evaluated with existent experimental methods, and (iii) be prepared in

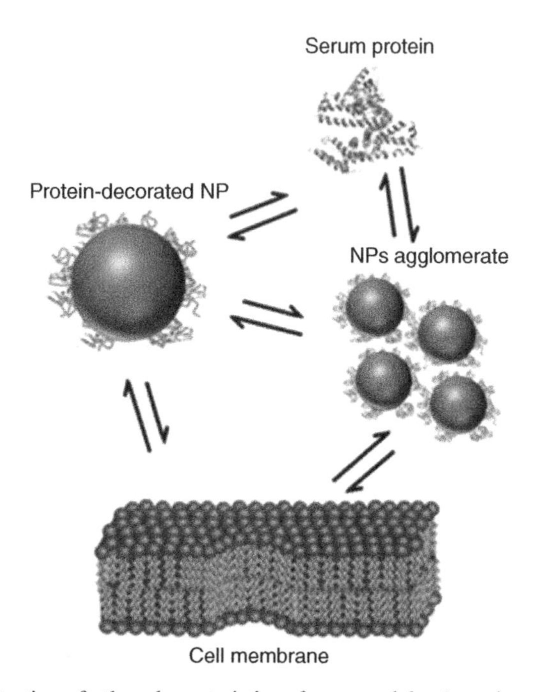

Figure 3.3 Illustration of other characteristics of nanoparticles (protein corona, agglomeration), which may influence nanoparticle toxicity. Source: Orts et al. [7], http://pubs.rsc.org/en/content/articlelanding/2013/ra/c3ra42112k#!divAbstract. Used under CC-BY 3.0 http://creativecommons.org/licenses/by/3.0/.

an adequate form to be used for toxicological studies. This means that parameters such as initial concentrations, pH, ionic strength, and agglomeration should be taken into account and be as close to the actual sample under analysis, for example, when the nanomaterial is subjected to *in vitro* studies. In fact, most of the existent RNMs are "not ready to use" in toxicological tests since they are not, for example, conceived to be compatible with isotonic solution at physiological pH [35].

One of the challenges in developing reference material for nanotoxicology is the need to take into account the dynamics of the nanomaterial with its immediate environment. When nanomaterials enter a biologically relevant medium, their characteristics can change, for example, protein can get adsorbed on to the surface resulting in protein corona. This is depicted in Figure 3.3. Moreover, most existing RNMs for nanotoxicology are monoparametric [7]; this means that only one characteristic of the nanoparticles is well defined, typically the primary particle size. The importance in the development of reference material in nanotoxicology has been stressed by several workers, and as pointed out by David Wahrheit, there is a need to question: "How Meaningful are the Results of Nanotoxicity Studies in the Absence of Adequate Material Characterization?" [36].

In the past, Stefaniak et al. [37] reported a systematic study about which properties are of priority for the risk assessment of ENPs. According to the literature from experts, the following properties were identified as being important (based on % frequency): surface area (100%), elemental composition (96%), surface chemistry (89%), particle size (86%), particle size distribution (86%), morphology-shape (86%), surface charge (86%), agglomeration state (71%), and crystal structure (61%). If we consider the development of a CRM, it has been shown that only particle size is a traceable property; this is the case for spherical particles such as silica and gold (see Table 3.1). In fact, the European Association of National Metrology Institutes (EURAMET) demonstrated the feasibility to determine nanoparticle size with an uncertainty of less than 1 nm. In the past, silica has been considered as a potential reference material by Bell et al. [38]. Their findings indicated the potential use of such nanoparticles as future reference material.

Since 2013, the traceability for other properties (besides particle size) such as composition and surface area are being evaluated for a series of different materials. Out of all the different properties that a nanomaterial can have, only particle size, surface area, and composition can be considered as traceable properties. Nevertheless, information about other particle properties is also considered to be very important. One such property is agglomeration. The importance of this stems from the fact that agglomeration has been shown to affect toxicity arising from silica [39, 40] silver, titanium dioxide[41], and gold [42] nanoparticles, among others. In fact, living cells react differently toward single nanoparticles and their agglomerates. However, monitoring of agglomeration still remains challenging, especially when NMs are in complex biological systems. The presence of proteins with concentrations typically exceeding the ones of nanoparticles by three orders of magnitude [43] will undoubtedly complicate the *in situ* measurement, and techniques such as dynamic light scattering (DLS) will fail to measure reliably if faced with such complex samples. Furthermore, agglomeration (in which particles have a tendency to flocculate) will result in an unstable sample, further posing challenges toward reliable measurement. As a result, errors can be incurred in the measurement resulting in unreliable findings. Overall, the development of CRM and even RM in nanotoxicology, with two or more well-defined characteristics still remains very challenging.

Since the toxicity associated with nanoparticles depends on experimental conditions which affect, for instance, agglomeration and surface properties (as depicted in Fig. 3.3), an alternative approach to RMs would be the use of functionalized nanoparticles as proposed in the past by several authors [44, 45]. Potentially, surface modification of nanoparticles would produce more robust RNMs in terms of colloidal stability [40, 46]. This would allow performing toxicological experiments without being forced to control the agglomeration state of their dispersions before and after *in vitro* testing.

3.3 SUMMARY

This chapter reviews relevant aspects of the still less-explored field of RNMs. In the first part, definition of RMs according to different metrological levels was presented as well as the general strategy on their development. In the second part, the importance of RNMs in material science and nanotoxicology is explained, and considerations on the use of silica nano-objects as potential future reference material are discussed.

Overall, several points should be highlighted from this chapter:

- Definition of a reference material depends on the level of standardization. CRM represents the highest level of traceability.
- Different types of reference materials, standard materials, and CRMs are available, as shown in Table 3.1.
- The preparation of nano-objects with well-defined mechanical properties is still an unsolved issue and thus further R&D activities are needed. Nevertheless, amorphous silica seems to be a promising candidate for reaching this objective.
- Traceable properties of nanomaterials are still limited to few cases such as primary particle size and thickness measurements of very thin films.
- In general, developed RMs are monoparametric thus, only one characteristic of the nano-objects is reported which might be not sufficient to correlate particle characteristics with mechanical properties or toxicological impact.
- In nanotoxicology, the challenge is in the development of suitable RNM that bears similarities to the measurements made on the sample analyzed.

ACKNOWLEDGMENTS

The research communicated in Section 3.2.1 is supported by the European Union by funding the European Metrology Research Programme (EMRP) project "Traceable measurement of mechanical properties of nano-objects (MechProNo)." TEM images were provided by Leonardo Agudo (Fig. 3.1) and Ines Häusler (Fig. 3.2b), both from BAM. The schematic presented in Figure 3.2a was conceptualized by Uwe Brand from PTB, Braunschweig, the leader of the MechProNO project.

REFERENCES

1. International Organization for Standardization ISO Guide 30. *Terms and Definitions Used in Connection with Reference Materials*. 2nd ed. Geneva: ISO; 1992.
2. Yan W, Pun CL, Wu Z, Simon GP. Some issues on nanoindentation method to measure the elastic modulus of particles in composites. Compos B 2011;42(8):2093–2097.
3. Stone V, Nowack B, Baun A, van den Brink N, von der Kammer F, Dusinska M, Handy R, Hankin S, Hassellöv M, Joner E, Fernandes TF. Nanomaterials for environmental studies: Classification, reference material issues, and strategies for physico-chemical characterisation. Sci Total Environ 2010;408(7):1745–1754.
4. The European Commission. Commission recommendation of 18 October 2011 on the definition of nanomaterial. Off J Eur Union 2011;L 275:38–41.
5. (SRU) GACotE. *Precautionary Strategies for Managing Nanomaterials*. Berlin: German Advisory Council on the Environment (SRU); 2011.

6. Tantra R, Shard A. We need answers. Nat Nanotechnol 2013;8:71.
7. Orts-Gil G, Natte K, Österle W. Multi-parametric reference nanomaterials for toxicology: state of the art, future challenges and potential candidates. RSC Adv 2013;3:18202–18215.
8. Roebben G, Ramirez-Garcia S, Hackley VA, Roesslein M, Klaessig F, Kestens V, Lynch I, Garner CM, Rawle A, Elder A, Colvin VL, Kreyling W, Krug HF, Lewicka ZA, McNeil S, Nel A, Patri A, Wick P, Wiesner M, Xia T, Oberdörster G. Interlaboratory comparison of size and surface charge measurements on nanoparticles prior to biological impact assessment. J Nanopart Res 2011;13(7):2675–2687.
9. Sigma-Aldrich, 2015. http://www.sigmaaldrich.com/analytical-chromatography/analytical-standards/application-area-technique/tracecert.html. Accessed 2015 Nov 19.
10. European Commission Joint Research Center. 2014. http://ihcp.jrc.ec.europa.eu/our_activities/nanotechnology/nanomaterials-repository. Accessed 2015 Oct 12.
11. NIST Office of Reference Materials, 2012. http://www.nist.gov/srm/index.cfm. Accessed 2015 Sep 30.
12. Pradel R. 2000. http://www.bam.de/en/fachthemen/referenzmaterialien/index.htm. Accessed 2015 Sep 03.
13. Basu S, Moseson A, Barsoum MW. On the determination of spherical nanoindentation stress–strain curves. J Mater Res 2006;21(10):2628–2637.
14. Gallego-Gómez F, Morales-Flórez V, Blanco Á, de la Rosa-Fox N, López C. Water-dependent micromechanical and rheological properties of silica colloidal crystals studied by nanoindentation. Nano Lett 2012;12(9):4920–4924.
15. Sun W, Zeng Q, Yu A, Kendall K. Calculation of normal contact forces between silica nanospheres. Langmuir 2013;29(25):7825–7837.
16. Kunz A, Pathak S, Greer JR. Size effects in Al nanopillars: Single crystalline vs. bicrystalline. Acta Mater 2011;59(11):4416–4424.
17. Dietiker M, Buzzi S, Pigozzi G, Löffler JF, Spolenak R. Deformation behavior of gold nano-pillars prepared by nanoimprinting and focused ion-beam milling. Acta Mater 2011;59(5):2180–2192.
18. Jang D, Greer JR. Size-induced weakening and grain boundary-assisted deformation in 60 nm grained Ni nanopillars. Scr Mater 2011;64(1):77–80.
19. Mutoh M, Nagoshi T, Mark Chang T-F, Sato T, Sone M. Micro-compression test using non-tapered micro-pillar of electrodeposited Cu. Microelectron Eng 2013;111:118–121.
20. Li Z, Gao S, Pohlenz F, Brand U, Koenders L, Erwin P. Determination of the mechanical properties of nano-pillars using the nanoindentation technique. Nanotechnol Precis Eng 2014;3:182–188.
21. Lopez T, Ortiz E, Alexander-Katz R, Basaldella E, Bokhimi X. Cortisol controlled release by mesoporous silica. Nanomedicine 2009;5(2):170–177.
22. Lin VSY. Nanomedicine: Veni, vidi, vici and then… vanished. Nat Mater 2009;8(4):252–253.
23. Jarvie HP, Al-Obaidi H, King SM, Bowes MJ, Lawrence MJ, Drake AF, Green MA, Dobson PJ. Fate of silica nanoparticles in simulated primary wastewater treatment. Environ Sci Technol 2009;43(22):8622–8628.
24. Zhang G, Sebastian R, Burkhart T, Friedrich K. Role of monodispersed nanoparticles on the tribological behavior of conventional epoxy composites filled with carbon fibers and graphite lubricants. Wear 2012;292–293:176–187.

25. Rahman I, Vejayakumaran P, Sipaut S, Ismail J, Chee K. Size-dependent physicochemical and optical properties of silica nanoparticles. Mater Chem Phys 2009;114:328–332.
26. Hori T. *Gate Dielectrics and MOS ULSIs: Principles, Technologies and Applications*. Berlin, Heidelberg: Springer-Verlag; 1997.
27. Brand U. 2012. http://www.ptb.de/emrp/mechprono-wp.html. Accessed 2015 Oct 12.
28. Plano catalogue No.11, chapter 1 p.23, printed in Wetzlar, Germany, 2013. www.plano-em.de. Accessed 2015 Oct 12.
29. Zhang G. Structure–tribological property relationship of nanoparticles and short carbon fibers reinforced PEEK hybrid composites. J Polym Sci B 2010;48(7):801–811.
30. Yu L, Misra S, Wang J, Qian S, Foldyna M, Xu J, Shi Y, Johnson E. Cabarrocas PRi: Understanding light harvesting in radial junction amorphous silicon thin film solar cells. Sci Rep 2014;4:4357
31. McFarland A, Colton J. Role of material microstructure in plate stiffness with relevance to microcantilever sensors. J Micromech Microeng 2005;15:1060.
32. Sundararajan S, Bhushan B. Development of AFM-based techniques to measure mechanical properties of nanoscale structures. Sens Actuators A 2002;101:338–351.
33. Peng LM, Chen Q, Liang XL, Gao S, Wang JY, Kleindiek S, Tai SW. Performing probe experiments in the SEM. Micron 2004;35(6):495–502.
34. Aitken RJ, Hankin SM, Lang Tran C, Donaldson K, Stone V, Cumpson P, Johnstone J, Chaudhry Q, Cash S, Garrod J. A multidisciplinary approach to the identification of reference materials for engineered nanoparticle toxicology. Nanotoxicology 2008;2(2):71–78.
35. Schulze C, Kroll A, Lehr C-M, SchÃ¤fer UF, Becker K, Schnekenburger J, Schulze Isfort C, Landsiedel R, Wohlleben W. Not ready to use â€“ overcoming pitfalls when dispersing nanoparticles in physiological media. Nanotoxicology 2008;2(2):51–61.
36. Wahrheit DB. How meaningful are the results of nanotoxicity studies in the absence of adequate material characterization? Toxicol Sci 2008;101:183–185.
37. Stefaniak AB, Hackley VA, Roebben G, Ehara K, Hankin S, Postek MT, Lynch I, Fu W-E, Linsinger TPJ, Thünemann AF. Nanoscale reference materials for environmental, health and safety measurements: needs, gaps and opportunities. Nanotoxicology 2013;7:1325–1337.
38. Bell NC, Minelli C, Tompkins J, Stevens MM, Shard AG. Emerging techniques for submicrometer particle sizing applied to Stöber silica. Langmuir 2012;28(29): 10860–10872.
39. Díaz B, Sánchez-Espinel C, Arruebo M, Faro J, de Miguel E, Magadán S, Yagüe C, Fernández-Pacheco R, Ibarra MR, Santamaría J, González-Fernández A. Assessing methods for blood cell cytotoxic responses to inorganic nanoparticles and nanoparticle aggregates. Small 2008;4(11):2025–2034.
40. Orts-Gil G, Natte K, Drescher D, Bresch H, Mantion A, Kneipp J, Österle W. Characterisation of silica nanoparticles prior to in vitro studies: from primary particles to agglomerates. J Nanopart Res 2011;13(4):1593–1604.
41. Lankoff A, Sandberg WJ, Wegierek-Ciuk A, Lisowska H, Refsnes M, Sartowska B, Schwarze PE, Meczynska-Wielgosz S, Wojewodzka M, Kruszewski M. The effect of agglomeration state of silver and titanium dioxide nanoparticles on cellular response of HepG2, A549 and THP-1 cells. Toxicol Lett 2012;208(3):197–213.
42. Albanese A, Tang PS, Chan WCW. The effect of nanoparticle size, shape, and surface chemistry on biological systems. Annu Rev Biomed Eng 2012;14(1):1–16.

43. Hondow N, Brydson R, Wang P, Holton M, Brown M, Rees P, Summers H, Brown A. Quantitative characterization of nanoparticle agglomeration within biological media. J Nanopart Res 2012;14(7):1–15.
44. Oesterle W. 2011. http://www.nano.bam.de/en/projekte/nanotox_refmat.htm. Accessed 2015 Aug 27.
45. Dean L. Size matters: Measurement helps solve nanoparticle toxicity challenges. Chem Int 2012;34(4):6–9
46. Drescher D, Orts-Gil G, Laube G, Natte K, Veh RW, Österle W, Kneipp J. Toxicity of amorphous silica nanoparticles on eukaryotic cell model is determined by particle agglomeration and serum protein adsorption effects. Anal Bioanal Chem 2011;400(5):1593–1604.

4

PARTICLE NUMBER SIZE DISTRIBUTION

D. BARTCZAK AND H. GOENAGA-INFANTE
LGC Limited, Middlesex TW11 0LY, UK

4.1 INTRODUCTION

Nanomaterial, according to the recently published European Commission (EC) Recommendation (2011/696/EU) [1], is

> (...) natural, incidental or manufactured material containing particles, in an unbound state or as an aggregate or as agglomerate and where, for 50% or more of the particles in the number size distribution, one or more external dimensions is in the size range 1 nm–100 nm.
>
> In specific cases and where warranted by concerns for the environment, health, safety or competitiveness the number size distribution threshold of 50% may be replaced by a threshold between 1 and 50%.

This is neither the first nor the only definition of nanomaterial encompassing the number-based qualifier. As summarized in a recent published JRC report (EUR 26744 EN, 2014) [2], European Trade Union Confederation states an 80% [3] particle number threshold, while Australian and Swiss indicate a 10% [4] and 1% [5] threshold, respectively. Nonetheless, the EC definition of nanomaterial has formed basis of the emerging European Union legislation, as manifested in Cosmetic Product (No 1223/2009) [6], Food Information to Consumer (No 1169/2011) [7],

Nanomaterial Characterization: An Introduction, First Edition. Edited by Ratna Tantra.

Biocidal Products (No 528/2012) [8], and Medical Devices (draft: COM(2012) 542 final) [9] Regulations.

In order to satisfy the EC definition of nanomaterial for regulatory purposes, a number-based concentration of particles falling into the 1–100 nm size range must be determined. Subsequently, this means that measurement of particle size based on number distribution is needed. This is different from mass, volume, or intensity-based measurements, which are considered to be ensemble (bulk analysis)-based techniques, such as dynamic light scattering (DLS). It is clear that regulation is one of the main driving forces behind developments toward more reliable methodology for particle number size distribution measurements. However, it is by no means the only reason as to why the determination of the number-based concentration is so important.

In occupational health, more adequate metrics for the measurement of dose in relation to nanomaterial exposure are needed. Instead of mass, the particle number has been identified as more relevant. According to Oberdörster et al., the number of surface molecules (i.e., reactive sites) increases exponentially when particle size decreases below 100 nm [10], and this higher surface-area-to-volume ratio results in higher reactivity per gram of substance. The inversely proportional relationship between nanoparticle (NP) size and surface area (hence reactivity) is not only of relevance to nanotoxicology [11] but also catalytic activity [12]. In a recent study, Ramachandran et al. compared data for three control groups that have been exposed on a regular basis to particles containing mixtures of diesel and gasoline exhaust and found that depending on the metric used (i.e., surface area, mass, or number), the exposure rankings changed significantly [13]. For those reasons, in nanotechnology, particle number size distribution or equivalent surface area per gram of NP with defined size, rather than mass or intensity-based measurements, often seem more appropriate [10–13].

Although various techniques for measuring particle size distribution are commercially available, establishing accurate and precise measurement system is not trivial. Unlike chemicals, particulate materials are not homogeneous in nature. Atoms are bound together in a particulate form and hence are not free to interact individually with the surrounding environment. The issue with size measurements of nanomaterial lies in the nanoscale regime and the associated inherent polydispersity. Even nanomaterials that are considered "monodisperse," actually consist of particles with varied size and shape [14–16]. Therefore, if particle size and/or size distributions are measured on the basis of mass or intensity, then the equivalent particle number reported will not be accurate, as algorithms used to convert mass or intensity to number will require exact information about particle size, shape, and density of all individual particles present. This is not the case for soluble chemicals or molecules, in which the number (i.e., molar) or amount is comparable to the mass-based concentration. The inability to measure particle size by number accurately, precisely, and reproducibly will lead to serious implications with regard to reliability of research findings in relation to understanding chemical, biological, and toxicological impact of nanomaterials.

This chapter provides an overview of the state of the art of currently available techniques capable of performing number-based measurements on nanomaterial suspensions. For each technique presented, the measurement principles, pre-requisite for reliable measurements, for example, sample preparation and advantages/limitations, are discussed. A case study is provided to establish whether silicon dioxide food additive found in commercially available coffee creamer should be classified as nanomaterial. The experimental study assesses the effects of different sample preparation strategies on data findings, for example, the effects of adding a separation step (involving asymmetric flow field-flow fractionation) in the method are evaluated. Transmission electron microscopy (TEM) is used to measure particle number size distribution. Finally, the authors' viewpoint on the expected instrumental developments in this area is discussed throughout.

4.2 MEASURING METHODS

4.2.1 Particle Tracking Analysis

Particle tracking analysis (PTA), also called nanoparticle tracking analysis (NTA) is commercially available and can directly measure particle number size distribution. The measurement technique leverages on two important properties of nano-objects (when suspended in a liquid matrix): ability to scatter light and to move under Brownian motion [17].

Measurement is conducted with the nano-object suspension placed in a glass chamber and illuminated with a laser beam. The light is scattered by the particles in the path of the beam and scattered light is detected by a highly sensitive camera, mounted on a conventional optical microscope equipped with a magnification objective. The camera then captures a video of particles moving under the Brownian motion within a field of view with known dimensions (Fig. 4.1). It is important to note, however, that the particles themselves are not being imaged as they are in the nanoscale size range with dimensions below the Rayleigh [19] or Abbé [20] limit, meaning that their structural information cannot be resolved. Instead, the particles act as points of scatter and Brownian motion of each particle is followed in real time

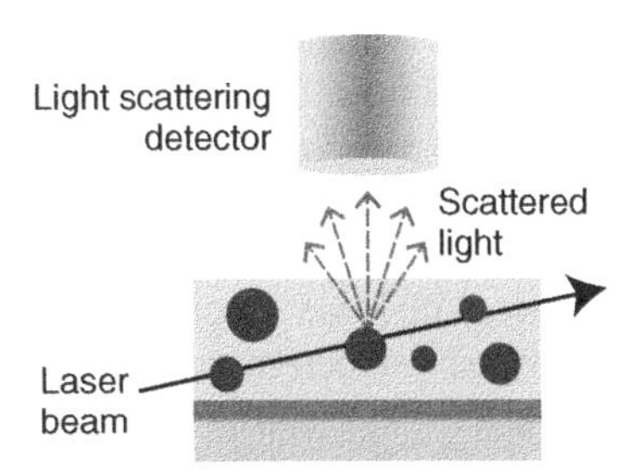

Figure 4.1 Schematic representation of the PTA (NTA) principle. Carr and Wright [18], Figure 4.1. Reproduced with permission of Wiley.

via video. The recorded movies are then analyzed by software, which locates and tracks individual particles frame-by-frame and calculates their number-based size distributions from the Stokes–Einstein equation [21].

In order to ensure statistically viable data sets, it is important that a sufficient number of particles are analyzed within the sample and over elapsed time, having in mind that NTA is not an ensemble technique, but each particle is analyzed individually irrespective of the others. By counting the particles within the known field of view (e.g., 100 μm by 80 μm [18], depending on the type of instrument) and known depth of the illuminating beam (in the order of 10 μm [18], depending on the instrument manufacturer), it is possible to estimate the equivalent particle concentration per milliliter. However, the effective scattering volume in which particles can be detected and counted varies as a function of particle size and refractive index as well as the characteristics of illuminating laser beam, such as power, wavelength, and dimensions. Furthermore, manual adjustment of the camera sensitivity and/or built-in software algorithms may affect the number of particles detected and tracked, which in turn can impact the overall concentration count. Ideally, in order to obtain unbiased number concentration and number size distribution values, PTA should be calibrated using a reference material of the same chemical composition, shape, and density.

The main advantage of the technique is its simplicity, requiring no input from the user other than the viscosity of the dispersant medium, if different from water. The chemical composition and density of the analyzed material may be unknown, though these strongly influence the lower limit of detection, with a minimum particle size detection limit of around 10 nm for high-density materials (such as gold or titania) and an upper limit of approximately 1000 nm. If the sample contains two or more types of particles of similar size but substantially different chemical composition (i.e., metal vs. polystyrene), intensity measurements allow differentiation between them and simultaneous characterization of the individual fractions. In addition to the number-based concentration, PTA also allows size and zeta-potential distribution determination. Such measurements can also be performed with complex samples, for example, sample containing different particle types and sizes and/or suspended in biological or environmental matrices. However, for such complex samples, the accuracy of particle counting is often compromised due to differences in the scattering volume and background scattering noise. It is important to note that for very small particles (with lower depth of the scattering volume), a broadening of size the distribution measured may occur. However, this limitation is possible to overcome with recently introduced, improved tracking algorithms. Improved tracking algorithms and detection might also have an effect on the particle concentration measurements especially in case of complex samples, resulting in software to software variability. The main drawback of the PTA is the limit of detection, not covering the entire nanoscale and dependent on the particle density, with lower density (e.g., silica or organic) particles smaller than 30–40 nm being difficult to detect.

Since PTA is not an ensemble technique and does not suffer from intensity weighting issues, it is suited for analysis of polydisperse samples (where the number of particles, present in each size fraction can be determined). Required sample preparation of suspensions is minimal and often involves just a simple dilution to an appropriate concentration. However, for samples containing significant portion of particles larger than 1000–2000 nm, removal of such large particles via filtration, centrifugation, or sedimentation is recommended before analysis. This is because such particles are too large to move freely under the Brownian motion and sediment fairly quickly, causing increased background scattering (noise), which could affect the concentration and size measurements of the sample. Finally, it is important to ensure that the diluent itself is free of particles and does not absorb light in a frequency of the laser source installed in the system (potentially leading to thermal convection).

Recent developments associated with this technique have been focused on software rather than the hardware, and this trend is expected to continue. Although improved tracking algorithms and user interface have increased the resolution and detection of complex samples, lower size limitation is still an issue.

4.2.2 Resistive Pulse Sensing

Resistive pulse sensing (RPS), also called scanning ion occlusion spectroscopy (SIOS), can be used for the direct measurement of particle size number distribution. The technique utilizes a nonoptical detection based on the Coulter principle [22], where a displacement of an electrolyte caused by a particle passing through a small aperture originates pulses of impedance in an electric current. Traditional Coulter counters use solid-state cylindrical aperture, connecting two compartments (Fig. 4.2). Particles suspended in an electrolyte are pulled from one container into another by voltage or pressure differences. Two electrodes located on both ends of the aperture continuously measure changes in the ion current, while individual particles pass through the aperture. In modern versions of the instrument, solid-state cylindrical aperture was replaced by a flexible membrane with a tunable size, conical-shaped pore, giving rise to a tunable resistive pulse sensing (TRPS) [23]. The size of the pore can be easily changed in real-time by axial stretching of the elastic membrane. In this way, the pore diameter is tuned to suit the size of objects being analyzed and to regulate the passage of the electrolyte and particles.

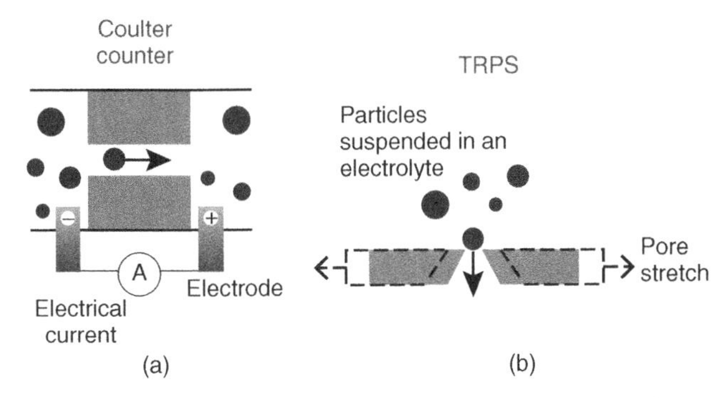

Figure 4.2 Schematic illustration of the Coulter counter (a) and TRPS principle (b), adapted from [23].

If a particle passing through a pore or aperture is of appreciable size compared with the width of the orifice, then the current flowing through the circuit changes a measurable amount. A discrete blockade of the event signal is recorded by the system. In cases where the particle is more conductive than the electrolyte, the measured resistance decreases, whereas if the particle is less conductive than the electrolyte, the signal increases. The magnitude of a measured blockade is a key indicator of the volume displaced by the particle as it flows through the sensing zone, which is equal to the particle volume, thus the size. Furthermore, the frequency of measured blockade events is directly related to the concentration of particles in a sample. As the concentration changes, the particle flow through the pore also changes in a linear fashion, with an increase in sample concentration resulting in an increase in measured blockade frequency. The duration of the blockade signal is proportional to the particle translocation velocity, which in the absence of particle–orifice interactions depends on the applied pressure, diffusion (e.g., thermal and for small objects also Brownian motions), electro-osmotic force, and electrophoretic force, with the latter used for the nanoparticle surface charge determination.

Modern (T)RPS instruments enable control over key parameters, such as the system pressure and applied voltage, which drive particle transport, resulting in the improved measurement sensitivity, dynamic range, and resolution. The relationship between the applied pressure and the measured frequency of blockade events is linear for the given type of particles, with the gradient depending on the particle concentration.

Main advantage of (T)RPS is that particle-by-particle measurement of size, number concentration, and relative surface charge distribution are feasible and can be determined simultaneously. The method is suitable to all types of particles but care must be taken when analyzing soft particles (such as liposomes or gels), for which deformation or compression might occur during translocations. The blockade event frequency can also show a temperature dependence, which altogether might lead to misinterpretation of results.

One drawback of (T)RPS is the lower limit of detection (typically around 50 nm), which covers only the top half of the nanoscale size range. Also, the requirement for high ionic strength environment (electrolyte) limits the possible applications, making the technique unsuitable for water (or dilute ionic) suspensions, organic solvents, or metal or semiconductor particles with weak (e.g., electrostatic) surface stabilization, for which the presence of an electrolyte could lead to agglomeration or aggregation and result in ambiguous readouts. Furthermore, for nanomaterials suspended in complex matrices, for example, of biological origin, interferences from the matrix (e.g., platelets) may lead to concentration overestimate and other misinterpretations. Finally, as sample preparation involves dilution to an appropriate concentration with high ionic strength buffer, care must be taken to ensure that the analyzed sample is stable in such environments and that the buffer used is free of particulate matter.

Since current commercial (T)RPS instruments suffer from a narrower dynamic range compared with other particle-counting instruments, the hardware improvements, aimed at developing smaller and tunable, that is, over broader size range (nano)pores, are expected in the future.

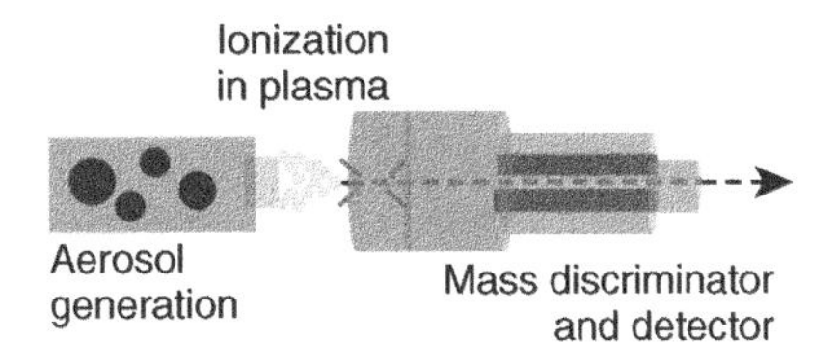

Figure 4.3 Schematic illustration of the ICP-MS principle.

4.2.3 Single Particle Inductively Coupled Plasma Mass Spectrometry

Single-particle inductively coupled plasma mass spectrometry (spICP-MS) is a recently emerged method, which leverages on traditional ICP-MS measurement principle (Fig. 4.3). In recent years, spICP-MS is being increasingly used for the determination of particle number size distribution. The approach was originally developed by Degueldre et al. [24] and the measurement principle has been described in detail elsewhere [25].

In spICP-MS, sample preparation (of the nanomaterial-containing sample) often requires the need to dilute to a concentration in the range of ng/l. The diluted sample is then introduced into the ICP-MS instrument at a certain flow rate. The technique works by acquiring individual intensity readings with very short dwell times (in the range of a few milliseconds). The signal corresponding to one particle can then be converted into particle mass (based on a calibration curve), which is obtained by measuring the signal intensity corresponding to the relevant ionic standard(s) or to particle reference material(s). The particle diameter can be calculated from the determined mass if a spherical shape and a solid par is assumed and the composition and density are known. The sum of masses of all particles divided by the measured volume of sample gives the mass concentration.

For dwell times above 10 ms, more than one particle may be registered by the detector in one event, which leads to a violation of a single particle rule. Therefore, with most current instrumentation, typically a dwell time between 1 and 3 ms is used, where in order to satisfy the single-particle rule, the number of peaks should not exceed 10% of the maximum possible to detected per minute (a typical run time, called a scan time), which in case of 3 ms dwell time equals to around 2000 events per minute [26]. The overall number of peaks detected per minute is directly proportional to the number of nanoparticles in the suspension, while the intensity of the peak corresponds to the particle mass and spherical equivalent diameter to the third power.

In practical terms, the determination of particle nebulization efficiency is important, since this number is required to determine the mass and number concentration of particles. For this, a well-known reference or certified particle standard should be used. Often, the availability of well-characterized or certified particle materials (of similar characteristics of the analyte) is rare. As such, certified particles of a different composition than the target particle can be used instead. The mass and size of individual particles are then calculated from an ionic (or dissolved elemental) standard of the same element. Similarly, the same ionic standard can be used to calculate the concentration of element ions in solution.

Depending on the particle diameter, dissolved ions (e.g., for silver, silica, or zinc oxide) can also be detected (which is shown as a constant signal) and, therefore, be distinguished from the particle signal.

Till date, most strategies so far have been based on external calibration. Internal standardization strategies (e.g., with isotopically enriched materials) are yet to be investigated and their performances (in terms of accuracy and measurement uncertainty) would have to be compared with the external calibration-based strategies.

The spICP-MS method is compatible with most of the aqueous suspensions of metal and metal(loid) oxide particles, since mass spectrometer can be tuned to measure a particular element of interest. However, due to the fast detection required for single-particle analysis, only one isotope (*m/z* value) can be monitored at a time (per run). Recent developments of ICP-MS instrumentation show features such as dwell time as low as 0.1 ms, offering a potential for multi-isotope/multielement approaches to be developed for spICP-MS. In the case of isotopes, which suffer from polyatomic interferences, a signal-to-noise ratio, critical for distinguishing particles from the background, can be improved by using interference-minimizing detection (as obtained by using collision reaction cell or sector field-based instrumentation).

The resolution of spICP-MS (in terms of particle size) depends on the background noise and sample concentration. Currently, this method has been successful in the analysis of gold and silver nanoparticles with a diameter of around 15 nm (when using a common single-quadrupole ICP-MS) and of TiO_2 nanoparticles with a diameter of 95 nm [27]. However, recent advances in data processing techniques, such as deconvolution algorithms [28], in combination with more powerful detectors (dwell time down to 0.1 ms) could potentially result in new capabilities extending beyond a current lower-end size limitation.

The lowest particle concentrations that can be detected with spICP-MS have been shown to range from ng/l (e.g., for gold and silver) to μg/l (e.g., for silica) [27]. Dynamic range of spICP-MS is limited to two orders of magnitude, since sample dilution is required.

A disadvantage of spICP-MS is in its restriction to aqueous suspensions, which limits its potential applications. To overcome this limitation, various extraction techniques (e.g., enzymatic or chemical) of particles from nonaqueous matrix can be used, but in order to relate the particle number in an aqueous suspension to the concentration in the original sample, an in-depth knowledge on the extraction efficiency and/or matrix effects is required.

As already mentioned, spICP-MS is only compatible with metal and metal(loid) oxides, and it cannot measure carbon-based nanomaterial. Furthermore, unless the sample contains an elemental tag that is visible to the ICP-MS, for example, sulfur, halides, or phosphorous (in, e.g., their capping layer formed of proteins or nucleic acids), the spICP-MS technique is considered unsuitable.

It is also important to highlight that since the equivalent (to the peak intensity) nanoparticle diameter is assumed to be spherical, the technique cannot easily deal with size characterization of anisotropic (e.g., rod-like) objects. This consideration might have an impact on the accuracy and uncertainty of the calculated particle diameter and has to be investigated in detail.

One of the key advantages of the technique is in its ability to determine quantitatively the chemical composition of the material, which no other counting technique listed in this chapter, can do. This, in conjunction with a potential for multi-isotope and multielement analysis, makes spICP-MS attractive for applications concerning more complex nanomaterial, e.g., capped or core–shell nanomaterials. As previously mentioned, in addition to the particle number, ionic concentrations (i.e., dissolved fraction) can also be determined (from a constant signal, which differs from the individual peaks representing particulate matter). Such analysis is of great importance, for example, in toxicological studies in which spICP-MS is the only counting technique that can potentially quantify the dissolved elemental fraction [29]. The technique is also suitable for samples suspended in complex (i.e., food or environmental) matrices. It has also been demonstrated by Telgmann et al. [29] that the matrix-dependent analyte responses can be compensated by isotope dilution analysis, where isotopically enriched spikes are used as internal calibrants.

4.2.4 Electron Microscopy

Electron microscopy (EM), as opposed to more traditional visible-light microscopy, uses a beam of electrons rather than a beam of light to visualize the sample. As proposed by Abbé [20], the ability to resolve detail in an object is limited by the wavelength of the source; hence, the use of electrons with a much shorter wavelength (de Broglie theory [30]) than the visible light results in a significantly higher resolution. In TEM (Fig. 4.4) high-energy (even above 100,000 eV) electrons interact with an ultrathin specimen, while passing through [31]. The elastic scattered electrons are then focused into an image by a sophisticated system of electromagnetic lenses. In scanning electron microscopy (SEM), an image is formed from low-energy (less than 50 eV) secondary electrons ejected from the specimen surface by inelastic scattering interactions with the electron beam [32]. As these electrons originate from the top of the sample surface, there is no restriction with regard to the thickness of the specimen being imaged.

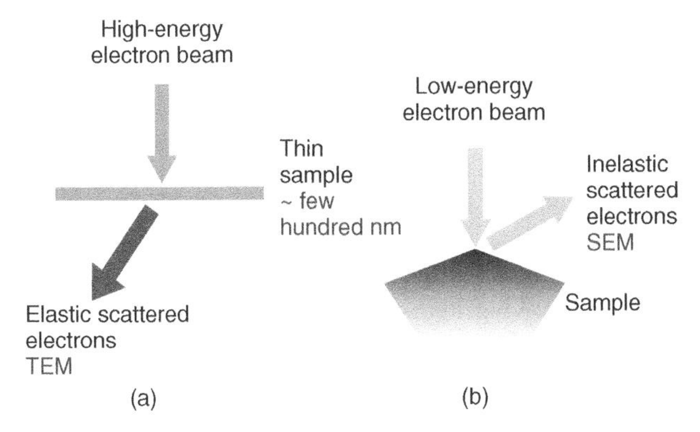

Figure 4.4 Schematic illustration of TEM (a) and SEM (b) principles.

The spatial resolution of an electron microscope generally is in the nanometer range, but it varies from instrument to instrument. In SEM, this depends on the size of an electron spot and a volume of the specimen interacting with the electron beam. However, because both of those parameters are larger than distances between atoms, SEM cannot image individual atoms and the typical resolution falls in the range of approximately 1–20 nm. Due to a shorter wavelength (hence higher energy electrons) TEM is often employed for nanomaterials. TEM can resolve individual atoms, and with recent advances in an aberration correction, a resolution even down to 0.5 Å can be achieved, which comfortably covers and exceeds the lower end of the nanoscale.

Compared with TEM, SEM has several advantages, such as the ability to image larger areas of a specimen and the ability to image relatively thick samples.

Both techniques in principal produce an image that is a two-dimensional (2D) projection of a three-dimensional (3D) object, even though scanning electron micrographs display topographical features of the sample surface giving an impression of a 3D appearance. This, to some extent, allows differentiation between flat (2D) and 3D objects, but with no ability to measure the length/dimension in the vertical plane; it, therefore, offers information about the size of material, only in the horizontal plane. While interpreting EM images, especially transmission electron micrographs, care must be taken in order to avoid incorrect classification of the material (e.g., circular nanoplates instead of nanospheres). In some cases and for very experienced users, differences in contrast (lower for 2D than a 3D object, of the same chemical composition and crystal structure) may give some useful indication on the class of particles. Alternatively, a technique called TEM tomography, allowing sequential imaging at incremental degrees of rotation around the centre of a sample and assembly of the obtained individual 2D images in one 3D image, can be used. This, however, is at the cost of resolution and time required for such analysis. Nonetheless, both TEM and SEM can be considered as counting methods, which provide size values of all objects present in the selected area of an image, as well as, often critical in nanoscience, information about the shape. For samples of well-dispersed materials (and those appearing (on the image) as a collection of individual objects), EM measurements (for counting as well as sizing) can be automated. Automation is only possible with such samples, as image processing softwares generally require significant differences in contrast between neighboring pixels to determine the edges of an object. In order to obtain the best estimate of size, using the diameter of a circle with the surface area equal to the 2D image of the object imaged [33], which is limited to near-spherical shapes, is recommended. For nonspherical objects, such as branched structures, manual or semiautomated analyses offer a better estimate.

One of the main advantages of EM over the other techniques listed in this chapter is the ability to resolve individual objects from agglomerates and aggregates, although with limitations, as this requires extensive operator input and interpretation.

In traditional EM involving high-vacuum chambers, sample preparation often involves deposition on a grid/substrate and drying before analysis. In some cases, this could induce aggregation or agglomeration of analyzed nanomaterial, thus leading to artifacts and/or misinterpretation. Potential measurement error can be further amplified by the high-energy electron beam, which may cause problems for samples containing organic material, for example, liposomes, or thick organic capping layer, which can be easily destroyed in this harsh environment. Therefore, for such samples, cryo or environmental chambers are better suited. Furthermore, the use of fixatives and heavy-element stains (osmium tetroxide, lead citrate, uranyl acetate, etc.) to preserve and increase contrast of the fragile organic components might also be considered, but yet again at a cost of spatial resolution.

In relation to future advances, there is a clear need to improve sample preparation, in order to minimize artifacts arising from inherent agglomeration of the particles on a substrate. As already mentioned, a promising alternative to the traditional EMs are instruments equipped with an environmental chamber, allowing sample analysis in wet/moist environment. However, spatial resolution of such environmental EMs requires further improvements.

4.2.5 Atomic Force Microscopy

Atomic force microscopy (AFM) is an imaging technique, which measures forces between a sample surface and a probe (flexible cantilever terminated with a sharp tip), at very short distances (0.1–20 nm) [34, 35]. The forces (van der Waals, electrostatic, etc.) cause a deflection of the cantilever, according to Hooke's law [36]. The deflection is typically measured with a laser beam reflected from the cantilever into a photodetector (Fig. 4.5). The probe scans the sample surface in a horizontal plane, as well as moving in the vertical direction; hence, AFM, unlike EM techniques, provides a 3D image. The image resolution obtained in the vertical plane is not the same as in the horizontal direction, where it is limited by the dimensions of the tip. Consequently, for very small objects the height information is often more reliable than the lateral plane readouts. Nonetheless, being an imaging technique, AFM allows counting particle-by-particle and offers additional information about the size and shape (with limitations) of the material.

AFM can operate in three different modes, namely contact, tapping, and noncontact, permitting analysis of a wide range of samples, including those of organic origin. In contact mode, the tip is kept close to the sample surface (closer than 0.5 nm) where the overall force is repulsive. As the tip moves across the sample, the cantilever bends,

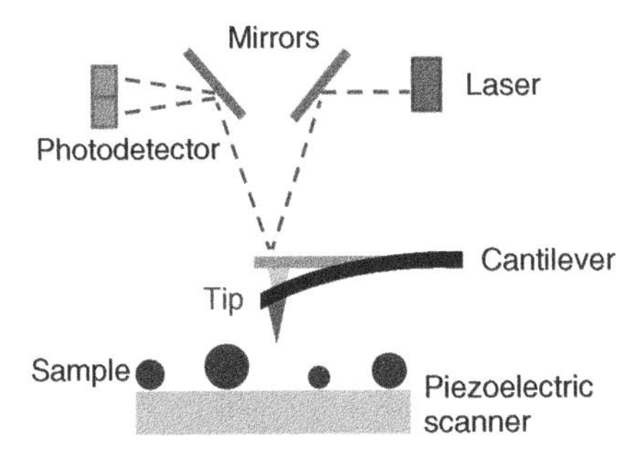

Figure 4.5 Schematic illustration of AFM principle.

while the instrument measures forces required to bring the cantilever back to a constant position. Due to a very strong interaction between the tip and the sample, this method is not suitable for fragile samples, for example, organic material.

For fragile samples, a tapping mode, in which the cantilever oscillates at a resonant frequency with an amplitude of approximately 20–100 nm, thus gently touching the surface, is often employed. When the tip comes close to the sample surface, the forces cause a decrease in the oscillation amplitude, which needs to be compensated for in order to maintain constant amplitude, and can be measured by the instrument. Tapping mode is quite gentle and allows imaging of delicate organic materials, such as nucleic acids, also in a liquid medium, provided the individual objects remain adsorbed to the substrate. Another option is to use the noncontact mode. In the noncontact mode, the tip does not touch the surface, but oscillates above, where the overall force is attractive. The forces decrease the oscillation amplitude of the cantilever. In order to maintain constant oscillation amplitude, the average tip-to-sample distance needs to be adjusted, which is then measured by the instrument. Noncontact AFM is recommended for the analysis of soft samples, such as organic films.

Main disadvantages of AFM compared with EM are the maximum size of an image and relatively slow scanning speed. Maximum height of an AFM image is in the order of 10–20 μm, with a maximum scanning area of about 150 by 150 μm. It takes several minutes to acquire a typical scan, which altogether makes it the most time-consuming counting technique described in this chapter. Due to the nature of AFM probes, they are not suited to analyze steep walls or overhangs. However, cantilevers can be specially designed to additionally move sideways to allow such analysis, but they are more expensive, offer lower lateral resolution, and may lead to artifacts.

Sample preparation for an AFM requires sample deposition on a surface, even for liquid samples, the material being imaged has to adhere to the surface. It is also important that the deposited sample is free of contamination. Since AFM resolution in the horizontal plane is limited by the probe dimensions, future advances in this technique are expected to come from developments in tip design and fabrication.

4.3 SUMMARY OF CAPABILITIES OF THE COUNTING TECHNIQUES

All currently available nanoparticle-counting techniques have some advantages as well as limitations, as detailed in the previous sections. The summary of key parameters, which should be taken under consideration when choosing the most appropriate technique, matched against the technique's capabilities is shown in Table 4.1.

4.4 EXPERIMENTAL CASE STUDY

4.4.1 Introduction

Fumed silica is an EU-approved food additive (E551) [37] containing particles with a primary crystal size below 15 nm, while titania (E171) [38] contains particles below

TABLE 4.1 Summary of Capabilities of the Counting Techniques

	PTA (NTA)	(T)RPS or SIOS	sp ICP-MS	EM	AFM
Covers the entire nanoscale?	–	–	–	+	+
Distinguishes between aggregates, agglomerates, and individual nanoparticles?	–	–	–	+	+
Counting in solution?	+	+	+	±	+
Compatible with water suspensions?	+	–	+	+	+
Compatible with buffered solutions?	+	+	+	+	+
Compatible with high ionic strength solutions?	+	+	–	–	–
Compatible with organic solvents?	+	–	–	+	+
Suitable for inorganic materials?	+	+	+	+	+
Suitable for carbon-based materials?	+	+	–	+	+
Suitable for organic materials?	+	+	±	+	+
Is it time efficient?	+	+	+	+	–

In the table "–"represents "no," "+" means "yes," whereas "±" depicts "yes, but only in some specific cases."

10 nm in diameter. Such materials added to food, upon interaction with the matrix components, may or may not retain their nanoscale size and properties. Therefore, for the purpose of toxicological assessments and to satisfy the requirements of recommended by EU definition of nanomaterial along with the upcoming EU regulations, it is absolutely vital to determine what percentage (if any) of such materials added to food remains at the nanoscale.

The aim here is to provide experimental data that allow the determination of whether the additive present in the sample is still in the nanoscale and thus labeled as nanomaterial product. According to regulation guidelines, this can only be determined through providing number-based size distribution data. Undoubtedly, sufficient number of data points must be collected in order to provide representative and statistically relevant data. Since food matrix is very complex, consisting of an array of organic and inorganic moieties, some of them forming naturally occurring nanoscale micelles (e.g., caseins in milk [39]), chemical composition of an analyte must also be determined, in order to confirm the presence of an element of interest (e.g., silicon nanomaterial) in the analyzed fraction.

As explained in the previous section, among currently available particle-counting techniques EM is ideal for the measurement of particle number size distribution. TEM is a particularly attractive choice as it is the only number-based size technique that can measure at the lower end of the nanoscale (i.e., sub-nanometer) and hence the choice of TEM for this particular study is justified. Also, since TEM is an imaging-based technique, the number size distribution of particles is determined directly, rather than indirectly via theoretical equations. Furthermore, it is possible to combine TEM with an additional energy-dispersive X-ray (EDX) detection to allow chemical characterization of a sample, thus providing confirmation that the objects being imaged contain the element of interest, for example, silicon, as found in the fumed silica additive.

(Note: EDX utilizes X-rays emitted from a specimen upon illumination with an electron beam. Emitted X-rays vary in energy, which is the characteristic feature of an element.)

4.4.2 Method

Samples for TEM/EDX were prepared by a droplet deposition method on carbon film coated 400 mesh copper grid (Agar Scientific Ltd.) and air dried. The grids were imaged with FEI Technai12 Transmission Electron Microscope operating at 80–120 kV bias voltage and equipped with TEAM™ EDS Apollo XLT X-ray microanalysis platform. The instrument was calibrated by the service engineer. Number-based size distributions were obtained by manual analysis of the collected images. However, available online free-of-charge image-processing software, such as ImageJ [40] or Pebbles [41], could be used to automate the procedure.

Two sample preparation steps were carried out in the study. For the first preparation, a real-life sample of commercially available coffee creamer [42] was suspended in water and mixed well for 1 min. An aliquot of 0.5 ml was placed in a centrifugal filter tube (Amicon Ultra, 3 kDa) and centrifuged for 30 min at 14 000 g, until the sample was concentrated to 50 µl. A droplet of the concentrated coffee creamer suspension was deposited on a TEM grid and air dried.

For the second sample preparation, the coffee creamer suspension underwent a sample separation step. This was carried out using a commercially available asymmetric-flow field-flow fractionation system (AF4), which was coupled online to an ICP-MS. A fraction containing the element of interest (i.e., silicon) was collected over 2 min timeframe (with total collected volume of 1 ml) and further processed in the same way as before (i.e., first preparation step), for subsequent analysis under TEM/EDX.

4.4.3 Results and Interpretation

Results are shown in Figure 4.6. TEM image of crude coffee creamer suspension (a) shows the presence of the food matrix, while a close-up (b) indicates the presence of nanoscale objects. However, due to the background of the food matrix, the image quality is undoubtedly compromised, which together with an uneven contrast makes it unsuitable for automated analysis.

To improve the image quality, sample was precleaned with AF4. An image obtained (c) clearly shows individual objects in the nanoscale. A total of 10 images were taken, in order to obtain sufficient number of data points (a total of at least 500 particles [43]) and provide size distribution histogram (d), indicating majority of particles (almost 50% of the total number of detected) in the size range of 7–11 nm, which agrees well with the size detection range reported for food grade or Aerosil silica particles [44].

Additional EDX spectrum (e) confirms the presence of silicon (Si) in the collected fraction. The peaks representing copper (Cu) and carbon (C) from the spectra come from the TEM grid.

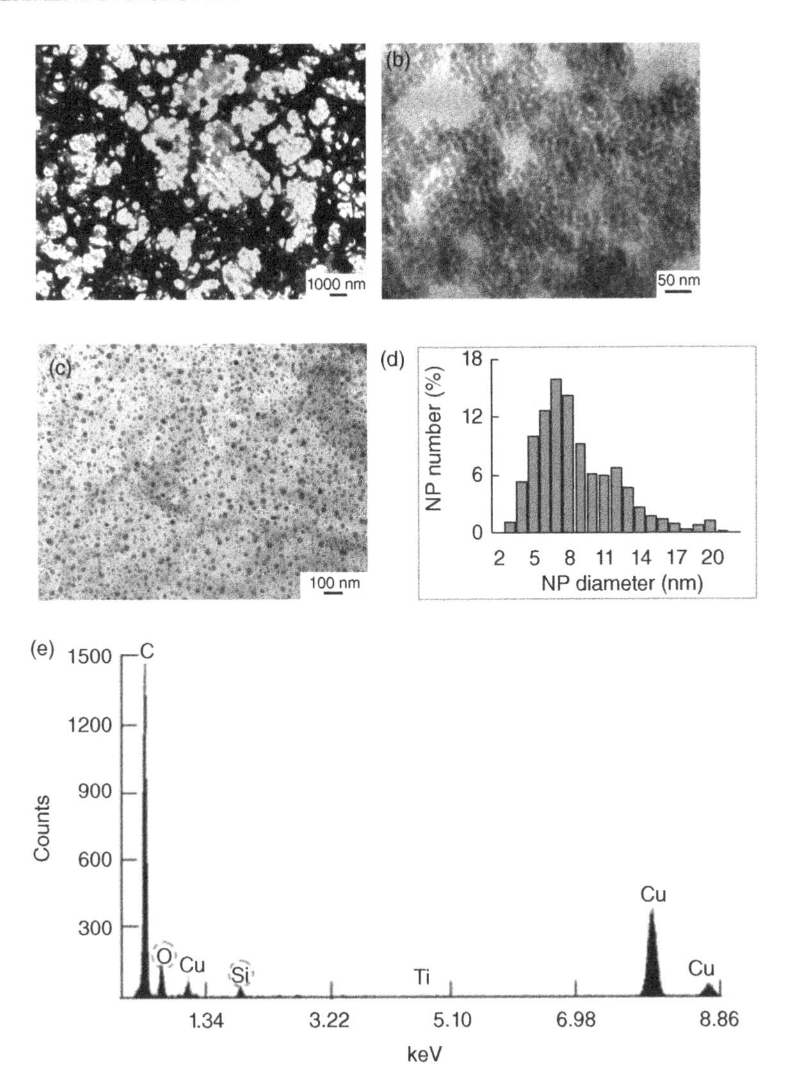

Figure 4.6 Representative TEM images of crude (a and b) and precleaned (c) coffee creamer suspension with corresponding number-based size distribution histogram (d) and EDX spectra (e).

4.4.4 Conclusion

Overall, TEM supported by EDX clearly indicates that all particles detected in the collected fraction are in the nanoscale. However, in order to assign prefix "nano" to this material, one would need to investigate whether all silicon present in the sample was accounted for and whether there is a portion of the material in the dissolved form (which would not be seen on TEM and would not be detected in a fraction separated by AF4/ICP-MS). The analysis performed with AF4/ICP-MS, supported by total ICP-MS showed that more than 50% of the total silicon present in the sample, eluted in the fraction collected for TEM analysis [42] and since TEM/EDX showed that all of the particles present in this fraction are in the nanoscale, the analysed material should be classed as 'nano'.

4.5 SUMMARY

All currently available nanoparticle-counting techniques have advantages, as well as limitations. It is clear that there is no ideal technique, which can deal will all kinds of nanomaterials and dispersant media. Some of the main measurement challenges associated with the techniques described in this chapter have been identified. For example, most techniques (apart from imaging-based techniques) struggle to measure at the lower end of the nanoscale, that is, below 10 nm. However, imaging-based techniques require the need to have a well-controlled sample preparation step (especially when the particles are deposited on the substrates).

The case study presented here demonstrated the potential use of TEM/EDX to analyze fumed silica additive in a coffee creamer sample. The study highlighted the importance of a sample separation step; the use AF4 to separate the nanomaterial from the complex matrix in the final method is shown to be the key in getting reliable measurements.

Finally, there is a clear potential for further advances and developments in many of the measurement systems discussed here. Future technology developments should concentrate on low dwell time detectors and deconvolution algorithms (for spICP-MS), tunable nanopore technology (for TRPS), and ultrasensitive light sensors (for PTA). Improving advances in technology will undoubtedly open new avenues, making these techniques more suitable for detection and characterization of particles, for example, ability to characterize particles with diameters lower than 10 nm, as well as real-life samples (such as environmental, food, cosmetics, and pharmaceutical, which often contain polydispersed material suspended in a complex matrix). Until these improvements are made, a multimethod and application-specific approach to number-based characterization of nanomaterials is advised.

REFERENCES

1. European Commission. European Commission Recommendation of 18 October 2011 on the definition of nanomaterial, 2011/696/EU. Off J Eur Union 2011;275:38–40.
2. G. Roebben, H. Rauscher, V. Amenta, K. Aschberger, A. Boix Sanfeliu, L. Calzolai, H. Emons, C. Gaillard, N. Gibson, U. Holzwarth, R. Koeber, T. Lisinger, K. Rasmussen, B. Sokull-Kluttgen and H. Stamm, JRC Scientific and Policy Report, Toward a review of the EC Recommendation for a definition of the term "nanomaterial" Part2: Assessment of collected information concerning the experience with the definition, EUR 267744 EN, Luxembourg: Publications Office of the European Union, 2014.

3. European Trade Union Confederation. 2010. ETUC concept of a regulatory definition of a substance in the nanoform, Brussels. Available at http://www.etuc.org/sites/www.etuc.org/files/REACH_nanosubstance_definition_ETUC_concept_1.pdf. Accessed 2014 Oct 02.
4. NICNAS Communications. NICNAS working definition for 'industrial nanomaterial'. Available at http://www.nicnas.gov.au/communications/issues/nanomaterials-nanotechnology/nicnas-working-definition-for-industrial-nanomaterial. Accessed 2014 Oct 02.
5. State Secretariat for Economic Affairs (SECO). *Safety Date Sheet (SDS): Guidelines for Synthetic Nanomaterials*. Zurich: Nano SDS Guidelines; 2012.
6. European Parliament and Council. Regulation (EC) No 1223/2009 of the European Parliament and of the Council of 30 Nov 2009 on cosmetic products. Off J Eur Union 2009;L342:59–209.
7. European Parliament and Council. Regulation (EU) No 1169/2011 of the European Parliament and of the Council of 25 October 2011 on the provision of food information to consumers, amending Regulations (EC) No 1924/2006 and (EC) No 1925/2006 of the European Parliament and of the Council, and repealing Commission Directive 87/250/EEC, Council Directive 90/496/EEC, Commission Directive 1999/10/EC, Directive 2000/13/EC of the European Parliament and of the Council, Commission Directives 2002/67/EC and 2008/5/EC and Commission Regulation (EC) No 608/2004. Off J Eur Union 2011;L304:18–63.
8. European Parliament and Council. Regulation (EU) No 528/2012 of the European Parliament and of the Council of 22 May 2012 concerning the making available on the market and use of biocidal products. Off J Eur Union 2012;L167/1(1):1–123.
9. European Parliament and Council. Proposal for a Regulation of the European Parliament and of the Council on medical devices, and amending Directive 2001/83/EC, Regulation (EC) No 178/202 and Regulation (EC) No 1223/2009, COM(2012) 542 Final, 2012.
10. Oberdörster G, Oberdörster E, Oberdörster J. Nanotoxicology: An emerging discipline evolving from studies of ultrafine particles. Environ Health Perspect 2005;113:823–839.
11. Oberdörster G, Finkelstein JN, Johnston C, Gelein R, Cox C, Baggs R, Elder AC. Acute pulmonary effects of ultrafine particles in rats and mice. Res Rep Health Eff Inst 2000;96:5–74.
12. Hayden BE, Pletcher D, Suchsland J-P. Enhanced activity for electrocatalytic oxidation of carbon monoxide on titania-supported gold nanoparticles. Ang Chem 2007;46:3530–3532.
13. Ramachandran G, Paulsen D, Watts W, Kittelso D. J Environ Monit 2005;7:728–735.
14. Hussain I, Graham S, Wang Z, Tan B, Sherrington DC, Rannard SP, Cooper AI, Brust M. J Am Chem Soc 2005;127:16398–16399.
15. O'Brien S, Brus L, Murray CB. J Am Chem Soc 2001;123:12085–12086.
16. Rogach AL, Talapin DV, Shevchenko EV, Koronowski A, Haase M, Weller H. Adv Funct Mater 2002;12:653–664.
17. Saveyn H, De Baets B, Thas O, Hole P, Smith J, Van der Meeren P. J Colloid Interface Sci 2010; 352:593–600.
18. Carr B, Wright M. *Nanoparticle Tracking Analysis: A Review of Applications and Usage 2010–2012*. NanoSight Ltd; 2013.
19. Lord FRS. Rayleigh. Philos Mag 1879;8:261–274.

20. Born M, Wolf E. *Principles of Optics*. 7th ed. Cambridge University Press; 1999, ISBN 0 521 642221.
21. Kholodenko AL, Douglas JF. Phys Rev E 1995;51:1081–1090.
22. W. H. Coulter, Means for counting particles suspended in a fluid, US Patent 2,656,508, 1953 Oct 20.
23. Kozak D, Anderson W, Vogel R, Trau M. Nano Today 2011;6:531–545.
24. Degueldre C, Favarger P-Y, Bitea C. Anal Chim Acta 2004;518:137–142.
25. Pace HE, Rogers NJ, Jarolimek C, Coleman VA, Higgins CP, Ranville JF. Anal Chem 2011;83:9361–9369.
26. Olesik JW, Gray PJ. J Anal At Spectrom 2012;27:1143–1155.
27. Lee S, Bi X, Reed RB, Ranville JF, Herckes P, Westerhoff P. Environ Sci Technol 2014;48:10291–10300.
28. Cornelis G, Hassellov M. J Anal At Spectrom 2014;29:134–144.
29. Telgmann L, Metcalfe CD, Hintelmann H. J Anal At Spectrom 2014;29:1265–1272.
30. de Broglie L. Philos Mag Ser 6 1924;47:446–458.
31. Williams DB, Barry CC. *The Transmission Electron Microscope*. US: Springer; 1996, ISBN 978 0 387 76500 6.
32. Reimer L. *Scanning Electron Microscopy: Physics of Image Formation and Microanalysis*. Springer Series in Optical Sciences vol. 45. 2nd ed. Springer; 1998, ISBN 978 3 540 38967 5.
33. ISO 9276-1 (to-6). *Representation of Results of Particle Size Analysis*. Geneva: International Organization for Standardization; 1998–2000.
34. Binnig G, Quate CF, Gerber C. Phys Rev Lett 1986;56:930–934.
35. Giessibl FJ. Rev Mod Phys 2003;75:949–983.
36. Giordano NJ. *College Physics: Reasoning and Relationships*. 2nd ed. Brooks/Cole: Cengage Learning; 2010.
37. European Parliament and Council. Regulation (EC) No 1333/2008 of the European Parliament and of the Council of 16 December 2008 on food additives. Off J Eur Union 2008;L354:16–33.
38. European Parliament and Council. European Parliament and Council Directive 94/36/EC of 30 June 1994 on colours in foodstuffs. Off J Eur Commun 1994;L237:13–29.
39. Glantz M, Hakansson A, Lindmark Mansson H, Paulsson M, Nilsson L. Langmuir 2010;26:12585–12591.
40. Pascau J, Mateos Pérez JM. *Image Processing with ImageJ*. PACKT Publishing; 2013. , ISBN 139781783283958.
41. Mondini S, Ferretti AM, Puglisi A, Ponti A. Nanoscale 2012;4:5356–5372.
42. Heroult J, Nischwitz V, Bartczak D, Goenaga-Infante H. Anal Bioanal Chem 2014;406:3919–3927.
43. Braun A, Kestens V, Franks K, Roebben G, Lamberty A, Linsinger TPJ. J Nanopart Res 2012;14:1012–1024.
44. Dekkers S, Krystek P, Peters RJB, Lankveld DPK, Bokkers BGH, van Hoeven-Arentzen PH, Bouwmeester H, Oomen AG. Nanotoxicology 2011;5:393–405.

5

SOLUBILITY PART 1: OVERVIEW

R. Tantra

Quantitative Surface Chemical Spectroscopy Group, Analytical Science, National Physical Laboratory, Teddington, TW11 0LW, UK

E. Bolea

Group of Analytical Spectroscopy and Sensors (GEAS), Institute of Environmental Sciences (IUCA), Universidad de Zaragoza, 50009, Zaragoza, Spain

H. Bouwmeester

Toxicology and Bioassays, RIKILT – Wageningen University & Research Center, 6708 WB, Wageningen, The Netherlands

C. Rey-Castro and C. A. David

Departament de Química and Agrotecnio, Universitat de Lleida, 25198, Lleida, Spain

J-M Dogné

Department of Pharmacy, University of Namur (UNamur), 5000, Namur, Belgium

F. Laborda

Group of Analytical Spectroscopy and Sensors (GEAS), Institute of Environmental Sciences (IUCA), Universidad de Zaragoza, 50009, Zaragoza, Spain

J. Laloy

Department of Pharmacy, University of Namur (UNamur), 5000, Namur, Belgium

K. N. Robinson

Quantitative Surface Chemical Spectroscopy Group, Analytical Science, National Physical Laboratory, Teddington, TW11 0LW, UK

A. K. Undas and M. van der Zande

Toxicology and Bioassays, RIKILT – Wageningen University & Research Center, 6708 WB, Wageningen, The Netherlands

Nanomaterial Characterization: An Introduction, First Edition. Edited by Ratna Tantra.

5.1 INTRODUCTION

Solubility is an important physicochemical parameter of relevance in many nano-specific scenarios. For example, let us take the case of sun-cream products often contain ZnO nanomaterial. The function of this nanomaterial in the cosmetic formulation is to filter out UV radiation. The ability to measure the solubility of ZnO in the formulation is important; if ZnO nanomaterial dissolves, then the product will no longer be able to give sufficient protection against harmful rays of the sun. In this sense, the ability to measure solubility is important in designing stable nanomaterial-based products. From a risk management point of view, the measurement of solubility is also important; if shown to be completely soluble, then disposal can be treated much in the same way as "ordinary" chemicals potentially simplifying testing and characterization regimes.

But what do we mean by solubility of nanomaterials?

Organisation for Economic Co-operation and Development (OECD) has defined solubility as "the degree to which material (the solute) can be dissolved in another material (the solvent) so that a single homogeneous phase results …" with water solubility being "the saturation mass concentration of the substance at a given temperature. Water solubility is expressed in mass of solute per volume of solution. The SI unit is kg/m^3 but g/l is commonly used" [1]. In relation to nanomaterials (of relevance to nanotoxicology), solubility has been defined by International Organization for Standardization (ISO) as "… the maximum mass or concentration of the solute that can be dissolved in a unit mass or volume of the solvent at specified (or standard) temperature and pressure, unit [kg/kg] or [kg/m^3] or [mole/mole]" [2]. In addition to the formal definitions of solubility, there is a need to understand differences between solubility and other similar terms, such as dissolution and dispersibility, which are often used interchangeably. Solubility can be differentiated from dissolution on the basis that solubility is assumed to be at equilibrium, whereas dissolution is considered to be a kinetic process [3]. Dispersibility, however, has an entirely different meaning and has been defined as the degree to which a particulate material can be uniformly distributed in another material (being the dispersing medium or continuous phase) [3].

Although the ISO definition for solubility is clear, its measurement is not trivial. There are several issues that can potentially complicate the measurement. Firstly, there is a need to have a method that is sufficiently selective to measure dissolved species, as opposed to particulate matter, for example, nanomaterials. Secondly, there is a need to appreciate different factors that can affect dissolution. For example, temperature increase can result in an unstable suspension, which leads to agglomeration and sedimentation of the nanomaterial. Potentially, this can lead to changes in solubility due to changes in particle size and hence the amount of particles exposed [4, 5] to the surrounding medium. Furthermore, the presence of dissolved species at different concentrations, such as "free" (hydrated) metal ions, soluble inorganic complexes (e.g., chlorides, hydroxides), soluble organic complexes (e.g., amino acids, proteins), can potentially affect dissolution.

In the case of ZnO nanomaterial, the dissolution process is dependent on external environmental conditions, such as pH [6, 7]. At low pH, the dissolve zinc is likely to exist as free ions [8]:

$$ZnO + 2H_3O^+ \rightarrow Zn^{2+} + 3H_2O \tag{5.1}$$

Although the Zn^{2+} species found in (Eq. 5.1) is referred to as "free," strictly speaking these ions will exist in hydrated forms, with the chemical formula of $[Zn(H_2O)_n]^{2+}$. At moderately to highly alkaline pH, different species of dissolve zinc exist [9, 10]:

$$Zn^{2+} + OH^- \rightarrow Zn(OH)^+ \tag{5.2}$$

$$Zn^{2+} + 2OH^- \rightarrow Zn(OH)_{2(aq)} \tag{5.3}$$

$$Zn^{2+} + 3OH^- \rightarrow Zn(OH)_3^-, \text{ and so on} \tag{5.4}$$

leading to the formation of soluble hydroxo complexes such as $Zn(OH)_3{}^-$ or $Zn(OH)_4{}^{2-}$ [8].

One main complication that can arise in the measurement of solubility is to know what to measure. According to Equations 5.1–5.4, dissolved Zn can be in various forms: Zn^{2+}, $Zn(OH)_3{}^-$, $Zn(OH)_4{}^{2-}$, and so on. The question as to which of the dissolved species to be measured will be dependent on the testing requirements. For example, according to the OECD and ISO solubility definition, solubility should be focused on measuring total dissolved nanomaterial, without taking into account species differentiation. However, it may be argued that this definition may not be entirely supportive in some applications such as nanoregulation. In fact, in 2008 European Chemicals Agency (ECHA) released guidance, recognizing the need to determine the distribution of the element among their dissolved species [11], a view shared among the trace metal toxicology community. This has been supported by past research, which underlines the importance of measuring the free ions, as they are often shown to be of relevance to toxicity [12–16], as exemplified by nano-Ag [17–19], and nano-ZnO [20–23].

The purpose of the chapter is to give an overview of different methods that can potentially be used to determine the solubility of nanomaterials. The chapter will start by presenting the different methods. As it is not the intent of this chapter to delve into the details of the individual techniques, only a brief technical background will be given. In general, the methods presented can be broadly divided into four categories: (i) separation methods, (ii) methods to quantify free ions, (iii) methods to quantify total dissolved species, and (iv) theoretical modeling/predictions. It is worthy to note that some of the separation methods, for example, filtration and centrifugation, do not measure solubility by themselves and are often integrated with detection methods. After presenting an overview of the different methods, a case study will be presented. The aim here is to illustrate nanomaterial solubility measurement in the context of a specific nano-specific application. In particular, this case study will explore the feasibility of microfluidics technology to measure solubility of ZnO. The device to be assessed is a miniaturized capillary electrophoresis (CE) with conductivity detection. The aim of the case study is to investigate as to whether such a device is fit for purpose and to determine if there is scope for further method development.

5.2 SEPARATION METHODS

An important requirement to measure the solubility of nanomaterials is the ability to differentiate signals arising from the dissolved component on top of the background, that is, interferents, such as particulates. If a technique cannot differentiate such signals, then it is important to remove the interferents prior to the detection and quantification of the dissolved species. This section assesses the different separation methods that can be employed as part of final method to determine the solubility of nanomaterials.

5.2.1 Filtration, Centrifugation, Dialysis, and Ultrafiltration

To date, there exists a published OECD document, which gives guidance on the measurement of solubility. However, this guidance document is for chemical and pure substances (not for nanomaterials) that are stable in water and not volatile [1]. This document specifies two types of methods that should be considered [1], which will be referred to as "column method" and "flask method." These methods have been developed in order to achieve conditions for solubility, that is, when at equilibrium conditions, for the tested material. The guideline also specified the need to have a separation method incorporated in the analysis. In the column method, the recommendation is to use a plug of glass wool at the column outlet to filter out particles. In the flask method, the recommended separation step is to use centrifugation. In both cases, the final liquid sample collected is assumed to be free from particulate matter, which can then be analyzed using an appropriate analytical method. Although the OECD column and flask method have been developed for chemicals, their relevance to nanomaterials is currently being considered and revisions to this guideline to adapt it to nanomaterials are still ongoing. To date, neither the column nor the flask method has been employed for nanomaterials.

Undoubtedly, one of the main advantages of using the OECD methods is accessibility of the technology along with ease of use and affordability. In this respect, other methods are currently available worth considering. The simplest is the use of filter membranes that will remove particulate matter, with 20-nm-pore-size filter membranes being commercially available. In addition to filters, samples have also been centrifuged in order to extract the resultant supernatant for analysis as reported by earlier studies [24, 25]. However, these methods may be unreliable, potentially due to nanoscale particulates, which can end up in the final analysis sample. In the case of filtration, subnanometer particulates can pass through the pores of filters. In centrifugation, care must be taken when extracting the resultant supernatant. In particular, the operator must ensure that the extraction process takes place without disturbing the particulates formed, that is, pellet formation after centrifugation has taken place; this is not easy to achieve. In fact, Xu and coworkers have reported that incomplete removal of nanomaterials from the supernatant can occur even after exhaustive centrifugation [26], confirming the potential unreliability of such methods.

In an effort to improve the reliability, alternative ways to remove particles in a suspension should be considered, such as equilibrium dialysis and ultrafiltration (UF). In equilibrium dialysis, the basic principle is that it relies on diffusion of analytes across a semipermeable membrane. In the past, several researchers have used equilibrium dialysis to measure the solubility of nanomaterials [24, 27]. A disadvantage in dialysis is that it can take a long time. Thus, a quicker method of ultrafiltration (UF) is often preferred to speed up the process. Similarly to dialysis, UF also uses a semipermeable membrane but it differs in that it is a pressure-driven process through the use of a vacuum/centrifugal force. However, separation still occurs across the semipermeable membrane [28]. UF is often used as a preparative method in biology, to simultaneously purify, preconcentrate, and fractionate macromolecules or particulate-based suspensions [29]. The use of UF has been extensively applied to nanomaterials and membranes with pore size in the range of 3–100 kDa; it has been used to study the solubility/dissolution of Ag [24, 30], ZnO [7, 24, 31, 32], CeO_2 [24, 30, 33], TiO_2 [24], CuO [34], quantum dots and Au [24]. One disadvantage in UF is potential interactions of nanomaterials and other dissolved species with the membrane. The choice of membrane is therefore crucial. Even though membrane pore sizes as low as 1–3 kDa (smaller than 1 nm) sizes are currently available, a complete removal of all particles may prove difficult to achieve if subnanometer particles below 1 nm exist. In a situation where a better separation is needed, other methods should be explored, which are described in the following sections.

5.2.2 Ion Exchange

Another promising separation method is ion exchange technology (IET), ion exchange resins being the most common method, and is mainly applicable in the separation of free metal ions. The method involves the equilibration of a known mass of resin in a column with a given volume of a sample [35, 36], and separation is based on the assumption that the resin/aqueous partition coefficient for the metal is constant with respect to the free metal ion concentration. Although a column resin is being specified here, binding the sample in a batch mode can also be done.

The IET requires a careful control of pH and percolation time to ensure that equilibrium conditions are attained. Once equilibrium is attained, steps of washing and elution follow and the metal ions contained in the eluted sample can then be quantified using a suitable analytical technique. IET has several advantages: (i) it can not only separate but also preconcentrate samples, (ii) large preconcentration factors are possible, which implies that very low detection limits can be reached (down to 10^{-9} mol/l)

[37, 38], and (iii) it is shown to be reliable, as shown through cross validation with electrochemical techniques [39]. Although the method allows primarily the determination of free ions, potentially it can be used for the simultaneous evaluation of free ions and total dissolved metal concentrations, as achieved through certain schemes [40]. Despite the promising features of IET, only few studies have been carried out in relation to the measurement of nanomaterial dissolution/solubility [38], suggesting that further method development is needed for nanomaterials.

Another method worthy of note is diffusion gradients in thin films (DGT) [41]. The DGT device is commercially available and also involves the use of ion exchange (or chelating) resins [42] ("DRT Research Ltd., Technical Documentation") [42]. The main component of the device is a passive sampler that consists of a three-layer assembly encased in a cylindrical plastic holder with a circular window that allows contact with the aqueous sample. The three-layer assembly is composed of (i) a membrane filter (0.45 μm pore size) that is in contact with the sample solution, (ii) a second hydrogel layer (of a known thickness, in which the diffusion layer is usually 800 μm thick), and (iii) a thin hydrogel layer (400 μm thick) containing embedded beads of a metal-binding resin. The hydrogels in the device are typically made of cross-linked agarose/polyacrylamide, which can be supplied with different porosities, for example, "open pore" and "restricted pore" gels. Typically, the binding resin Chelex® 100 is used for most metal cations, but other binding resins are also available for inorganic anions, for example, titanium dioxide or ferrihydrite [43]. Past studies have shown that the Chelex chelating resins have been successfully applied to determine more than 30 elements including alkaline earth and transition metals, metalloids, lanthanides, and actinides [44].

The DGT involves deploying the device in an aqueous sample under (ideally) constant convective conditions, for a given period of time, typically from 1 h to a few days (depending on the sensitivity required). During this time, the chemical species in the sample diffuse through the membrane filter, then to the hydrogel layer before reaching the innermost film, where the ionic species of interest bind to the ion exchange resin. The physical principles of DGT thus lie on the interplay between chemical kinetics and diffusive transport of the analytes within the three-layer assembly, and, thus, separation is achieved on the basis of charge, size, and chemical affinity. After a suitable deployment period, the DGT device can be retrieved from the sample solution, disassembled, and analytes desorbed/eluted from the binding layer with, for example, nitric acid solutions. The eluted sample can then be quantified using a suitable analytical technique, typically atomic spectrometry. The concentration of analytes measured by DGT is usually regarded as a labile fraction of the dissolved element. Generally, the value of the DGT-measured concentration lies between the free ion concentration and the total dissolved metal concentration. Further theoretical details on DGT will not be covered here and can be found elsewhere [43, 45, 46]. In relation to nanomaterials and the measurement of solubility, past studies have used DGT in the context of ecotoxicological studies and have been applied to several nanomaterials such as silver [19, 47–51], copper oxide [50, 52, 53], and zinc oxide [50].

DGT shares many of the advantages with IET: simple, robust, can be used over a wide range of pH, capable of multielement analysis, has good reproducibility and sensitivity. In addition, large preconcentration factors and consequently low detection limits, and high elemental selectivity can be achieved in combination with inductively coupled plasma mass spectroscopy (ICP-MS) for the analysis of eluted samples. However, they have their limitations. As with IET, their selectivity and primary use are toward the quantification of a certain fraction (free or labile) rather than total dissolved species. DGT requires careful control in temperature to optimize performance, although theoretical corrections can be applied to account for temperature variations [44]. DGT devices, however, are relatively nonexpensive, easy to use, and can be deployed directly in situ (for ecotoxicological study purposes). In fact, several studies have shown its suitability for *in situ* speciation technique for the analytical measurement of labile inorganic species in aqueous media [40, 41, 43, 54, 55].

A disadvantage of both DGT and IET is associated with the large sample volume needed for analysis, that is, in the order of hundreds of milliliters. In relation to their use for analyzing nanomaterial samples, there are several issues that are still associated with DGT and IET. In IET, the direct contact of the exchange resin with the sample means that adsorption of particles can take place on the resin beads. In DGT, the effect of partial penetration of nanoparticles into the diffusion gel is still not well understood.

5.2.3 High-Performance Liquid Chromatography, Electrophoresis, Field Flow Fractionation

High-performance liquid chromatography (HPLC) employs a chromatography column (stationary phase) to separate different components in a sample. When the sample enters the column, the sample interacts with the stationary phase. A solvent (eluent) is then added to the column; this is (flowing) mobile phase and separation is achieved due to differences in partitioning behavior between the flowing mobile phase and the stationary (often reverse) phase. The composition and temperature of the eluent play a role in the separation process as they govern the interactions between sample components and the stationary phase. However, the nature of the stationary phase, as well as the partition mechanism, is more important than the mobile phase itself.

HPLC is not just a separation method, as an HPLC system will have a detection system integrated, to detect/quantify the separated analytes. Different detection modes can be attached to an HPLC system, including fluorescence, UV–Vis, and electrochemical detectors. In relation to nanomaterial characterization, HPLC methods have been used to determine particle size/concentration of gold nanoparticles [56] and fullerenes [57, 58]. To date, HPLC has not been used in the measurement of nanomaterial solubility.

Another powerful separation method is capillary zone electrophoresis (CZE). This electrokinetic-separation-based method is particularly powerful and widely used in separation science. In CZE, high voltages are applied in order to separate charged molecules when moving in an electric field. The CZE setup typically consists of a buffer-filled capillary (small-diameter tubes) placed between two buffer reservoirs, with a potential field being applied across the capillary. The velocity of a given analyte in the capillary is governed by electrophoretic mobility as dictated by the charge to mass ratio of the analyte. In addition, the velocity is also affected by electroosmotic flow (EOF), as it is governed by zeta-potential of the surface of the capillary wall. The EOF can be defined as an induced bulk liquid flow that is moving with the same velocity as the counter ions, that is, these are the cations in the diffuse layer at the walls of the capillary. The electrophoretically separated species are often detected near the outlet end of the capillary, where various detection schemes can be employed such as fluorescence, absorbance, electrochemical, and refractive index. The final choice of detection method will however be dependent on the analytical criteria or requirements. These requirements often involves taking into account factors such as: selectivity, sensitivity, sample type and so on.

The use of CZE-based methods have been developed and refined over decades, widely used for separation of different analytes found in complex mixtures. These methods have been used for analysis of inorganic anions and cations, as well as organic ions such as carboxylic acids, amines, amino acids, peptides, proteins, DNA fragments, antibiotics, and many other pharmaceutical compounds. In relation to nanomaterial characterization, electrophoresis has been shown to be able to separate nanomaterials of different sizes, shapes, and compositions [59–64]. In particular, CZE has been used to separate carbon nanotubes [65–71], gold [72–76], iron oxide [77–81], silica [82–83], quantum dots [84, 85, 86], polystyrene [87, 88], core shell particles [81, 89–91], and others [92–96]. CZE has been shown to be particularly useful in the study of nanoparticle–protein interactions [97–99]. Although these studies have shown the potential use of CZE in relation to nanomaterial analysis, much fundamental work is still required. Unlike molecules, the theoretical understanding of separation mechanism and electrokinetic behavior of nanomaterials are not completely understood. For example, electrophoretic motion can no longer be estimated by considering molecular weight and expected charge as particles possess "electrophoretic heterogeneity," that is, exhibiting a wide distribution of charge, size, and shape, all of which can vary with experimental conditions and time. In relation to using CZE-based methods to measure the solubility of nanomaterials, its use has been tested for ZnO nanomaterials [100].

Finally, a reliable separation method worth considering is field flow fractionation (FFF). FFF is a family of techniques designed to separate analytes with different physicochemical properties. The separation principle is based on the differential movement of analytes in a fluid flowing in laminar regime inside a thin channel, generally directed toward a detector. Separation is not directly caused by the flow itself, but by a generated field perpendicular to the direction of the flow. The applied field determines the properties on which the analytes will be separated. The field may be generated by sedimentation, electrical, or magnetic forces, thermal gradients, or a cross flow [101]. The latter, called flow FFF, is the most commonly used FFF technique in which the analytes are driven toward the boundary layer, that is, the so-called accumulation wall. This wall consists of a semipermeable membrane lying at the top of a supporting frit. The lower size limit of this technique is thus set by the molecular weight or size cut off of the membrane. The most critical factor for flow FFF is thus the choice of membrane since interactions of nanomaterials with the membrane could occur, depending on the type of membrane [102]. Separation in flow FFF is driven by differences in diffusion of the analytes, which is inversely correlated with the hydrodynamic diameter of the analyte, that is, the lower the analyte diameter, the larger the diffusion. Smaller analytes have higher diffusion rates and will therefore be transported at a higher flow speed in the parabolic flow profile than larger ones. The original flow FFF design includes two permeable walls, but the more recent asymmetrical design (with one permeable wall) is the most commonly used and this design results in asymmetrical flow FFF or AF4 [103]. A variety of detectors may be coupled to FFF and the choice is dependent on the type of analyte to be detected. Commonly used detectors for the detection of nanomaterials include multiangle light scattering (MALS), UV absorbance, and elemental detectors such as ICP-MS or inductively coupled plasma optical emission spectrometry (ICP-OES) [101, 103].

In relation to the separation of nanomaterials, AF4 has several advantages over other separation techniques. Firstly, it does not comprise a stationary phase, which is in contrast to traditional liquid chromatography techniques such as size exclusion and ion exchange chromatography. A stationary phase is often not desirable for the separation of nanomaterials since unpredictable interactions between the stationary phase and nanomaterials could occur possibly, for example, irreversible binding of the nanomaterial to the stationary phase. Secondly, the carrier solution may be changed with respect to pH and ionic strength in order to match the carrier solution with the sample composition. Finally, the outcome of AF4 is easy to interpret. Separation is driven only by size, so the retention time is directly proportional to physical properties of nanoparticles, whereas techniques such as ion exchange chromatography and capillary electrophoresis are driven by both size and charge. AF4 is also particularly advantageous when nanomaterial is dispersed in complex biological matrices and applicable to the analysis of heterogeneous samples (containing a broad mass/size ranging from ~1 nm up to 100 μm), thus reduces the need for sample preprocessing [104]. The selectivity and speed of the technique are also tunable. Furthermore, the low shear rates provide the possibility to handle samples in which weak forces prevail and thus the "soft" fractionation of this technique compared to other techniques [101]. Compared to CZE, AF4 has a greater sensitivity due to larger injection volumes that can be used. Nevertheless, the AF4 channel can easily be overloaded, requiring dilution of the sample and a sensitive detection system.

Over the last decades, asymmetrical flow FFF or AF4 [102–107] has been extensively refined with respect to separation and characterization of various nanomaterials, in particular Au [108–110], Ag [111–115], SiO_2 [116], TiO_2 [117, 118], CNTs [119–122], quantum dots [123], liposomes [124–128], and hybrid nanomaterials [129, 130]. Most of these studies, however, have been in relation to characterization of specific physicochemical properties of nanomaterials, such as size and mass. However, regarding solubility, there is little information on using AF4 as a method. This is mainly due to the fact that soluble (dissolved) material will mostly be transported through the membrane by the cross flow. The cross flow is used as a means to separate the nanomaterial and is collected separately, often regarded as waste. Therefore, this flow is typically not on-line and not connected to a detection system. Nevertheless, the cross flow may be collected and measured as a separate fraction, theoretically rendering this technique to be suitable for solubility measurements [131]. In that case, it should be pointed out that, although possible, the high dilution of the species collected will make its detection challenging.

5.3 QUANTIFICATION METHODS: FREE IONS (AND LABILE FRACTIONS)

For species characterization, there is a need to have methods that can quantify free ions. In this section, an overview of some of these methods is given. Electrochemical method and colorimetric method are presented.

5.3.1 Electrochemical Methods

Perhaps the simplest and cheapest method to measure the presence of ions is measuring electrical conductivity, that is, a solution's ability to carry a current and conduct electricity. A typical setup for a conductivity meter consists of two metal electrodes in contact with the solution and then applying alternating voltage to the electrodes, which causes the ions to move back and forth to produce a current. An analyzer then measures the current and using Ohm's law, resistance of the solution is calculated and hence conductance (the reciprocal of resistance). The measurement of conductivity has several advantages such as it is rugged, easy to use, with quick analysis time. Furthermore, the meter can be calibrated against a solution of known conductivity, that is, solutions of potassium chloride. Therefore, it is not surprising that measurement of conductivity has been widely accepted and has been shown to be particularly useful for the assessment of water quality [132, 133]. Strictly speaking, conductivity is governed by the concentration of all ionic species. Hence, one limitation in conductivity measurement is that it is unspecific, and without a separation technique, the conductivity signal can be easily masked by background ions. Hence, in order to use it for nanomaterial characterization, a reliable separation method is needed. Integration with CZE, as previously discussed, is one such possibility [100].

Electrochemical methods that offer better selectivity than conductivity methods are potentiometry, that is, ion selective electrodes (ISE), and stripping voltammetry [134–136]. In potentiometry, measurement of free ions involves an electrochemical cell that consists of a reference electrode (with an electric potential independent of the measuring conditions) and an ISE. The ISE is a membrane-based electrochemical sensor that responds specifically to the activity of a particular ionic species. There are different kinds of ISEs, depending on the nature of the membrane (glass, liquid, and solid state). When the ISE is immersed in solution, an electric potential is created across the electrodes and measured by a millivolt meter under (ideally) zero-current conditions [135, 136]. Theoretically, the measured electric potential in these cells is a linear function of the logarithm of the free ion activity, as described by Nernst equation. ISEs usually show a linear Nernstian behavior over at least 5–7 orders of magnitude in the analyte concentration, offering a very wide analytical window. The theoretical value of the Nernst slope at 25 °C is 59.1 mV per pH unit, which means that a 1 mV change in potential corresponds to a 4–8% change in the activity of a monovalent/divalent ion. Therefore, the analytical sensitivity is relatively low, and, accordingly, small uncertainties in the measured potential can lead to relatively large errors (uncertainties are generally in the range of ±0.1–1 mV). The main source of uncertainty is often associated with instability of the reference electrode junction potential [136]. ISE electrodes must be calibrated either by using concentration standards or by performing serial dilutions of a standard stock solution in the same medium used for the solubility test [135, 137].

ISE potentiometry has several advantages. ISE is considered to be a nondestructive technique, with fast response (often less than 1 min), not affected by sample color or turbidity, relatively inexpensive, and does not entail extensive operator training. Furthermore, only a small amount of sample (in the order of 10 ml) is required for analysis, and specially designed flow cells can be used with much smaller volumes [136]. However, several limitations are associated with ISEs. Firstly, the detection/quantification limit is not much lower than 10^{-6} M, although in well-buffered systems (with excess of a metal binding ligand) values as low as 10^{-14} M have been reported by some authors [138, 139]. Secondly, they are potentially sensitive to certain interferences/contaminations, with most ISEs showing a response for other ions in solution apart from the target ion, to a greater or lesser extent [140]. Interference effects are mainly due to the presence of other ions of the same charge as the target ion. Lastly, commercial ISEs offer only a limited variety of the ions that can be analyzed, such as Ca^{2+}, Cd^{2+}, Cu^{2+}, Ag^{+}, and Pb^{2+}. To date, the vast majority of past studies surrounding the use of ISE to measure dissolution/solubility have been associated with ecotoxicology investigations using limited types of nanomaterials: silver [24, 47, 138, 139, 141] and copper [142].

Similar to ISE, stripping voltammetric methods have primarily been used for the measurement of free ions. These methods are dynamic (nonzero current) techniques that measure the electric current or potential in a three-electrode cell as a function of time, while an independent variable (usually the electric potential or the current) is controlled [134]. The measured current (or potential) signal is used to obtain quantitative information on the electroactive species of the sample concentration (but often also kinetic and transport parameters). Promising voltammetric methods are those based on some kind of preconcentration step, which are found in methods such as anodic stripping, adsorptive cathodic stripping voltammetry, and absence of gradients and Nernstian equilibrium stripping (AGNES). These will be explained in more detail below.

Anodic stripping voltammetry (ASV) consists of two stages. In the first stage, a small amount of electroactive metal species (negligible in comparison with the total content of the sample) is reduced and preconcentrated at an electrode, under controlled conditions of time, stirring, and deposition potential. This stage is what is referred to as an accumulation step. Usually, mercury is used as a working electrode (in thin film or hanging drop configurations), so that the reduced metal forms an amalgam although this is limited to a few metal ions such as Zn^{2+}, Cd^{2+}, Pb^{2+}, Cu^{2+}, In^{3+}, Sn^{2+}, or Tl^{+}. However, instead of mercury, bare or modified solid electrodes of Au, Ag, or Pt can also be used. The second stage is what is referred to as the "stripping stage," that is, the accumulated metal is reoxidized (or "stripped") as the potential is scanned anodically (toward more positive potentials) either in a linear way or, usually, in a pulse waveform (such as the differential pulse mode, DPASV) [136]. The resulting signal measured is a peak current, whose profile is dependent on characteristics of metal ions (such as diffusion coefficient, charge number), the geometry of the electrode, the hydrodynamic conditions during the deposition step, and so on and the medium composition (due to possible contributions from the dissociation of metal complexes during deposition).

In adsorptive cathodic stripping voltammetry (AdCSV), an indirect method of determination of dissolved metal concentrations is employed. This is based on the competitive ligand complexation with a suitable chelating agent [35, 143], resulting in the formation of a surface-active, electroactive metal complex. The metal complex is then accumulated by adsorption on the electrode surface at a controlled potential (during the accumulation step). In the stripping step, a negative-going potential scan is applied, and the current due to the reduction of the adsorbed metal complex (proportional to its electrode surface concentration) is recorded.

AGNES is also a stripping-based technique, which employs a mercury working electrode. Unlike ASV and AdCSV, it has been designed to measure exclusively the free ion concentration, regardless of the formation of labile metal complexes. The fundamental difference in AGNES (in relation to other stripping methods) is that the accumulation stage lasts until a special equilibrium state is reached where (i) there is no concentration gradient of any species within the mercury electrode or the surrounding solution and (ii) the redox couple is in Nernstian equilibrium at the mercury interphase [144]. Under these conditions, the applied deposition potential determines the gain or preconcentration factor in the amalgam, with respect to the free ion concentration in the solution, through Nernst equation. The stripping stage in AGNES allows the quantification of the metal accumulated in the amalgam by measuring the current under diffusion-limited conditions or the charge as the response function [144, 145]. The faradaic current measured at a fixed time (after subtraction of a suitable blank) is proportional to the free metal ion concentration.

Overall, stripping-based methods can potentially have remarkably high sensitivities, mostly due to large preconcentration factors (of the order of 10^2–10^3) being achieved prior to stripping. These methods are usually able to reach sub-ppb detection limits (as low as 10^{-10}–10^{-12} mol/l), which compares favorably with other techniques such as inductively coupled plasma detection techniques. Time of analysis is short (typically, a few minutes) and relies on relatively inexpensive equipment. Moreover, in many cases, these techniques do not require a solid–liquid separation step, but having said this, there is potential for the adsorption of dissolved organic matter and particles on the working electrode, which may interfere with the measurement.

A disadvantage of stripping-based methods is the need for some level of expertise of operators and the current degree of automation is probably not so well developed. However, AGNES avoids typical complications in the interpretation of dissociation kinetics and mass transport of metal complexes, and so on, so that the results are much simpler to interpret. Regarding the measurement of nanomaterial solubility/dissolution, ASV has been used in solubility measurements for quantum dots [146], silver [147, 148], and gold [149]. The use of AdCSV has been limited so far to the study of solubility/dissolution of titanium dioxide nanoparticles [150, 151]. AGNES has been used for the solubility and ion binding measurement for various nanomaterials: latex [152], zinc oxide [22, 23, 153, 154], quantum dots [155], and clays [156].

5.3.2 Colorimetric Methods

In addition to electrochemical-based methods, colorimetric (including fluorimetric) assays, in conjunction with chelating reagents, can also be used to measure free metal ions [157, 158]. These methods rely on the interaction of the metal with the chelating agent to result in a colored complex, which can be monitored using appropriate spectrometers, for example, through measuring a change in absorbance or fluorescence signal. As with electrochemical-based methods, these methods are inexpensive [159] and do not require extensive sample preparation. Hence, they are not so labor intensive and thus viable alternatives to those methods that require more sophisticated instrumentation, such as atomic absorption spectrometry (AAS) or inductively coupled plasma mass spectrometry. A number of chelating agents already exist commercially, which offer selectivity to specific metal ions. For example, Zincon, that is, 2-carboxy-2′-hydroxy-5′-sulphoformazylbenzene, has been used for the determination of free zinc ions as well as copper and cobalt ions. Although the actual sensitivity of such assays is dependent on the molar absorptivity of the complex, a method using Zincon has been reported to achieve a detection limit of ~200 nM [159]. As their selectivity is not exclusive to one type of ion, analysis using Zincon may be problematic if other ions are also present in solution, which may interfere or mask the ions of interest. To overcome this, method development to improve selectivity to a particular ion is key, but this may not be straightforward. Säbel and coworkers have used Zincon for dual-metal quantification [159], that is, for Cu^{2+} and Zn^{2+}. This involves the assessment of total metal ion concentration before using EDTA to mask the Zn^{2+} ions in order to measure the Cu^{2+}. For the dual quantification of Zn^{2+} and Co^{2+}, on the other hand, PAR (4-(2-pyridylazo) resorcinol), instead of Zincon, has been shown to be more appropriate, when it comes to accuracy and precision. Other commonly used chelating agents include 5-Bromo-PAPS (2-(5-bromo-2-pyridylazo)-5-(*N*-propyl-*N*-sulfopropylamino)-phenol) and Nitroso-PSAP (2-Nitroso-5-(*N*-propyl-*N*-sulfopropylamino)phenol) for the determination of zinc and iron ions, respectively [160]. One advantage of using Nitroso-PSAP is that wavelength emission is considerably longer than that of the other metal complexes, which implies that free iron can be easily detected without interference from other metal ions. However, it has been reported that Nitroso-PSAP is affected by bilirubin

[161]. Another example of a chelating agent is bathocuproine (BC), which have been shown to be highly selective for Cu(I) ions in solution [162–164]. Although chelating agents have been used extensively in the past [165, 166], their use in relation to nanomaterial characterization is limited. An important point to highlight is the need to remove particulate matter, in particular, nanomaterials before adding relevant chelating agents into the sample solution [167].

5.4 QUANTIFICATION METHODS TO MEASURE TOTAL DISSOLVED SPECIES

5.4.1 Indirect Measurements

Although the methods discussed in the previous section are mainly suitable for the measurement of free (hydrated) ions, rather than total concentration of dissolved species, it is important to note that the total dissolved species can be sometimes estimated indirectly by titration experiments. This has already been reported in the cases of AGNES [23] and ISE potentiometry [138].

For example, consider the dissolution of ZnO in a test solution (with a mixture of different inorganic and organic compounds). The total solubility will be the sum of the free Zn^{2+} concentration plus the concentration of all complexes containing Zn:

$$S = c_{Zn^{2+}} + c_{Zn(OH)^+} + c_{ZnL^{2+}} + \dots \quad (5.5)$$

where $Zn(OH)^+$ and ZnL^{2+} are examples of the inorganic and organic soluble complexes (L is a representative organic ligand), respectively. The concentrations of each complex can be calculated from the free metal and ligand concentrations, using the corresponding conditional stability constants K_i^{cond}, which are referred to the actual background conditions (pH, ionic strength, temperature, etc.):

$$S = c_{Zn^{2+}}(1 + K_1^{cond} c_{(OH)^-} + K_2^{cond} c_L + \dots) \quad (5.6)$$

The value between brackets is characteristic of the medium composition and represents the ratio between the total solubility and the free metal concentration:

$$S = c_{Zn^{2+}} K_{medium} \quad (5.7)$$

In those cases where pH is constant, and the concentration of dissolved zinc is very low (so that the concentrations of free ligand are much larger than those of the metal complexes), the value of K_{medium} is approximately constant with respect to the free metal ion concentration. In this case, it can be calculated from independent experiments, in which the test medium (in the absence of ZnO) can be titrated against known additions of a soluble Zn standard solution. Subsequently, this allows the estimation of the total solubility of ZnO from measurements of the free ion concentration [23].

5.4.2 Direct Measurements

Although the possibility of measuring total dissolved species indirectly is an option worthy of note, a direct measurement for multielemental analysis can yield far more accurate results. As a result, methods such as atomic spectrometry, for example, ICP-OES, ICP-MS, and AAS will be discussed in this section.

Both ICP-OES and ICP-MS use inductively coupled plasma (ICP). An ICP is a discharge maintained by the interaction of a radiofrequency field and a partially ionized gas, usually argon. These plasmas can reach temperatures as high as 10,000 K, allowing the atomization and ionization of the elements in a sample and minimizing potential chemical interferences. In the case of ICP-OES, the plasma works as an excitation source for atoms and ions, whereas in ICP-MS it is a source of ions. Samples are introduced as solutions or suspensions through a nebulization system, consisting of a nebulizer and a spray chamber, which produces an aerosol of droplets. Once the droplets are into plasma, solvent evaporates, forming solid particles, which in turn are vaporized and their elements atomized and ionized. In ICP-OES, the UV-Vis radiation emitted by excited atoms and ions is collected by an optical spectrometer, which is used to separate the individual wavelengths of radiation and focus them onto a detector. In ICP-MS, ions are extracted through an interface into a mass spectrometer, where they are separated according to their mass/charge ratio and detected.

A promising use of ICP-MS is when the technique is being used in single-particle mode. Single-particle ICP-MS (SP-ICP-MS) is increasingly being used in nanomaterial analysis, primarily as it provides a means to detect individual nanoparticles. The SP-ICP-MS technique is mainly employed for the determination of particle size and number concentrations [114, 168–171], although it has been shown to be able to differentiate directly the dissolved and particulate forms of the analyte [38, 171]. The technique works by acquiring thousands of individual intensity readings with a very short dwell time (i.e., 1–10 ms) and at very low nanoparticle concentrations (i.e., thousands of nanoparticles per milliliter or nanogram per liter). Dwell times and concentrations are chosen to detect just one nanoparticle per reading. Intensity readings are collected as a function of time, and the number of pulses above the continuous background is proportional to the number concentration of nanoparticles. The intensity of each pulse is proportional to the mass of analyte in the particle, from which the particle size can be calculated (if the composition, shape, and density are known). This method requires limited sample preparation and no separation before measurement. Current limitations of this technique are that only one element can be measured at a time and the relatively large size limit of detection attainable (i.e., ~20 nm or larger). On the other hand, no distinction can be made between pristine and surface-modified nanoparticles because just the inorganic core is detected.

The use of SP-ICP-MS for solubility testing is based on the constant signal produced by the dissolved analyte, which induces a shift of the continuous background to higher values, maintaining the pulses due to the particles. By plotting the number of readings for each intensity, two distributions are obtained: the first one, at lower intensities, is due to the dissolved analyte, whereas the second one is due to the nanoparticles. By integrating the distributions and using dissolved standards, the mass concentration of the analyte in the dissolved and nanoparticle forms can be obtained [38, 114, 171]. It should be noted that the presence of dissolved species in SP-ICP-MS measurements, and its accompanying increase in background intensity, has a direct effect on the nanomaterial size detection limit of the technique [171].

As in the ICP techniques, AAS is a quantitative technique for the determination of the total element content in a sample. The technique involves the atomization of the analyte, in which atoms (in the ground state) are promoted to a higher excitation state by absorption of radiation at specific wavelengths. The amount of absorbed radiation is thus a quantitative measure for the concentration of the element analyzed. There are two commonly used atomizers: flame and graphite furnace (GF) atomizers. Flame atomizers are suitable for liquids and less sensitive than furnace atomizers. GF-AAS are used in order to quantify elements when in solutions or in solid samples; gas samples are uncommon in GF-AAS.

Most atomic spectrometry techniques often require the need to undergo conventional acid digestion as part of the sample preparation step. In the case of suspensions, this is usually carried out to prevent nebulizer blockage and coating of the spray chamber. For example, Fabricius et al. [24] have recommended the use of microwave-assisted acid digestion as the optimal strategy for reliable routine analysis of any kind of metallic nanoparticles.

The choice of ICP-OES, ICP-MS, or AAS for determining the total element content depends on the sample concentration level. ICP-MS is the most sensitive, with limits of detection below 1 ng/l, whereas in ICP-OES detection limits are 2–3 orders of magnitude higher. Detection limits of AAS vary greatly, with GF-AAS being in between ICP-OES and ICP-MS but flame-AAS being 3–4 orders of magnitude higher (i.e., worse) compared to ICP-MS. Due to the fact that the most sensitive method out of the three is ICP-MS, it has been widely used for the analysis of inorganic nanomaterials in various types of matrices, varying from *in vitro* cell culture media [172–174] to cells [175, 176] and tissues [177, 178]. Atomic spectrometry techniques measure the total element content and thus cannot distinguish between different forms of an element, that is, particulate versus dissolved forms; this is true apart from when ICP-MS is used in a single-particle mode. Hence, these techniques are often coupled to separation techniques such as FFF, hydrodynamic chromatography, HPLC, ultracentrifugation, dialysis, or ultrafiltration.

5.5 THEORETICAL MODELING USING SPECIATION SOFTWARE

Although not a measurement tool, predictive analytical models should be considered as simulated data can lead toward a better understanding of the solubility or the dissolution process. An important theoretical calculation to consider surrounds the equilibrium concentration of the different chemical species formed as a result of the dissolution of a bulk solid material, which can be carried out using a thermodynamic speciation software such as Visual MINTEQ (v. 3.0, downloadable from http://vminteq.lwr.kth.se/), MINEQL+ (v. 4.6, see http://www.mineql.com/), or

WHAM (v. 7, see http://www.ceh.ac.uk/products/software /wham/). All these programs have been originally designed for geochemistry and environmental chemistry applications. They combine a built-in thermodynamic database (with values of equilibrium stability constants of complexes, solubility products, standard redox potential values, reaction enthalpies, activity coefficients, etc., which can also be updated by the user) with a numerical algorithm for solving the set of nonlinear equations associated with simultaneous multiple equilibria in aqueous solution. In many cases, they also integrate models for the description of ion adsorption on surfaces or binding to natural colloids. Some of them also allow implementing mass transport and kinetic models, such as ORCHESTRA (see http://www.macaulay.ac.uk/ORCHESTRA/) or PHREEQC (v. 3, see http://wwwbrr.cr.usgs.gov/projects/GWC_coupled/phreeqc/index.html). Another application of the thermodynamic speciation software is calculation of the distribution of species resulting from the dissolution of a pure solid phase. For example, it can predict the precipitation (or oversaturation) of new solids in ZnO dispersions, the equilibrium of some common redox couples (e.g., Ag^+/Ag), and the distribution of a metal among inorganic (e.g., hydroxides, chlorides) and organic complexes (e.g., with amino acids, buffers, chelating agents). Calculations can take into account the effect of environmental variables such as pH, ionic strength, temperature, partial pressure of O_2 or CO_2, and so on.

In the case of nanomaterial dispersions, these models can be very useful to calculate the expected values of solubility and free ion concentrations, and several researchers have reported on the usefulness of such theoretical modeling. Most studies reported have focused on the assessment of exposure and fate of nanomaterials in ecotoxicity and *in vitro* testing, with particular reference to a number of nanomaterials such as silver [47, 179–182], zinc oxide [22, 23, 31, 153, 183], copper oxide [50, 142]. Despite immense potential, users must be aware of several limitations that exist: (i) an assumption of the existence of an equilibrium situation; (ii) results that are strongly dependent on the reliability and accuracy of the reference thermodynamic data included in the software; (iii) calculations that rely on thermodynamic data available from bulk materials, unless the database is updated manually by the user.

5.6 WHICH METHOD?

The selection of appropriate methods to measure solubility is not trivial as the selection of the technique is highly dependent on various factors such as the type of nanomaterials and corresponding matrix. However, whatever analytical method chosen, an assessment must be made in relation to whether it fulfils analytical requirements being set. As previously stated in Chapter 1, Eurochem states that "analytical measurements should be made to satisfy an agreed requirement i.e. to a defined objective and should be made using methods and equipment which have been tested to ensure that they are fit for purpose" [184]. "Fit for purpose" here implies that the technique/method must be sufficiently reliable (in relation to the level of specificity or selectivity, accuracy, precision, detectability, and sensitivity) and robust [185, 186].

Analytical requirements often considered include the need for a method to have the following characteristics:

- Highly selective, in order to differentiate signal from background, for example, other ionic/dissolved species from nanoparticles (if it is the case).
- Highly accurate (close to true value). This can only be achieved if suitable reference materials are available.
- Highly repeatable and reproducible as assessed by round robin exercise to measure laboratory-to-laboratory variability.
- Highly sensitive, often requiring low limit of quantification, for example, ppb.
- Commercially available.
- Robust/rugged, for example, not being sensitive to operator, day-to-day variability, precise concentrations of reagents, and so on.
- Having fast analysis time (to reflect the cost of the analysis). This is especially useful if multiple measurements are required or there is a need to monitor solubility close to real time.

Overall, there is no universal method to determine nanomaterial solubility, as this will be heavily dependent on the analytical requirements for each nano-specific scenarios. For example, if the measurement of total dissolved species is needed, then the use of atomic spectroscopy is an attractive choice. However, it is likely that researchers should look into the integration of a separation method prior to the detection/quantification. For example, ICP-MS is a popular detection technique that has been coupled with AF4, ultracentrifugation [187, 188], and ultrafiltration (for Ag [24, 30, 38, 47, 178, 189–191], CeO_2 [24, 30, 33], Be [192], Au, ZnO, and TiO_2 [24]). Unlike the popularity of ICP-MS as a detection technique, several studies have employed only ICP-OES to study dissolution of different nanomaterials. They are often coupled to ultrafiltration (for separation), and UF-ICP-OES has been used to study Ag [193] and ZnO [7, 31, 32, 194]. The use of AAS to study the nanomaterial solubility has been reported, but this has been somewhat limited. AAS has been used (in conjunction with ultrafiltration) to determine the solubility of ZnO [195], Ag [179, 196], and CuO [34].

If total elemental analysis is not needed, then other methods should be considered. Although atomic spectroscopy-based techniques such as ICP-MS are highly selective and sensitive, this is done at the expense of a higher instrumental cost if compared to other techniques, as exemplified by electrochemical-based methods. AGNES, for example, is highly promising and much more affordable, but it works well with certain metals only such as Zn, whereas it is not suitable for many other metals such as Ag. Unlike atomic spectrometry–based techniques, the main issue with electrochemical-based techniques is that it is applicable to limited types of ionic species that can be analysed. It is worthy to note that colorimetric based methods can also be associated to the same limitation.

Whatever method is chosen, the important thing is to have robust and reliable data. For this to happen, methods must be validated, which can be done by conducting well-controlled round robin studies. If they are not validated, then this can lead to a situation in which experimental data gets reported without proper understanding of the associated errors and propagation of such errors. Finally, due to the dynamics of the dissolution process, any protocol used to measure solubility must be reported in detail; this includes the reporting of experimental conditions under which the data were collected such as pH, temperature, composition, timeframe, and so on.

5.7 CASE STUDY: MINIATURIZED CAPILLARY ELECTROPHORESIS WITH CONDUCTIVITY DETECTION TO DETERMINE NANOMATERIAL SOLUBILITY

5.7.1 Introduction

In the early 1990s, the idea that a multifunctional chemical laboratory process can be integrated into a single microfabricated device was presented, subsequently dubbed as "lab-on-chip." Since then the miniaturization for chemistry, biology, and engineering became increasingly popular, with several advantages of faster analysis, lower cost, less sample consumption, and high-throughput analysis [197] having been pointed out.

In the past few years, the drive for smaller devices and much faster analysis time has led to the commercialization of miniaturized capillary electrophoresis (CE)-conductivity microchip [198]. This microchip device consists of fabricated channels comprising a capillary electrophoretic microcolumn suitable for the separation of ions. At the end of the column, a contactless capacitively coupled conductivity detection (C4D) detection system is integrated, suitable to quantify the separated ionic species. The theory of the CE-conductivity microchip has been previously reported, and the device has been shown to identify ions in highly complex medium such as blood [199], which suggests its usefulness to determine nanomaterial solubility.

In this case study, the performance of the CE-conductivity microchip will be evaluated to measure the solubility of ZnO nanomaterial. ZnO is chosen as the nanomaterial of interest as it is found in many nano applications, for example, cosmetic formulations. Initial tests will be carried out to establish if the device can satisfy two important analytical criteria: (i) it can differentiate the signal of interest above the background, (ii) it is sufficiently repeatable, in particular when different microchips are used. Hence, the case study can be subdivided into two parts. The first part investigates the ability of the device to detect dissolved Zn^{2+} produced as a result of dissolution of ZnO nanomaterial when in a complex medium; here, reconstituted fish medium, often employed in ecotoxicology studies, will be used as an example. The second part aims at assessing the degree of data variability arising from individual microchips and its ability to reliably quantify the amount of $[Zn^{2+}]$. A test material consisting a premade preparation of Li^+, Na^+, and K^+ in de-ionized (DI) water will be used to assess the degree of chip-to-chip variability.

5.7.2 Method

5.7.2.1 Materials Unless specified elsewhere, all chemicals were purchased from Sigma-Aldrich, UK.

All chemicals were used as received, and appropriate solution concentrations were made up with deionized (DI) water; DI water with a resistivity value of 18.2 MΩ cm was used throughout the experiments. All solutions were filtered through a 20-nm syringe filter prior to introduction into the microchip to prevent potential blockages in the microchannels.

The components of the reconstituted fish medium are shown in Table 5.1. The final pH of the fish medium was ~7.

The nanomaterial ZnO (NM-110) was supplied in glass vials, packed under argon atmosphere, by the JRC (Joint Research Centre). Once the vials were opened, the contents were used immediately and remainder discarded. This is the same material that has been used in the case study of Surface Area Chapter 7 (of this book) and the corresponding physicochemical properties have been reported in that chapter.

The standard test solutions used for the CE-conductivity microchip were purchased from eDAQ Europe. These consisted of a solution composed of LiCl (1 mM), KNO_3 (1 mM), and Na_2SO_4 (1 mM) in deionized water (supplied with associated certified weight reports as provided by Absolute Standards, Inc.).

Background electrolyte (BGE) of acetic acid (0.5 M) was used when the test solutions were analyzed, in accordance with the recommendations of the device supplier. BGE of 5 mM His and 3 mM HIBA (adjusted to pH 4.5 with acetic acid) were used for the zinc analysis. Analytical grade NaOH (0.5 M) was used as the wetting agent for the microchannels.

5.7.2.2 Dispersion Protocol The protocol employed for the dispersion of the ZnO nanomaterial in the fish medium was based on the recommended protocol used for the OECD nanomaterial testing and has been described elsewhere [25]. Overall, the dispersion involved mixing the nanomaterial powder into a paste before adding more liquid and sonicating using a sonicating probe (Cole Palmer® 130-Watt Ultrasonic Processors (50/60 Hz, VAC 220)) for 20 s.

5.7.2.3 Instrumentation: CE-Conductivity Device The CE-conductivity instrument was purchased from eDAQ Europe and consisted of the following units:

TABLE 5.1 Components of the Fish (Ecotox) Medium

Component	Concentration (mM)
Calcium chloride [$CaCl_2$]	4.87
Magnesium sulfate [$MgSO_4$]	1.92
Sodium bicarbonate [$NaHCO_3$]	1.54
Potassium chloride [KCl]	0.15

Tantra et al. [200], http://www.hindawi.com/journals/jt/2012/270651/.

a) Two high-voltage sequencers (HVS, Model ER230), each HVS unit having dual-channel high-voltage power supplies that are especially designed for EOF applications.
b) Capacitively coupled contactless conductivity detection (C4D) data system (ER225); this comprised a conductivity detector and signal recording functions.
c) C4D Micronit Chip Electrophoresis Platform (ET225). This unit has high-voltage cables that allowed the microchip to be connected to the HVS units. The unit included a safety interlock, to switch off the HVS if the cover plate had been accidentally lifted during the experiment. The platform had a ground connector to minimize noise in the C4D signal.

Fabricated borosilicate glass microchips were purchased from Micronit, the Netherlands (Model ET145-4). The chip contains a manifold in a double-T geometry and integrated contactless conductivity detection. The chip has a separation channel of 33 mm long, with a cross section of 100 μm wide and 10 μm deep. The arrangement of the detector consisted of two parallel electrodes measuring 200 μm by 500 μm and 200 nm thick, separated by 2.9 mm.

Figure 5.1 shows a schematic of the microchip and the allocated reservoir numbers. It also depicts the equivalent circuit model of how conductivity is measured. PowerChrom® software purchased from eDAQ Europe (ES280) was used to collect, display, and analyze the data.

5.7.2.4 CE-Conductivity Microchip: Measurement Protocol The capillary electrophoresis measurement was performed using a BGE that is optimal for the analyte of interest, that is, either zinc or test analytes. The instrumental setting is governed

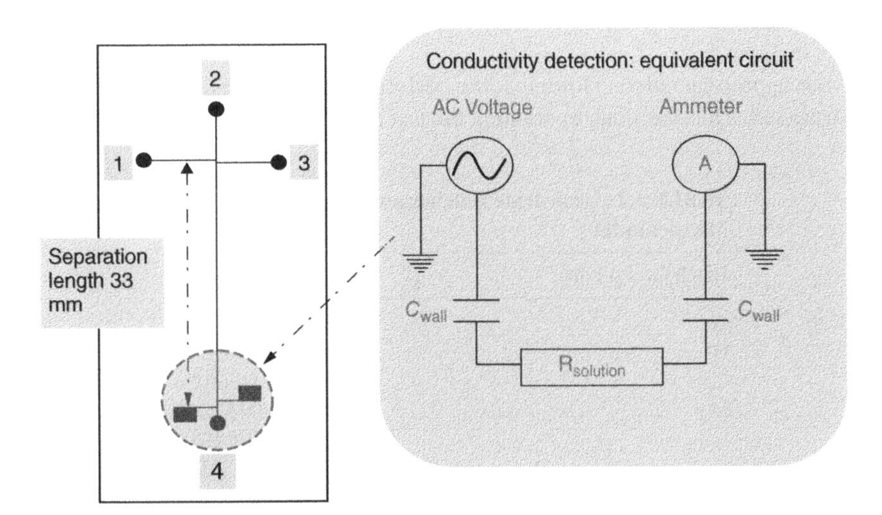

Figure 5.1 CE-conductivity microchip analysis. Adapted from [100].

by the type of BGE employed, which is established by following the manufacturer's instructions. The instrument settings for the test standard BGE (0.5 M acetic acid) were as follows: frequency 800 kHz, amplitude 100%, high gain off, 5 Hz low pass. The instrument settings for the BGE employed for zinc analysis, that is, 5 mM His, 3 mM HIBA, pH 4.5 were as follows: frequency 1150 kHz, amplitude 20%, high gain on, 5 Hz low pass).

For electrophoretic analysis, the microchip was preconditioned prior to use. This was achieved by flushing the microchannels with the following solutions under vacuum: 0.5 M NaOH for 2 min, DI water for 10 min, and finally BGE for 5 min. The microchip was mounted on the C4D Micronit Chip Electrophoresis Platform, and electrophoresis was carried out by applying the appropriate voltages to the four reservoirs. The voltage program was set so as to achieve sample plug formation, sample injection, separation, and detection.

5.7.2.5 Protocol: To Assess the Feasibility of Measuring the Zn^{2+} (from ZnO Nanomaterial) Signal above the Fish Medium Background A ZnO dispersion of 500 mg/l concentration (0.5 l total volume) was prepared and stored, at ambient temperature, in a 1-l storage media bottle. A corresponding "control" bottle containing just the fish medium, which contained no ZnO nanomaterial, was also prepared. Throughout a 3-week period, 2 ml samples were aliquoted from each bottle; prior to extracting the sample, each bottle was shaken gently by hand to aid homogeneity. Each 2 ml aliquot was passed through a 20-nm Anatop syringe filter (Fisher Scientific, UK) and the filtrate was collected for further analysis using the CE-conductivity microchip device.

5.7.2.6 Protocol: To Assess Data Variability between Different Microchips Six sets of four microchips were assessed for this study, each microchip set coming from a different glass substrate batch. The type of glass that was employed to make the substrate was the same for all batches. A total of 24 microchips were analyzed, as shown in Table 5.2.

After appropriate preconditioning, the performance of the individual microchips was assessed. This was done by running the test solution (diluted to 0.5 mM with DI

TABLE 5.2 Substrate Batches and Corresponding Microchip ID

Batch Group Label	Microchip Label
A	A1, A2, A3, A4
B	B5, B6, B7, B8
C	C9, C10, C11, C12
D	D13, D14, D15, D16
E	E17, E18, E19, E20
F	F21, F22, F23, F24

Tantra et.al. [201], Table 5.1. Reproduced with permission from Wiley-VCH Verlag GmbH & Co. KGaA, Weinheim.

water) using a BGE of 0.5 M acetic acid. After running a sample, the microchip was then electrophoretically cleaned for 2 min to flush the microchannels. The process of running 0.5 mM test solution followed by electrophoretic cleaning was repeated until six electropherograms of the test material were acquired. All solutions were filtered through a 20-nm syringe filter prior to introduction into the microchip to prevent potential blockages in the microchannels.

5.7.3 Results and Interpretation

5.7.3.1 Study 1: Assessing Feasibility of the CE-Conductivity Microchip to Detect Free Zn^{2+} Arising from Dispersion of ZnO in Fish Medium Figure 5.2 shows the results from the ZnO dissolution study, when a sample is taken at the end of the 3-week study. Two peaks are apparent: the first being a blank peak (i.e., fish medium only) and the second being the Zn^{2+} peak. Although it is not shown, the identity of the zinc peak has been confirmed and reported elsewhere [100].

The size of the zinc peak observed here is very small with a peak height <5 mV. This zinc peak was monitored throughout the 3 weeks of the study and was shown to first emerge after the first week. After this, the zinc peak seemed to appear and disappear within the 3-week time frame. Even though the zinc peak was not always present after the first week, this peak was never present in the corresponding blank samples throughout the study. The appearance/disappearance of the zinc peak in this case can only be explained by the fact that the magnitude of the measured signal was extremely small and perhaps close to the detection limit (~ppm level) of the CE-conductivity microchip device.

5.7.3.2 Study 2: Assessing Performance of Microchips Using Reference Test Material Only a summary of the findings will be reported here as the results of this study have been discussed in great detail elsewhere [201]:

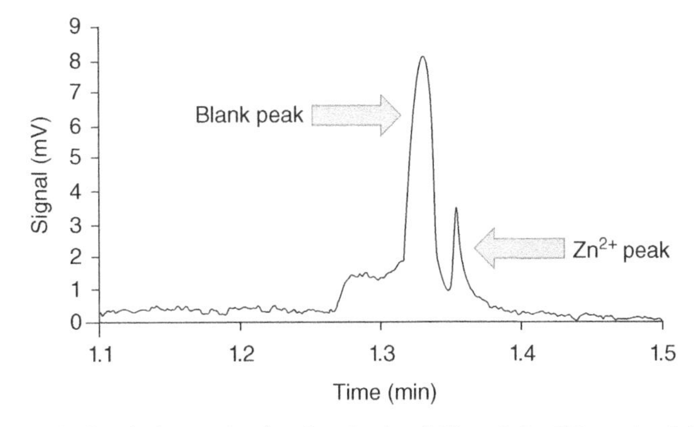

Figure 5.2 ZnO dissolution study showing the feasibility of the CE-conductivity device to detect free zinc from a dispersion of nano-ZnO in fish medium. Adapted from [100].

a) *Results associated with data variability in relation to the retention time, t_R and hence repeatability of detection.* The relative standard deviation (RSD) fluctuation for t_R was analyzed for the 24 (4×6 batches) of microchips tested. Note that each batch relates to the individual glass substrate from which a microchip has been cut. The findings show that the RSD in t_R was typically 2% or less, with the exception of four microchips (B6, D13, D14, and D16), in which the RSD in t_R was noticeably higher, although still within acceptable levels (<10%). The observation associated with chip-to-chip variation, in which peaks are eluting at slightly different times, is an interesting one and does not seem to be attributed to the glass substrate from which the microchip has been cut. There are several possible explanations as to why peaks elute faster or slower: differences associated with geometrical variations of the microchannels arising from manufacturing quality, slight variability in the actual injection time, broadening effect, and estimated shift of peak maxima.

b) *Results associated with data variability in relation to the peak area and hence repeatability of quantification.* The RSD fluctuation in relation to the peak area for the 24 microchips tested showed a much larger variation than the corresponding t_R data. Overall, this indicated that measurements of the peak area were less reliable (compared to t_R). Only two microchips had a RSD of 10% or less, with most tending to fluctuate between 10% and 40% and with three microchips having an RSD of 48, 51, and 53%. Out of the three ions analyzed, the largest RSD was associated with lithium, indicating that it is least reliable for quantification through peak area measurements. Offering an explanation for this observation is not straightforward, and we can only theorize on potential causes. One major difference between lithium and the other cations is that lithium ions possess a large solvation shell. The electrophoretic mobility of a much larger solvation shell surrounding a cation will mean that its signal will be more sensitive to fluctuations under applied electric field. This will ultimately affect the amount of the analyte that is present in sample plug and hence injected into the separation column. Interestingly, results showed no indication that data variability was due to substrate batch. The highly irregularly scattered data suggested that the source was either random or due to parameters whose values are inconsistent between the individual microchips. There are several possible sources of variability in relation to peak area measurements. First, this may be caused by voltage fluctuations from the high-voltage power supplies, which control the liquid flow within the microchannel and separation performance. This can control characteristics of the sample plug (size, shape) that is being created and subsequently sent down the separation column. Second potential reason may be associated with the variations in surface properties of the microchannel and, in particular, the microchannel structure, that is, the double T-junction shown in Figure 5.1. This part of the microchip is important, as this is the location where the sample plug is formed and the plug subsequently injected into the separation column. Variations in surface properties such as surface roughness around the T-junction area will undoubtedly affect the quality (size and shape) of the sample plug formed. This in turn will affect how much of analyte is being sent down the separation column and thus the analyte concentration subsequently measured.

5.7.4 Conclusion

The case study reports the findings from the preliminary study to assess the feasibility of using CE-conductivity microchip device to determine the nanomaterial solubility, potentially in nanotoxicological investigations. The first part of the study showed that the device can be highly selective at differentiating free ions in solution within a complex medium. The second part of the study assessed the extent of data variability arising from 24 CE-conductivity microchips (of six different batches). Huge variability (with some microchips having RSD of $\sim 50\%$) was observed for peak area measurements. The variability was not attributed to the glass substrate from which the microchip has been cut during fabrication. Several explanations were given to explain the potential sources of variations. In summary, the microchip technology assessed here may be suitable for identification but not for the quantification of ions.

5.8 SUMMARY

This chapter serves as a starting point for researchers to review the different analytical methods available to measure the solubility of nanomaterials. Our findings indicate the wide variety of methods available, with most being capable of measuring either total dissolved species or free ions, but not both. Electrochemical- and colorimetric-based detection methods are potentially suitable to measure free ions (as well as labile fractions), while atomic spectrometry–based techniques are more suited to measure total dissolved species. In some cases, a combination of a separation method with a quantification analysis platform is needed. The final choice on what method to use, however, will be dependent on the nanomaterial sample and the specific scenario that the researcher is faced with. However, an important point that has been highlighted is to ensure that methods are fit for purpose.

The case study presented in the second half of the chapter investigates the feasibility of state-of-the-art technology to determine the nanomaterial solubility. CE-conductivity microchip was assessed and our findings found its suitability for the detection but not for quantification purposes.

ACKNOWLEDGMENTS

The work leading to these results has received support from the European Union Seventh Framework Programme (FP7-NMP.2012.1.3-3) under grant agreement No. 310584 (NANoREG). This work was supported by the Service Public de Wallonie (SPW) – Direction générale opérationnelle – Economie, Emploi et Recherche (DGO6), Département des Programmes de Recherche (NANoREG, SPW/UNamur research convention No. 1317938). CRC and CAD thank Joseph Galceran and Jaume Puy (U. of Lleida) for their useful suggestions and comments.

REFERENCES

1. OECD. *Test No. 105: Water Solubility*. OECD Publishing; 1995.
2. ISO. *ISO/TR 13014:2012: Nanotechnologies – Guidance on Physico-Chemical Characterization of Engineered Nanoscale Materials for Toxicologic Assessment*. ISO; 2012.
3. European Chemicals Agency. *ECHA-12-G-03-EN. Appendix R7-1 Recommendations for Nanomaterials Applicable to Chapter R7a Endpoint Specific Guidance in Guidance on Information Requirements and Chemical Safety Assessment*. European Chemicals Agency; 2012.
4. Tantra, R., Jing S., Gohil D., Technical issues surrounding the preparation, characterisation and testing of nanoparticles for ecotoxicological studies. Edited Contributions Presented at the Third International Conference on Environmental Toxicology (Environmental Toxicology III); 2010; Cyprus. WIT Press; 2010.
5. Tantra R, Jing S, Pichaimuthu SK, Walker N, Noble J, Hackley VA. Dispersion stability of nanoparticles in ecotoxicological investigations: the need for adequate measurement tools. J Nanoparticle Res 2011;13(9):3765–3780.
6. Bian SW, Mudunkotuwa IA, Rupasinghe T, Grassian VH. Aggregation and dissolution of 4 nm ZnO nanoparticles in aqueous environments: Influence of pH, ionic strength, size, and adsorption of humic acid. Langmuir 2011;27(10):6059–6068.
7. Reed RB et al. Solubility of nano-zinc oxide in environmentally and biologically important matrices. Environ Toxicol Chem 2011;31:93–99.
8. Zhang Y, Muhammed M. Critical evaluation of thermodynamics of complex formation of metal ions in aqueous solutions VI. Hydrolysis and hydroxo-complexes of Zn^{2+} at 298.15 K. Hydrometallurgy 2001;60(3):215–236.
9. Reichle RA, McCurdy KG, Hepler LG. Zinc hydroxide: Solubility product and hydroxy-complex stability constants from 12.5–75 degrees C. Can J Chem 1975; 53:3841–3845.
10. Degen A, Kosec M. Effect of pH and impurities on the surface charge of zinc oxide in aqueous solution. J Euro Ceram Soc 2000;20(6):667–673.
11. European Chemicals Agency. *Guidance on Information Requirements and Chemical Safety Assessment. Appendix R.7.13-2: Environmental Risk Assessment for Metals and Metal Compounds*. European Chemicals Agency; 2008.
12. Misra SK, Dybowska A, Berhanu D, Luoma SN, Valsami-Jones E. The complexity of nanoparticle dissolution and its importance in nanotoxicological studies. Sci Total Environ 2012;438:225–232.
13. Ivask A et al. Metal-containing nano-antimicrobials: Differentiating the impact of solubilized metals and particles. In: Cioffi N, Rai M, editors. *Nano-Antimicrobials*. Berlin, Heidelberg: Springer; 2012. p 253–290.
14. Batley GE, Kirby JK, McLaughlin MJ. Fate and risks of nanomaterials in aquatic and terrestrial environments. Acc Chem Res 2012;46(3):854–862.
15. Pettitt ME, Lead JR. Minimum physicochemical characterisation requirements for nanomaterial regulation. Environ Int 2013;52:41–50.

16. Tiede K et al. Considerations for environmental fate and ecotoxicity testing to support environmental risk assessments for engineered nanoparticles. J Chromatogr A 2009;1216(3):503–509.
17. Levard C et al. Environmental transformations of silver nanoparticles: Impact on stability and toxicity. Environ Sci Technol 2012;46(13):6900–6914.
18. Miao AJ et al. The algal toxicity of silver engineered nanoparticles and detoxification by exopolymeric substances. Environ Pollut 2009;157(11):3034–3041.
19. Schlich K et al. Effects of silver nanoparticles and silver nitrate in the earthworm reproduction test. Environ Toxicol Chem 2013;32(1):181–188.
20. Li M, Lin D, Zhu L. Effects of water chemistry on the dissolution of ZnO nanoparticles and their toxicity to Escherichia coli. Environ Pollut 2013;173:97–102.
21. Li M, Zhu L, Lin D. Toxicity of ZnO nanoparticles to *Escherichia coli*: Mechanism and the influence of medium components. Environ Sci Technol 2011;45(5):1977–1983.
22. Adam N et al. The chronic toxicity of ZnO nanoparticles and ZnCl2 to Daphnia magna and the use of different methods to assess nanoparticle aggregation and dissolution. Nanotoxicology 2014;8(7):709–717.
23. Mu Q et al. Systematic investigation of the physicochemical factors that contribute to the toxicity of ZnO nanoparticles. Chem Res Toxicol 2014;27(4):558–567.
24. Fabricius A-L et al. ICP-MS-based characterization of inorganic nanoparticles – sample preparation and off-line fractionation strategies. Anal Bioanal Chem 2014;406:467–479.
25. Tantra R, Boyd R, Cackett A, Fry AT, Gohil DD, Goldberg S, Lee JLS, Minelli C, Peck R, Quincey P, Smith S, Snowden J, Spencer S, Tompkins J, Wang J, Yang L. *NPL Report: Final Report on the Physico-Chemical Characterisation of PROSPEcT Engineered Nanomaterials*. National Physical Laboratory; 2012. http://publications.npl.co.uk/dbtw-wpd/exec/dbtwpub.dll?&QB0=AND&QF0=ID&QI0=%206281%20&TN=NPLPUBS&RF=WFullRecordDetails&DL=0&RL=0&NP=4&AC=QBE_QUERY.
26. Xu M et al. Challenge to assess the toxic contribution of metal cation released from nanomaterials for nanotoxicology – The case of ZnO nanoparticles. Nanoscale 2013;5(11):4763–4769.
27. Fabrega J et al. Sequestration of zinc from zinc oxide nanoparticles and life cycle effects in the sediment dweller amphipod corsophium volutator. Environ Sci Technol 2012;46(2):1128–1135.
28. Cheryan M. *Ultrafiltration and Microfiltration Handbook*. CRC Press; 1998.
29. Gheldof D et al. Thrombin generation assay and transmission electron microscopy: a useful combination to study tissue factor-bearing microvesicles. J Extracell Vesicles 2013;2.
30. Cornelis G et al. A method for determination of retention of silver and cerium oxide manufactured nanoparticles in soils. Environ Chem 2010;7:298–308.
31. Sivry Y et al. Behavior and fate of industrial zinc oxide nanoparticles in a carbonate-rich river water. Chemosphere 2014;95:519–526.
32. Mudunkotuwa IA et al. Dissolution of ZnO nanoparticles at circumneutral pH: a study of size effects in the presence and absence of citric acid. Langmuir 2012;28:396–403.
33. Cornelis G et al. Solubility and batch retention of CeO_2 nanoparticles in soils. Environ Sci Technol 2011;45:2777–2782.
34. Wang Z et al. CuO nanoparticle interaction with human epithelial cells: Cellular uptake, location, export, and genotoxicity. Chem Res Toxicol 2012;25:1512–1521.

35. Pesavento M, Alberti G, Biesuz R. Analytical methods for determination of free metal ion concentration, labile species fraction and metal complexation capacity of environmental waters: A review. Anal Chim Acta 2009;631(2):129–141.
36. Fortin C, Campbell PG. An ion-exchange technique for free-metal ion measurements (Cd^{2+} Zn^{2+}): Applications to complex aqueous media. Int J Environ Anal Chem 1998;72(3):173–194.
37. Hadioui M, Peyrot C, Wilkinson KJ. Improvements to single particle ICPMS by the online coupling of ion exchange resins. Anal Chem 2014;86(10):4668–4674.
38. Hadioui M, Leclerc S, Wilkinson KJ. Multimethod quantification of Ag^+ release from nanosilver. Talanta 2013;105:15–19.
39. Alberti G et al. A comparison between the determination of free Pb(II) by two techniques: Absence of gradients and Nernstian equilibrium stripping and resin titration. Anal Chim Acta 2007;599(1):41–50.
40. Pesavento M, Biesuz R. Simultaneous determination of total and free metal-ion concentration in solution by sorption on iminodiacetate resin. Anal Chem 1995; 67(19):3558–3563.
41. Zhang H, Davison W. Performance-characteristics of diffusion gradients in thin-films for the in-situ measurement of trace-metals in aqueous-solution. Anal Chem 1995; 67(19):3391–3400.
42. DRT Research Ltd. 2003, Technical documentation. Available at www.dgtresearch.com/dgtresearch/dgtresearch.pdf. Accessed 2015 Oct 15.
43. Davison W, Zhang H. Progress in understanding the use of diffusive gradients in thin films (DGT) – Back to basics. Environ Chem 2012;9(1):1–13.
44. Garmo OA et al. Performance study of diffusive gradients in thin films for 55 elements. Anal Chem 2003;75(14):3573–3580.
45. Uribe R et al. Contribution of partially labile complexes to the DGT metal flux. Environ Sci Technol 2011;45(12):5317–5322.
46. Mongin S et al. Limits of the linear accumulation regime of DGT sensors. Environ Sci Technol 2013;47(18):10438–10445.
47. Navarro E et al. Toxicity of silver nanoparticles to *Chlamydomonas reinhardtii*. Environ Sci Technol 2008;42(23):8959–8964.
48. Buffet P-E et al. Biochemical and behavioural responses of the endobenthic bivalve *Scrobicularia plana* to silver nanoparticles in seawater and microalgal food. Ecotoxicol Environ Saf 2013;89:117–124.
49. Buffet PE et al. A marine mesocosm study on the environmental fate of silver nanoparticles and toxicity effects on two endobenthic species: The ragworm *Hediste diversicolor* and the bivalve mollusc *Scrobicularia plana*. Sci Total Environ 2014;470–471:1151–1159.
50. Odzak N et al. Dissolution of metal and metal oxide nanoparticles in aqueous media. Environ Pollut 2014;191:132–138.
51. Kunniger T et al. Release and environmental impact of silver nanoparticles and conventional organic biocides from coated wooden facades. Environ Pollut 2014;184:464–471.
52. Buffet PE et al. Behavioural and biochemical responses of two marine invertebrates *Scrobicularia plana* and *Hediste diversicolor* to copper oxide nanoparticles. Chemosphere 2011;84(1):166–174.

53. Buffet P-E et al. A mesocosm study of fate and effects of CuO nanoparticles on *Endobenthic* species (*Scrobicularia plana*, *Hediste diversicolor*). Environ Sci Technol 2012;47(3):1620–1628.
54. Davison W, Zhang H. In-situ speciation measurements of trace components in natural-waters using thin-film gels. Nature 1994;367(6463):546–548.
55. Sigg L et al. Comparison of analytical techniques for dynamic trace metal speciation in natural freshwaters. Environ Sci Technol 2006;40(6):1934–1941.
56. Song Y et al. Estimation of size for 1–2 nm nanoparticles using an HPLC electrochemical detector of double layer charging. Anal Chem 2003;75(19):5088–5096.
57. Baati T et al. The prolongation of the lifespan of rats by repeated oral administration of [60]fullerene. Biomaterials 2012;33(19):4936–4946.
58. Braun T et al. Olive oil as a biocompatible solvent for pristine C60. Fullerenes Nanotubes Carbon Nanostruct 2007;15(4):311–314.
59. Rodriguez MA, Armstrong DW. Separation and analysis of colloidal/nano-particles including microorganisms by capillary electrophoresis: a fundamental review. J Chromatogr B 2004;800(1–2):7–25.
60. Muralidharan, S., K. Du, and P. Amaratunga, Fundamental characterization of nanoparticles and their interactions by capillary electrophoresis. Abstracts of Papers of the American Chemical Society, 2007. p 233.
61. Keeney, R.A., M.R. Ivanov, and A.J. Haes, Nanoparticle synthesis for separation optimization with capillary electrophoresis. Abstracts of Papers of the American Chemical Society, 2009. p 237.
62. Hwang WM et al. Separation of nanoparticles in different sizes and compositions by capillary electrophoresis. Bull Korean Chem Soc 2003;24(5):684–686.
63. Lin KH, Chu TC, Liu FK. On-line enhancement and separation of nanoparticles using capillary electrophoresis. J Chromatogr A 2007;1161(1–2):314–321.
64. Xue Y, Yang HY, Yang YT. Determination of size distribution of nano-particles by capillary zone electrophoresis. Chin Chem Lett 2005;16(1):67–70.
65. Doorn SK et al. High resolution capillary electrophoresis of carbon nanotubes. J Am Chem Soc 2002;124(12):3169–3174.
66. Doorn SK et al. Capillary electrophoresis separations of bundled and individual carbon nanotubes. J Phys Chem B 2003;107(25):6063–6069.
67. Lopez-Pastor M et al. Separation of single-walled carbon nanotubes by use of ionic liquid-aided capillary electrophoresis. Anal Chem 2008;80(8):2672–2679.
68. Jimenez-Soto JM et al. Evaluation of the performance of single-walled carbon nanohorns in capillary electrophoresis. Electrophoresis 2010;31(10):1681–1688.
69. Pauwels J, Van Schepdael A. Carbon nanotubes in capillary electrophoresis, capillary electrochromatography and microchip electrophoresis. Cent Eur J Chem 2012;10(3):785–801.
70. Suarez B et al. Separation of carbon nanotubes in aqueous medium by capillary electrophoresis. J Chromatogr A 2006;1128(1–2):282–289.
71. Yamamoto T et al. An analytical system for single nanomaterials: Combination of capillary electrophoresis with raman spectroscopy or with scanning probe microscopy for individual single-walled carbon nanotube analysis. Anal Chem 2009;81(17):7336–7341.
72. Cao W et al. A highly efficient and versatile microchip capillary electrophoresis method for DNA separation using gold nanoparticle as a tag. J Sep Sci 2011;34(8):939–946.

73. Du, K., H.L. Ma, and S. Muralidharan, Synthesis of gold nanoparticle clusters and their characterization by capillary electrophoresis, light scattering, and transmission electron microscopy. Abstracts of Papers of the American Chemical Society, 2006. p 231.
74. Liu YM et al. Sensitive chemiluminescence immunoassay by capillary electrophoresis with gold nanoparticles. Anal Chem 2011;83(3):1137–1143.
75. Liu FK, Lin YY, Wu CH. Highly efficient approach for characterizing nanometer-sized gold particles by capillary electrophoresis. Anal Chim Acta 2005;528(2):249–254.
76. Wei GT et al. Separation of nanostructured gold particles by capillary zone electrophoresis. J Chin Chem Soc 1998;45(1):47–52.
77. d'Orlye F, Varenne A, Gareil P. Determination of nanoparticle diffusion coefficients by Taylor dispersion analysis using a capillary electrophoresis instrument. J Chromatogr A 2008;1204(2):226–232.
78. d'Orlye F, Varenne A, Gareil P. Size-based characterization of nanometric cationic maghemite particles using capillary zone electrophoresis. Electrophoresis 2008;29(18):3768–3778.
79. Vanifatova NG et al. Investigation of iron oxide nanoparticles by capillary zone electrophoresis. Talanta 2005;66(3):605–610.
80. Baharifar H et al. Influence of polymeric coating on capillary electrophoresis of iron oxide nanoparticles. J Iran Chem Soc 2014;11(1):279–284.
81. Petr J et al. Separation of alpha-lactalbumin grafted- and non-grafted maghemite core/silica shell nanoparticles by capillary zone electrophoresis. Electrophoresis 2010;31(16):2754–2761.
82. Vanifatova NG, Spivakov BY, Kamyshny A. Comparison of potential of capillary zone electrophoresis and Malvern's improved laser Doppler velocimetry for characterisation of silica nanomaterials in aqueous media. Int J Nanopart 2011;4(4):369–380.
83. Vanifatova NG et al. Size separation of silica nanospheres by means of capillary zone electrophoresis. Talanta 2003;59(2):345–353.
84. Wang Q et al. General solution-based route to V–VI semiconductors nanorods from hydrolysate. J Nanopart Res 2007;9(2):269–274.
85. Celiz MD et al. Study on the effects of humic and fulvic acids on quantum dot nanoparticles using capillary electrophoresis with laser-induced fluorescence detection. Environ Sci Technol 2011;45(7):2917–2924.
86. Oszwaldowski S, Zawistowska-Gibula K, Roberts KP. Characterization of CdSe nanocrystals coated with amphiphiles. A capillary electrophoresis study. Microchim Acta 2012;176(3–4):345–358.
87. Liu P et al. Separation and detection of monodisperse polystyrene nanospheres by capillary electrophoresis. In: Cong HL, editor. *Advanced in Nanoscience and Technology*. Stafa-Zurich: Trans Tech Publications Ltd; 2012. p 51–55.
88. Vanifatova NG et al. Separation of unmodified polystyrene nanosphere standards by capillary zone electrophoresis. J Chromatogr A 2000;898(2):257–263.
89. Wu TM, Lin SH. Characterization and electrical properties of polypyrrole/multiwalled carbon nanotube composites synthesized by *in situ* chemical oxidative polymerization. J Polym Sci B 2006;44(10):1413–1418.
90. Liu F-K et al. Analytical separation of Au/Ag core/shell nanoparticles by capillary electrophoresis. J Chromatogr A 2006;1133(1–2):340–346.

91. Vanifatova NG et al. Study of properties of silicone-silica crystalline nanospheres in aqueous solutions by capillary zone electrophoresis. Int J Nanopart 2010;3(1):65–76.
92. Abdel-Haq H, Bossù E. Capillary electrophoresis as a tool for the characterization of pentosan nanoparticles. J Chromatogr A 2012;1257:125–130.
93. Helle A et al. Quantitative determination of drug encapsulation in poly (lactic acid) nanoparticles by capillary electrophoresis. J Chromatogr A 2008;1178(1):248–255.
94. Oukacine F, Morel A, Cottet H. Characterization of carboxylated nanolatexes by capillary electrophoresis. Langmuir 2011;27(7):4040–4047.
95. Rudnev A et al. Characterization of calcium hydroxyapatite polycrystalline nanoparticles by capillary zone electrophoresis and scanning electron microscopy. J Anal Chem 2012;67(6):565–571.
96. Eckhoff DA et al. Capillary electrophoresis of ultrasmall carboxylate functionalized silicon nanoparticles. J Chem Phys 2006;125(8):081103.
97. Li N et al. Probing nanoparticle– protein interaction by capillary electrophoresis. Anal Chem 2010;82(17):7460–7466.
98. Zhong, W.W., Capillary electrophoresis in the study of nanoparticle-protein interactions. Abstracts of Papers of the American Chemical Society, 2012. p 243.
99. Xie MY, Guo ZP, Chen Y. Inspection into the interaction of bovine serum albumin with gold nanoparticles by capillary electrophoresis. Chem J Chin Universities 2010;31(11):2162–2166.
100. Tantra R, Jarman J. *μTAS (Micro Total Analysis Systems) for the High-Throughput Measurement Of Nanomaterial Solubility. In Journal of Physics: Conference Series.* IOP Publishing; 2013.
101. Giddings JC. Field-flow fractionation: analysis of macromolecular, colloidal, and particulate materials. Science 1993;260(5113):1456–1465.
102. Hassellov M et al. Nanoparticle analysis and characterization methodologies in environmental risk assessment of engineered nanoparticles. Ecotoxicology 2008;17(5):344–361.
103. Williams SK, Runyon JR, Ashames AA. Field-flow fractionation: addressing the nano challenge. Anal Chem 2011;83(3):634–642.
104. von der Kammer F et al. Analysis of engineered nanomaterials in complex matrices (environment and biota): general considerations and conceptual case studies. Environ Toxicol Chem 2012;31(1):32–49.
105. Luykx DM et al. A review of analytical methods for the identification and characterization of nano delivery systems in food. J Agric Food Chem 2008;56(18):8231–8247.
106. Bednar AJ et al. Comparison of on-line detectors for field flow fractionation analysis of nanomaterials. Talanta 2013;104:140–148.
107. Mitrano D et al. Field-flow fractionation coupled with ICP-MS for the analysis of engineered nanoparticles in environmental samples. Spectroscopy 2012;27(9):36–44.
108. Schmidt B et al. Quantitative characterization of gold nanoparticles by field-flow fractionation coupled online with light scattering detection and inductively coupled plasma mass spectrometry. Anal Chem 2011;83(7):2461–2468.
109. Calzolai L et al. Separation and characterization of gold nanoparticle mixtures by flow-field-flow fractionation. J Chromatogr A 2011;1218(27):4234–4239.
110. Cho TJ, Hackley VA. Fractionation and characterization of gold nanoparticles in aqueous solution: asymmetric-flow field flow fractionation with MALS, DLS, and UV–vis detection. Anal Bioanal Chem 2010;398(5):2003–2018.

111. Poda AR et al. Characterization of silver nanoparticles using flow-field flow fractionation interfaced to inductively coupled plasma mass spectrometry. J Chromatogr A 2011;1218(27):4219–4225.

112. Cumberland SA, Lead JR. Particle size distributions of silver nanoparticles at environmentally relevant conditions. J Chromatogr A 2009;1216(52):9099–9105.

113. Bolea E et al. Size characterization and quantification of silver nanoparticles by asymmetric flow field-flow fractionation coupled with inductively coupled plasma mass spectrometry. Anal Bioanal Chem 2011;401(9):2723–2732.

114. Mitrano DM et al. Silver nanoparticle characterization using single particle ICP-MS (SP-ICP-MS) and asymmetrical flow field flow fractionation ICP-MS (AF4-ICP-MS). J Anal Atom Spectrom 2012;27(7):1131–1142.

115. Hagendorfer H et al. Characterization of silver nanoparticle products using asymmetric flow field flow fractionation with a multidetector approach – A comparison to transmission electron microscopy and batch dynamic light scattering. Anal Chem 2012;84(6):2678–2685.

116. Tadjiki S et al. Detection, separation, and quantification of unlabeled silica nanoparticles in biological media using sedimentation field-flow fractionation. J Nanopart Res 2009;11(4):981–988.

117. Contado C, Pagnoni A. TiO_2 in commercial sunscreen lotion: Flow field-flow fractionation and ICP-AES together for size analysis. Anal Chem 2008;80(19):7594–7608.

118. Contado C, Pagnoni A. TiO_2 nano- and micro-particles in commercial foundation creams: Field Flow-Fractionation techniques together with ICP-AES and SQW Voltammetry for their characterization. Anal Methods 2010;2(8):1112–1124.

119. Chen BL, Selegue JP. Separation and characterization of single-walled and multi-walled carbon nanotubes by using flow field-flow fractionation. Anal Chem 2002; 74(18):4774–4780.

120. Chun J et al. Size separation of single-wall carbon nanotubes by flow-field flow fractionation. Anal Chem 2008;80(7):2514–2523.

121. Peng HQ et al. Dielectrophoresis field flow fractionation of single-walled carbon nanotubes. J Am Chem Soc 2006;128(26):8396–8397.

122. Tagmatarchis N et al. Separation and purification of functionalised water-soluble rnulti-walled carbon nanotubes by flow field-flow fractionation. Carbon 2005; 43(9):1984–1989.

123. Rameshwar T et al. Determination of the size of water-soluble nanoparticles and quantum dots by field-flow fractionation. J Nanosci Nanotechnol 2006;6(8):2461–2467.

124. Arifin DR, Palmer AF. Determination of size distribution and encapsulation efficiency of liposome-encapsulated hemoglobin blood substitutes using asymmetric flow field-flow fractionation coupled with multi-angle static light scattering. Biotechnol Prog 2003;19(6):1798–1811.

125. Korgel BA, van Zanten JH, Monbouquette HG. Vesicle size distributions measured by flow field-flow fractionation coupled with multiangle light scattering. Biophys J 1998;74(6):3264–3272.

126. Kirkland JJ, Yau WW, Szoka FC. Sedimentation field flow fractionation of liposomes. Science 1982;215(4530):296–298.

127. Hupfeld S et al. Liposome size analysis by dynamic/static light scattering upon size exclusion-/field flow-fractionation. J Nanosci Nanotechnol 2006;6(9–10):3025–3031.

128. Hupfeld S et al. Liposome fractionation and size analysis by asymmetrical flow field-flow fractionation/multi-angle light scattering: influence of ionic strength and osmotic pressure of the carrier liquid. Chem Phys Lipids 2010;163(2):141–147.
129. Lee NS et al. Application of carbon nanotubes to field emission displays. Diamond Relat Mater 2001;10(2):265–270.
130. Colfen H, Volkel A. Hybrid colloid analysis combining analytical ultracentrifugation and flow-field flow fractionation. Eur Biophys J 2003;32(5):432–436.
131. Kanzer J et al. *In situ* formation of nanoparticles upon dispersion of melt extrudate formulations in aqueous medium assessed by asymmetrical flow field-flow fractionation. J Pharm Biomed Anal 2010;53(3):359–365.
132. Thompson MY, Brandes D, Kney AD. Using electronic conductivity and hardness data for rapid assessment of stream water quality. J Environ Manage 2012;104:152–157.
133. Kney AD, Brandes D. A graphical screening method for assessing stream water quality using specific conductivity and alkalinity data. J Environ Manage 2007;82(4):519–528.
134. Bard AJ, Faulkner LR. *Electrochemical Methods: Fundamentals and Applications*. 2nd ed. New York: Wiley; 2001. p xxi, 833.
135. Bailey PL. *Analysis with Ion-Selective Electrodes*. Heyden international topics in science. 2nd ed. London; Philadelphia: Heyden; 1980. p xv, 247.
136. Wang J. *Analytical Electrochemistry*. 2nd ed. New York: Wiley-VCH; 2000. p xvi, 209.
137. Buck RP, Cosofret VV. Recommended procedures for calibration of ion-selective electrodes. Pure Appl Chem 1993;65(8):1849–1858.
138. Koch M et al. Use of a silver ion selective electrode to assess mechanisms responsible for biological effects of silver nanoparticles. J Nanopart Res 2012;14(2).
139. Benoit R, Wilkinson KJ, Sauvé S. Partitioning of silver and chemical speciation of free Ag in soils amended with nanoparticles. Chem Cent Journal 2013;7(1).
140. Umezawa Y et al. Potentiometric selectivity coefficients of ion-selective electrodes Part I. Inorganic cations – (Technical report). Pure Appl Chem 2000;72(10):1851–2082.
141. Benn TM, Westerhoff P. Nanoparticle silver released into water from commercially available sock fabrics. Environ Sci Technol 2008;42(11):4133–4139.
142. Kakinen A et al. The effect of composition of different ecotoxicological test media on free and bioavailable copper from $CuSO_4$ and CuO nanoparticles: comparative evidence from a Cu-selective electrode and a Cu-biosensor. Sensors 2011;11(11):10502–10521.
143. van den Berg CM. Potentials and potentialities of cathodic stripping voltammetry of trace elements in natural waters. Anal Chim Acta 1991;250:265–276.
144. Galceran J et al. AGNES: A new electroanalytical technique for measuring free metal ion concentration. J Electroanal Chem 2004;566(1):95–109.
145. Parat C et al. Determination of free metal ion concentrations using screen-printed electrodes and AGNES with the charge as response function. Electroanalysis 2011;23(3):619–627.
146. Kokkinos C et al. Microfabricated tin-film electrodes for protein and DNA sensing based on stripping voltammetric detection of Cd(II) released from quantum dots labels. Anal Chem 2013;85(22):10686–10691.
147. Szymanski M, Turner APF, Porter R. Electrochemical dissolution of silver nanoparticles and its application in metalloimmunoassay. Electroanalysis 2010;22(2):191–198.
148. Morones JR et al. The bactericidal effect of silver nanoparticles. Nanotechnology 2005;16(10):2346–2353.

149. Pinijsuwan S et al. Sub-femtomolar electrochemical detection of DNA hybridization based on latex/gold nanoparticle-assisted signal amplification. Anal Chem 2008;80(17):6779–6784.

150. Schmidt J, Vogelsberger W. Dissolution kinetics of titanium dioxide nanoparticles: The observation of an unusual kinetic size effect. J Phys Chem B 2006;110(9):3955–3963.

151. Schmidt J, Vogelsberger W. Aqueous long-term solubility of titania nanoparticles and titanium(IV) hydrolysis in a sodium chloride system studied by adsorptive stripping voltammetry. J Solution Chem 2009;38(10):1267–1282.

152. Domingos RF et al. Comparison of AGNES (absence of gradients and Nernstian equilibrium stripping) and SSCP (scanned stripping chronopotentiometry) for trace metal speciation analysis. J Electroanal Chem 2008;617(2):141–148.

153. David CA et al. Dissolution kinetics and solubility of ZnO nanoparticles followed by AGNES. J Phys Chem C 2012;116(21):11758–11767.

154. Galceran J et al. The impact of electrodic adsorption on Zn, Cd and Pb speciation measurements with AGNES. J Electroanal Chem 2014;722–723(0):110–118.

155. Domingos RF, Franco C, Pinheiro JP. Stability of core/shell quantum dots-role of pH and small organic ligands. Environ Sci Pollut Res 2013;20(7):4872–4880.

156. Rotureau E. Analysis of metal speciation dynamics in clay minerals dispersion by stripping chronopotentiometry techniques. Colloids Surf A 2014;441:291–297.

157. Crouch SS, Douglas A, West DM, Holler FJ. *Fundamentals of Analytical Chemistry*. 9 ed. Brooks Cole; 2013.

158. Vogel A. *Vogel's Textbook of Quantitative Inorganic Analysis, Including Elementary Instrumental Analysis*. Longman Sc & Tech; 1980.

159. Säbel CE, Neureuther JM, Siemann S. A spectrophotometric method for the determination of zinc, copper, and cobalt ions in metalloproteins using Zincon. Anal Biochem 2010;397(2):218–226.

160. Noelia Corral-Gallego M, Ramis-Fossas J, Aulesa-Martinez C. Evaluation of two colorimetric methods for the determination of zinc in seminal plasma. Revista Internacional De Andrologia 2012;10(2):57–62.

161. Fujita T et al. New enzymatic assay of iron in serum. Clin Chem 1994;40(5):763–767.

162. Cherny RA et al. Chelation and intercalation: Complementary properties in a compound for the treatment of Alzheimer's disease. J Struct Biol 2000;130(2–3):209–216.

163. Syed AA, Syeda A. Neocuproine and bathocuproine as new reagents for the spectrophotometric determination of certain proton pump inhibitors. Bull Chem Soc Ethiop 2007;21(3):315–321.

164. Horstkotte B et al. Automatic determination of copper by in-syringe dispersive liquid-liquid microextraction of its bathocuproine-complex using long path-length spectrophotometric detection. Talanta 2012;99:349–356.

165. Abe A, Yamashita S, Noma A. Sensitive, direct colorimetric assay for copper in serum. Clin Chem 1989;35(4):552–554.

166. Mukherjee AK. Use of some new chelating agents for the colorimetric determination of iron: Ethylenediamine-bis-sulphosalicylaldehyde. Anal Chim Acta 1955;13:268–272.

167. Singh C, Friedrichs S, Levin M, Birkedal R, Jensen KA, Pojana G, Wohlleben W, Schulte S, Wiench K, Turney T, Koulaeva O, Marshall D, Hund-Rinke K, Kördel W, Van Doren E, De Temmerman P-J, Abi Daoud Francisco M, Mast J, Gibson N, Koeber R, Linsinger T, Klein CL. *NM-Series of Representative Manufactured Nanomaterials – Zinc*

Oxide NM-110, NM-111, NM-112, NM-113 Characterisation and Test Item Preparation. J.R.C.I.f.R.M.a. Measurements; 2011.

168. Peters RJB et al. Development and validation of single particle ICP-MS for sizing and quantitative determination of nano-silver in chicken meat. Anal Bioanal Chem 2014;406(16):3875–3885.
169. Mitrano DM et al. Detecting nanoparticulate silver using single-particle inductively coupled plasma-mass spectrometry. Environ Toxicol Chem 2012;31(1):115–121.
170. Pace HE et al. Determining transport efficiency for the purpose of counting and sizing nanoparticles via single particle inductively coupled plasma mass spectrometry (vol 83, pg 9361, 2011). Anal Chem 2012;84(10):4633.
171. Laborda F et al. Selective identification, characterization and determination of dissolved silver(I) and silver nanoparticles based on single particle detection by inductively coupled plasma mass spectrometry. J Anal Atom Spectrom 2011;26(7):1362–1371.
172. Takenaka S et al. A morphologic study on the fate of ultrafine silver particles: Distribution pattern of phagocytized metallic silver in vitro and in vivo. Inhal Toxicol 2000;12:291–299.
173. Tang JL et al. Silver Nanoparticles Crossing Through and Distribution in the Blood–brain Barrier In Vitro. J Nanosci Nanotechnol 2010;10(10):6313–6317.
174. Bouwmeester H et al. Characterization of translocation of silver nanoparticles and effects on whole-genome gene expression using an in vitro intestinal epithelium coculture model. ACS Nano 2011;5(5):4091–4103.
175. Liu W et al. Impact of silver nanoparticles on human cells: Effect of particle size. Nanotoxicology 2010;4(3):319–330.
176. Alkilany AM et al. Cellular uptake and cytotoxicity of gold nanorods: molecular origin of cytotoxicity and surface effects. Small 2009;5(6):701–708.
177. van der Zande M et al. Sub-chronic toxicity study in rats orally exposed to nanostructured silica. Part Fibre Toxicol 2014;11.
178. Loeschner K et al. Distribution of silver in rats following 28 days of repeated oral exposure to silver nanoparticles or silver acetate. Part Fibre Toxicol 2011;8.
179. Liu JY, Hurt RH. Ion release kinetics and particle persistence in aqueous nano-silver colloids. Environ Sci Technol 2010;44(6):2169–2175.
180. Xiu ZM et al. Negligible particle-specific antibacterial activity of silver nanoparticles. Nano Lett 2012;12(8):4271–4275.
181. Behra R et al. Bioavailability of silver nanoparticles and ions: From a chemical and biochemical perspective. J R Soc Interface 2013;10(87).
182. Loza K et al. The dissolution and biological effects of silver nanoparticles in biological media. J Mater Chem B 2014;2(12):1634–1643.
183. Ma HB et al. Toxicity of manufactured zinc oxide nanoparticles in the nematode caenorhabditis elegans. Environ Toxicol Chem 2009;28(6):1324–1330.
184. Eurachem. *The fitness for purpose of analytical methods: A laboratory guide to method validation and related topics.* Laboratory of the Government Chemist; 1998.
185. Büttner J et al. Provisional recommendation on quality of control in clinical chemistry part 1. General principles and terminology. Clin Chim Acta 1975;63(1):F25–F38.
186. Horwitz W, Cohen S, Hankin L, Krett J, Perrin CH, Thornburg W. Quality assurance practices for health laboratories. In: Inhorn S, editor. *Analytical Food Chemistry.* Washington, DC; 1978. p 545–646.

187. Kittler S et al. Toxicity of silver nanoparticles increases during storage because of slow dissolution under release of silver ions. Chem Mater 2010;22(16):4548–4554.
188. Ma R et al. Size-controlled dissolution of organic-coated silver nanoparticles. Environ Sci Technol 2012;46(2):752–759.
189. Lee YJ et al. Ion-release kinetics and ecotoxicity effects of silver nanoparticles. Environ Toxicol Chem 2012;31(1):155–159.
190. Farmen E et al. Acute and sub-lethal effects in juvenile Atlantic salmon exposed to low μg/L concentrations of Ag nanoparticles. Aquat Toxicol 2012;108:78–84.
191. Unrine JM et al. Biotic and abiotic interactions in aquatic microcosms determine fate and toxicity of Ag nanoparticles. Part 1. Aggregation and dissolution. Environ Sci Technol 2012;46(13):6915–6924.
192. Huang W et al. Dissolution and nanoparticle generation behavior of Be-associated materials in synthetic lung fluid using inductively coupled plasma mass spectroscopy and flow field-flow fractionation. J Chromatogr A 2011;1218(27):4149–4159.
193. Loeschner K et al. Detection and characterization of silver nanoparticles in chicken meat by asymmetric flow field flow fractionation with detection by conventional or single particle ICP-MS. Anal Bioanal Chem 2013;405:8185–8195.
194. Liu J et al. Degradation products from consumer nanocomposites: a case study on quantum dot lighting. Environ Sci Technol 2012;46:3220–3227.
195. Prach M, Stone V, Proudfoot L. Zinc oxide nanoparticles and monocytes: Impact of size, charge and solubility on activation status. Toxicol Appl Pharmacol 2013;266(1):19–26.
196. van der Zande M et al. Distribution, elimination, and toxicity of silver nanoparticles and silver ions in rats after 28-day oral exposure. ACS Nano 2012;6(8):7427–7442.
197. Weigl BH, Bardell RL, Cabrera CR. Lab-on-a-chip for drug development. Adv Drug Deliv Rev 2003;55(3):349–377.
198. Elbashir AA, Aboul-Enein HY. Recent advances in applications of capillary electrophoresis with capacitively coupled contactless conductivity detection (CE-C4D): an update. Biomed Chromatogr 2012;26(8):990–1000.
199. Vrouwe EX et al. Microchip capillary electrophoresis for point-of-care analysis of lithium. Clin Chem 2007;53(1):117–123.
200. Tantra R et al. Dispersion stability of nanoparticles in ecotoxicological investigations: the need for adequate measurement tools. J Nanopart Res 2011;13(9):3765–3780.
201. Tantra R, Robinson K, Sikora A. Variability of microchip capillary electrophoresis with conductivity detection. Electrophoresis 2014;35(2–3):263–270.

6

SOLUBILITY PART 2: COLORIMETRY

R. TANTRA, K. N. ROBINSON, AND J. C. JARMAN

Quantitative Surface Chemical Spectroscopy Group, Analytical Science, National Physical Laboratory, Teddington TW11 0LW, UK

D. GOHIL

Advanced Engineered Materials Group, Materials, National Physical Laboratory, Teddington TW11 0LW, UK

6.1 INTRODUCTION

As discussed in Chapter 5, the property of solubility is of interest to nanoregulation, particularly in the context of hazard and risk management. Hence, the measurement of solubility has been highlighted as a topic for further investigation in a currently running pan-European project, Framework Programme (FP) 7 NANoREG; one of the goals of the project is to develop methods to determine the solubility of nanomaterials [1].

In Chapter 5, a small case study was presented, which investigated the feasibility of a microfluidics-based device, to quantify free zinc (Zn^{2+}) that arises from dissolution of ZnO nanomaterial when dispersed in a reconstituted fish medium. Our findings indicate that this microfluidic device is still immature, due to high levels of uncertainty associated with lack of quantification repeatability. Overall, such a state-of-the-art technology still requires further development, in order to make the device reliable and robust. Yet, in nanotoxicology research, the need exists for not only reliable and robust methods but also affordable and accessible.

Of the many methods described in Chapter 5, approaches based on electrochemical and colorimetric assays are worthy of note. In relation to electrochemical

Nanomaterial Characterization: An Introduction, First Edition. Edited by Ratna Tantra.

methods, commercially available ion selective electrodes (ISEs) are particularly attractive as such techniques are considered to be well established [2]. ISE measurements are simple and fast, which involve the immersion of electrodes into a sample solution in order to acquire a reading. Although many ISEs for different ions are available, to date no ISE for the quantification of Zn^{2+} is commercially available. Yet, such an ISE will be particularly useful to measure the solubility of nanomaterials that contain zinc such as ZnO. In this case, colorimetry-based assays capable of detecting Zn^{2+} should be considered.

Colorimetry has many advantages associated with ISE, in that it is affordable and simple to use for the quantification of free ions. It is particularly attractive as it requires the use of an ultraviolet–visible (UV–Vis) spectrometer, an instrument that nearly every laboratory has. In this chapter, as part of FP7 NANoREG project's output, colorimetry will be used to quantify the amount of free zinc ions generated from ZnO nanomaterial (powder) dissolution. The main goal of this study is to employ a colorimetric assay to estimate the amount of dissolved free zinc (Zn^{2+}) that arises when ZnO nanomaterial is exposed to digestive juices, to be used in *in vitro* digestion model that will simulate digestion from the mouth to the intestine. Generally, using a digestive model has its advantages, in that it can reveal the state of the nanomaterial at each stage of digestion, which can potentially be analyzed; this is not always possible in an *in vivo* experiment. The purpose of using the digestive juice model here is to understand exposure implications of the nanomaterial should there be an accidental ingestion in humans [3]. One objective of the study is to establish if there is a correlation between the amount of free zinc that arises from the dissolution of ZnO nanomaterial with respect to particle concentration. In addition to colorimetry, scanning electron microscope (SEM) will be used to offer complementary information. The aim here is to positively identify if the ZnO nanomaterial is present at the different stages of digestion, namely, after addition of saliva juice and at the end of digestion.

In order to develop a suitable colorimetry method, it is important to identify the dyes that are suitable for the intended application. In this study, it is important for the dyes to be water soluble and commercially available. There are about 10 commercial dyes to choose from, all having the potential capability to complex and thus measure the concentration of free zinc ions [Zn^{2+}]. From this list, not all can be classed as water soluble. Insoluble dyes include PAN (1-(2-Pyridylazo)-2-naphthol and Cu-PAN. Partially soluble dyes include Xylenol orange, Eriochrome Black T, Nitro-PAPS, 1-1 phenanthroline, PAR (4-(2-Pyridylazo) resorcinol. Only three commercially available dyes are worthy of note as they are water soluble: 5-Bromo-PAPS (chemical name = 2-(5-Bromo-2-pyridiylazo)-5-[*N*-*n*-propyl-*N*-(3-sulfopropyl)amino]phenol), Methylthymol blue (chemical name = 3,3′-Bis[*N*,*N*-di(carboxymethyl)aminomethyl] thymolsulfonephthalein sodium salt, MTB, Thymolsulfonphthalein-3′,3″-bis(methyliminodiacetic acid sodium salt)), and Zincon (chemical name = 2-[5-(2-Hydroxy-5-sulfophenyl)-3-phenyl-1-formazyl] benzoic acid monosodium salt, 2-Carboxy-2′-hydroxy-5′-sulfoformazyl-benzene monosodium salt). According to literature, all three dyes have been used to measure [Zn^{2+}] in different samples. Zincon has been used for the measurement of zinc in granular fertilizers [4], plant tissues [5,6], metalloproteins [7], beverages [8], and water [9]. Methylthymol blue has been employed for alloy samples [10], binary water–methanol mixtures [11]. Unlike the other two, 5-Bromo-PAPS has been frequently employed for the analysing biological fluids as well as environmental samples, such as serum [12–15], seminal plasma [16], and environmental water samples [17]. As the study here deals with the measurement of [Zn^{2+}] in digestive juices and, hence biological fluids, 5-Bromo-PAPS dye will be used.

It is worthy to note that what differentiates the FP7 NANoREG project from other projects is the need for participating research laboratories to follow a set of defined protocols. This is considered necessary to establish some level of quality control and assurance. Most of the protocols presented in this chapter (apart from colorimetry-based measurements) have been developed by other partners within the FP7 NANoREG consortium; these are mandatory protocols, and all relevant consortium partners are required to follow such protocols. Here, the mandatory protocols include: calibration of ultrasonic probes, benchmark activities and following a digestion protocol.

In relation to nanomaterials, there were restrictions with regards to what nanomaterials can be used for the study, which must be part of the "NM series". Here, SiO_2 and ZnO nanomaterials used are part of the NM series of representative test nanomaterials hosted by JRC Nanomaterials (Joint Research Centre). Test material JRCNM02000a (SiO_2), JRCNM01100a (ZnO) will be used. For the sake of simplicity, the common names of NM 200 (SiO_2) and NM 110 (ZnO), respectively, will be employed for the remainder of the chapter.

This chapter thus presents the findings obtained from studies on colorimetry of ZnO nanomaterial. The methods and results associated with other mandatory tests that are supplementary to the colorimetry tests will be reported in Appendix A6.

6.2 MATERIALS AND METHOD

6.2.1 Materials

DI water (Elga Purelab; resistivity = 18.2 MΩ cm) was used throughout the study. Unless stated otherwise, all chemicals used were of analytical grade or better, as supplied from Sigma Aldrich, UK. Nanomaterial powders JRCNM01100a (ZnO), also commonly referred to as ZnO (NM-110), were supplied sealed under argon in small amber glass vials and used as received from the European Commission Joint Research Centre, Institute for Health and Consumer Protection JRC (JRC-IHCP, Ispra, Italy), through FP7 NANoREG project. These vials were used within a few hours after first opening; the remaining material was disposed. For any given experiment conducted, the ID numbers on the vials were recorded.

6.2.2 Mandatory Protocol: NanoGenotox Dispersion for Nanomaterials

The NanoGenotox dispersion protocol was used to disperse the nanomaterial. This involved the prewetting of the nanomaterial powder using 0.5 vol% ethanol before dispersing in 0.05 wt.% BSA (in DI water) to result in a final particle concentration of 2.56 mg/ml. Probe sonication was carried out for 8 min at 5% amplitude, using a Misonix Sonicator 3000 (with a maximum power output of 400 W and operated at a frequency of 20 kHz). The diameter of the ultrasonication probe tip was 13 mm. The settings for amplitude and sonication time were determined when the probe was calibrated by following a few basic procedures and subsequently benchmarked against reference SiO_2 (NM 200) (see Appendix A6 for further details). Overall, the probe was calibrated so that a total energy of 7056 J was delivered. During sonication, the vial that held the nanomaterial dispersion was immersed in an ice water bath to minimize thermal effects generated during the ultrasonic procedure. Unless specified otherwise, freshly made dispersions were used within the first two hours of preparation. It was important to establish that there were no visible precipitates in the dispersion prior to use.

6.2.3 Mandatory Protocol: Simulated *In Vitro* Digestion Model

The day before carrying out the actual digestion steps, four types of media (saliva, gastric juice, duodenal juice, and bile juice) were prepared in accordance to the recipe composition tabulated in Table 6.1. These media were made up on Day 1 and stored

TABLE 6.1 Composition of Four Different Juices for the Fed *In Vitro* Digestion [18], made up on Day 1

Saliva	Gastric Juice	Duodenal Juice	Bile Juice
• 896 mg/l KCl • 200 mg/l KSCN • 1021 mg/l $NaH_2PO_4H_2O$ • 570 mg/l Na_2SO_4 • 298 mg/l NaCl • 1694 mg/l $NaHCO_3$ • 200 mg/l urea • 290 mg/l amylase • 15 mg/l uric acid • 25 mg/l mucin • • milli-Q water (~1 l)	• 2752 mg/l NaCl • 306 mg/l $NaH_2PO_4H_2O$ • 824 mg/l KCl • 302 mg/l $CaCl_2$ • 306 mg/l NH_4Cl • 6.5 ml/l 37% HCl • 650 mg/l glucose • 20 mg/l glucuronic acid • 85 mg/l urea • 330 mg/l glucosaminehy-drochloride • 1 g/l BSA • 2.5 g/l pepsin • 3 g/l mucin • milli-Q water (~1 l)	• 7012 mg/l NaCl • 3388 mg/l $NaHCO_3$ • 80 mg/l KH_2PO_4 • 564 mg/l KCl • 50 mg/l $MgCl_2 \cdot 6H_2O$ • 180 µl/l HCl (37%) • 100 mg/l urea • 151 mg/l $CaCl_2$ • 1 g/l BSA • 9 g/l pancreatin • 1.5 g/l lipase • milli-Q water (~1 l)	• 7012 mg/l NaCl • 3388 mg/l $NaHCO_3$ • 80 mg/l KH_2PO_4 • 564 mg/l KCl • 50 mg/l $MgCl_2.6H_2O$ • 180 µl/l HCl (37%) • 100 mg/l urea • 151 mg/l $CaCl_2$ • 1 g/l BSA • 9 g/l pancreatin • 1.5 g/l lipase • milli-Q water (~1 l)

overnight at room temperature in the incubator, prior to their use. The actual digestion was carried out on Day 2.

On Day 2, about 2 h before carrying out the digestion protocol, an incubator was set at $37 \pm 2\,^{\circ}C$. Before conducting the digestion experiment, it is important to do a quick check on the digestive solutions made on Day 1. This was performed by making up the following solution, consisting of 1 ml saliva, 2 ml gastric juice, 2 ml duodenal juice, 1 ml bile juice, and 28 mg (26.5–29.5 mg) $NaHCO_3$. The pH was checked, which had to be within the range 6.5 ± 0.5, otherwise the liquid juices could not be used in the digestion experiment that follows.

The digestion protocol was conducted on NM 110 (ZnO). The first step was to prepare the batch dispersion in accordance to the NanoGenotox protocol as detailed above. One milliliter of the nanomaterial dispersion was then transferred to a 50-ml Greiner tube, after which 6 ml saliva juice were added and incubated "head-over-heels" (or under tumbling motion) for 5 min at $37 \pm 2\,^{\circ}C$. Afterwards, 12 ml gastric juice was added to the tube and pH measured. The pH at this point was adjusted to 2.5 ± 0.5. The tube was then further incubated head-over-heels for 2 h at $37 \pm 2\,^{\circ}C$. Then, 12 ml of duodenal juice was added to the mixture, followed by 6 ml bile juice and 2 ml sodium carbonate solution ($NaHCO_3$; 84.7 g/l). The pH at this point was adjusted to 6.5 ± 0.5. The tube was then incubated head-over-heels for another 2 h at $37 \pm 2\,^{\circ}C$. For the pH measurements, a pH-meter was used in which the glass electrode was thoroughly cleaned and calibrated between measurements. pH adjustments were made using either hydrochloric acid (HCl) or sodium hydroxide (NaOH), as appropriate.

The digestion protocol was also conducted without NM 110 (ZnO), resulting in the corresponding collection of "digestive blanks," to act as controls for the experiment. For the purpose of SEM analysis, a separate digestion was carried out in parallel, so that the appropriate sized aliquots can be extracted for the SEM analysis. The protocol for this is detailed in Section 6.2.5. Samples were taken at specific time intervals during the digestion experiment: after conducting the NanoGenotox protocol, after saliva addition and at the end of the digestion protocol.

6.2.4 Colorimetry Analysis

The following reagents were used in the colorimetric zinc determination:

a) 5-Bromo-PAPS solution (100 mg/l) in DI water.
b) HEPES (4-(2-hydroxyethyl)-1-piperazineethanesulfonic acid) buffer (1M), pH 8. Zinc atomic spectroscopy standard stock concentrate was diluted with DI water to give a final concentration of 10 g/l. The corresponding zinc standard concentration series solutions were made up by subsequently diluting the stock with appropriate volume of DI water.

Prior to analysis, samples from the digestion experiment had to be processed in order to remove particulate matter. This was achieved in two steps: centrifugation of the sample for 30 min at 1000 rcf (using Centrifuge 5430, Eppendorf, UK) to extract the supernatant, then filtration of the resultant supernatant through a hydrophilic poly(vinylidene fluoride) (PVDF) membrane filter (pore size of 0.45 μm), to remove any remaining particulate matter. Colorimetric assays were then performed on the resultant solutions.

For colorimetry assay, 1 ml of 5-Bromo-PAPS solution was thoroughly mixed with 0.3 ml of sample solution and allowed to stand for 5 min. Then, 200 μl of this mixture was diluted with 600 μl of HEPES buffer before being pipetted into a clean quartz UV cell (Hellma Analytics, UK). UV–Visible absorbance spectrum was then subsequently acquired using a Lambda 850 UV–Vis spectrometer supported by UV Winlab software [Version 5.1.5] (Perkin Elmer, USA). The instrument wavelength calibration was checked using a 15246-Ho Holmium glass standard (Serial # 9392, Starna Scientific, UK). For the reference channel of the spectrophotometer, a matched cell containing the corresponding dispersing media was used. Absorbance spectra were recorded in the range 350–700 nm using a slit width of 2 nm and a scan rate of 50 nm/min. Three replicate measurements were carried out for each sample analyzed.

6.2.5 SEM Analysis

As previously mentioned, SEM analysis was carried out on the samples extracted at various stages of the digestion protocol: (i) the dispersion at the start of the experiment, that is, after following the NanoGenotox protocol, (ii) after saliva addition, and (iii) at the end of the digestion.

Sample preparation for SEM analysis on the extracted samples involved fixing the particles onto poly-L-lysine-coated microscope glass slides (Sigma Aldrich, UK). This was achieved by pipetting 1 ml aliquots of the sample to functionalized glass slides and incubating for 5 min; this step was conducted in a fume hood at room temperature. The slides were then dip rinsed in DI water to remove excess material and left to dry under cover in a fume hood. Sample slides were stored in sealed boxes at room temperature, ready for analysis.

SEM images were obtained using a Supra 40 field emission SEM from Carl Zeiss (Welwyn Garden City, Hertfordshire, UK), in which the optimal spatial resolution of the microscope was 1.2 nm. The in-lens detector images were used to obtain the images, an accelerating voltage of 15 kV, a working distance of ≈3 mm, and a tilt angle 0° was used. The SEM was calibrated using a SIRA grid calibration set (SIRA, Chislehurst, Kent, UK). These are metal replicas of cross-ruled gratings of area of 60 mm^2 with 19.7 lines/mm for low-magnification and 2160 lines/mm for high-magnification calibrations, accurate to 0.2%. For the image acquisition, an adequate magnification was chosen so that the shape and particle boundary of the primary particles became apparent.

To avoid charging effects, slides were thinly sputtered with gold using an Edwards S150B sputter coater unit (BOC Edwards, UK). Sputtering was conducted under vacuum (≈7 mbar or 0.7 mPa), while passing pure, dry argon into the coating chamber. Typical plate voltage and current were 1200 V and 15 mA, respectively. Sputtering time was approximately 10 s, which resulted in an estimated gold thickness of no more than 2 nm being deposited on top of the substrate.

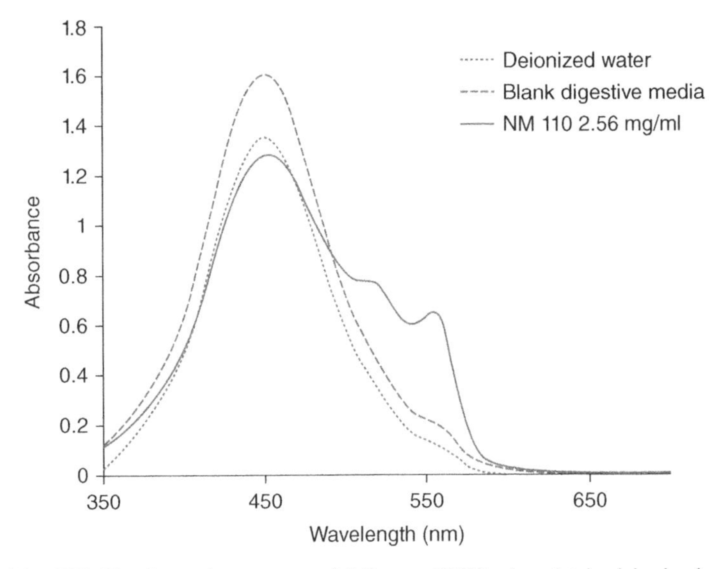

Figure 6.1 UV–Vis absorption spectra of 5-Bromo PAPS when (a) in deionized water, (b) digestive blank media, and (c) extracted supernatant after 2.56 mg/ml of ZnO (NM 110) has been dispersed and exposed to the digestion protocol.

6.3 RESULTS AND INTERPRETATION

A preliminary investigation was conducted to evaluate the signature wavelength of the dye used, that is, 5-Bromo-PAPS, above background (i.e., the digestive blank) and how this changes in the presence of dissolved zinc that arises from the dissolution of ZnO nanomaterial. Thus, Figure 6.1 shows the UV–Vis absorption spectra of 5-Bromo PAPS when in three different media: deionized water, digestive blank, and extracted supernatant from a nanomaterial suspension (2.56 mg/ml) after undergoing the digestion protocol. Results show similar spectra in the case of DI water and digestive blank but not for the extracted supernatant case. In the case of extracted supernatant, there is an apparent change in the UV–Vis spectral profile. Results show the presence of λ_{max} (absorbance (Abs) peak) at 556 nm on top of the blank (of 0.65 absorbance unit), which indicates the presence of $[Zn^{2+}]$-dye complex. The experiment was repeated but with the 5-Bromo-PAPS plus HEPES reagent (as detailed in the Method section) and similar result was acquired. Hence, this preliminary study confirms that the 556 peak maxima indicates the presence of a zinc complex and thus will be used to identify and measure $[Zn^{2+}]$. Note that absorbance $= -\log_{10} T$, where $T = I/I_o$; I_o and I are the intensities of the incident light and transmitted light, respectively.

Figure 6.2 shows a calibration plot, using dilution of standard stock solutions, of net absorbance with respect to $[Zn^{2+}]$. Results show a linear response within the concentration range of 0–10 mg/l. The corresponding regression equation is shown in

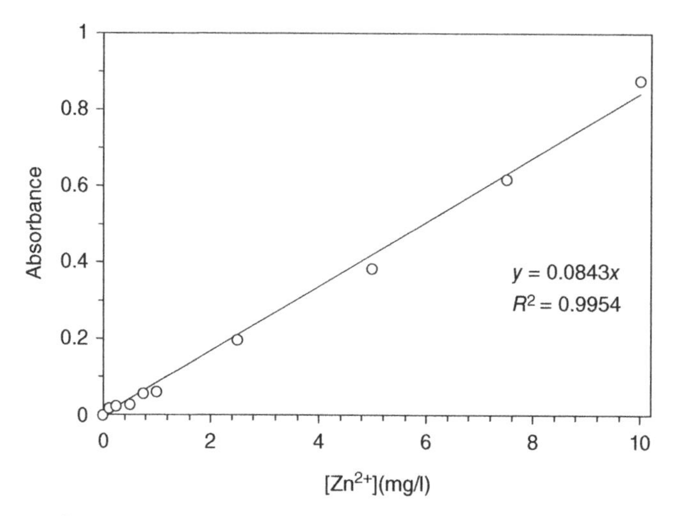

Figure 6.2 $[Zn^{2+}]$ concentration calibration plot using 5-Bromo-PAPS (λ_{max} = abs peak at 556 nm). Each data point is the mean of triplicate measurements; note that the standard deviation is too small to be visible.

Figure 6.2, which was used to estimate unknown $[Zn^{2+}]$ concentrations arising from the digestive juice experiment.

Table 6.2 shows the estimated $[Zn^{2+}]$ arising from dissolution of ZnO (NM 110) at various particle concentrations following the digestion protocol. The table also shows the corresponding UV–Vis Abs max values. Note that at zero ZnO concentration, the signal here is reported as zero; this is the net signal, which is obtained when the

TABLE 6.2 A Summary of UV–Vis Absorbance Signal (λ_{max} = abs peak at 556 nm) and Corresponding Estimated $[Zn^{2+}]$ Found in the Extracted Supernatant from ZnO (NM 110) Digestive Juice Experiment

Mass of ZnO [NM 110] (mg/ml)	5-Bromo-PAPS		
	Net Absorbance @ 556 nm (Corrected Relative to the Background)		Estimated $[Zn^{2+}]$ (μg/ml)
	Mean	RSD (%)	
0	0	0	0
0.00256	0.0112	3.6284	0.1331
0.0256	0.0137	1.3362	0.1622
0.128	0.0420	0.6704	0.4979
0.256	0.0406	0.7436	0.4816
1.28	0.1377	0.2191	1.6335
2.56	0.4425	0.0809	5.2489

The absorbance data reported are from three replicate measurements.

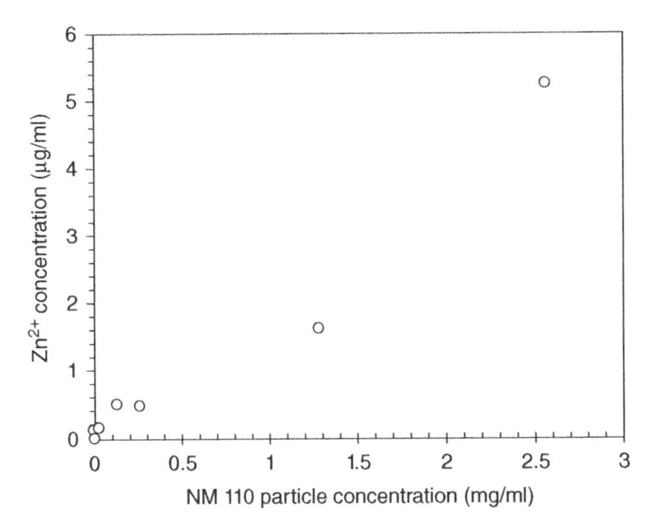

Figure 6.3 Estimated $[Zn^{2+}]$ as a function of ZnO (NM 110) particle concentration. The $[Zn^{2+}]$ reported arises from the dissolution of ZnO (NM 110) particle concentration as a result of digestive juice experiment.

background signal arising from the digestive blank is subtracted from the final signal. The digestive blank signal in this case is the signal arising from the sample in which the digestion experiment was carried out in the absence of ZnO.

Figure 6.3 shows the corresponding plot, indicating a positive correlation between particle concentration and estimated $[Zn^{2+}]$. This is expected as an increase in the nanomaterial concentration tends to result in an increase in dissolution and thus the measured $[Zn^{2+}]$.

Figure 6.4 shows SEM images of the ZnO (NM 110) nanomaterial dispersion, before running the digestion experiment. SEM results indicate that particles do not exist as primary particles but clusters of particles of different size agglomerates. Results show that the primary particles themselves are inhomogeneous, with respect to the particle size and shape, in which rods and spheres exist. The results coincided with past SEM images on the as-received powders of ZnO (NM 110), thus confirming that the ZnO NM 110 particles have distinct features that can be observed from the SEM images.

Figure 6.5 shows the SEM results after addition of saliva following the digestion protocol; images indicate the presence of ZnO (NM 110) particles. There is only a slight difference between these images compared to those of Figure 6.4, that is, pure NM dispersion. Images show an increase of background material in the corresponding blank associated with saliva case (as indicated in the figure). Although, the nanomaterial is clearly visible above the background, the overall quality of the images associated with the saliva case is poor. By this, we mean that particle boundaries are less visible, due to other matter found in the corresponding blank that can interfere with the quality of the SEM images.

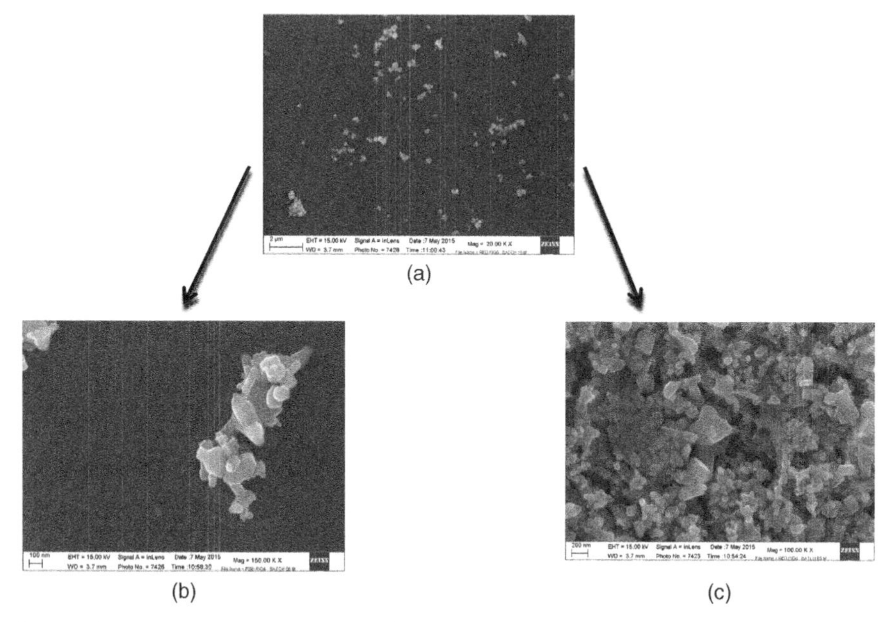

Figure 6.4 SEM images showing ZnO (NM 110) nanomaterials after dispersion using NanoGenotox protocol. Three different images were acquired, taken at different magnifications, showing (a) the presence of particle agglomerates, (b) size of the smallest agglomerates, and (c) polydispersity in primary particle size.

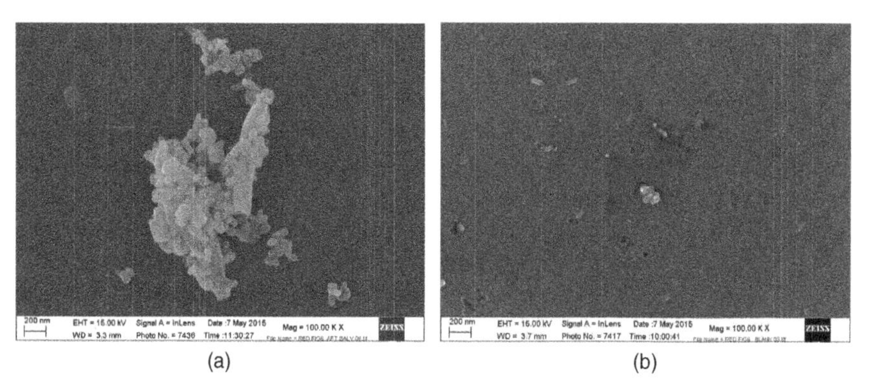

Figure 6.5 SEM images showing ZnO (NM 110) nanomaterial (a) after saliva addition. Results also show (b) the corresponding digestive blank after saliva addition.

Figure 6.6 shows the SEM images of particles observed at the end of digestion protocol; results also show the corresponding digestive blank control, that is, without addition of ZnO (NM 110). Results show no signs of the ZnO (NM 110) particles present at the end of the digestion. However, other particulate species are visible in the corresponding digestive blank. In addition to particles that are oblong in shape,

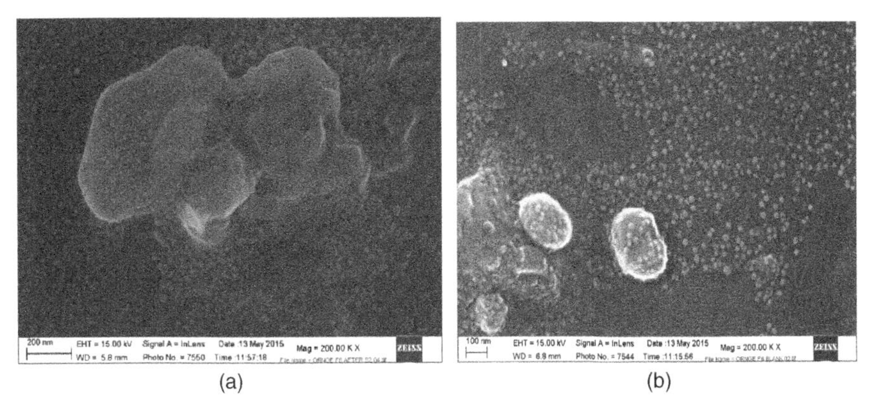

Figure 6.6 SEM images showing ZnO (NM 110) nanomaterial at the end of digestion (a). Results also show the corresponding digestive blank (b) at the end of the digestion protocol.

much smaller particles, are also apparent. Such particles are likely to originate from the background contribution as the corresponding SEM blank also show the presence of such particles. Again, the overall quality of SEM images are poorer compared to the batch dispersion before digestion; again, this can be attributed to a much higher level of background contributions arising from the sample.

Results from Figure 6.6 indicate that the original ZnO (NM 110) particles are no longer present at the end of the digestion. This is not surprising as part of the digestion protocol involved the addition of gastric juice and adjusting the pH to 2.5 ± 0.5. At this low pH, high solubility of the ZnO (NM 110) nanomaterial [19, 20] is expected. However, the colorimetry results show that only a fraction of the dissolved zinc exists as free ions. This is not surprising as the dissolved zinc arising from the nanomaterial dissolution is likely to complex with other species within the digestive solutions. It is well known that species in digestive media can potentially bind to zinc free ions; an example of such species is bovine serum albumin [21].

6.4 CONCLUSION

This chapter presented a colorimetric method and demonstrated its suitability to measure [Zn^{2+}]. The method was used to quantify [Zn^{2+}] arising from the dissolution of ZnO (NM 110) following a digestion protocol. Some of the attractive features in using colorimetry-based methods surround its simple operation, potentially suitable for routine analysis. 5-Bromo-PAPS was used as the colorimetric dye due to its commercial availability and high water solubility. Results showed a positive correlation between ZnO (NM 110) particle concentration and [Zn^{2+}] released as a result of the digestive juice experiment. The SEM images indicated complete dissolution of the ZnO (NM 110) material at the end of the digestion; this was attributed to a low pH environment associated with part of the digestion protocol, that is, addition of gastric juice. The colorimetry results, however, indicated that the dissolved zinc exists as complexes rather than free ions. Lastly, as part of the FP7 NANoREG project, it was mandatory for research partners to supply a set of supporting characterization data. The purpose here was to provide some level of assurance about data quality reported; this supplementary data is presented in the Appendix section.

ACKNOWLEDGMENTS

The work leading to these results has received support from the European Union Seventh Framework Programme (FP7-NMP.2012.1.3-3) under grant agreement no. 310584 (NANoREG). Many thanks to colleagues within the NANoREG project for the supply of protocols, through the WP co-ordination by Dr Keld Alstrup Jensen from NRCWE, Denmark. Special thanks to Dr Meike van der Zande from RIKILT, the Netherlands, for providing the digestion protocol and supplying digestion media solutions for use in the experiments. Many thanks to JRC, ISPRA, for providing useful feedback/comments and the supply of nanomaterials.

APPENDIX A6

PROBE CALIBRATION AND BENCHMARKING EXERCISES

A6.1 MATERIALS AND METHOD

A6.1.1 Materials

In addition to the materials already reported in Section 6.2, a nanomaterial powder, JRCNM02000a (SiO_2), referred to as SiO_2 (NM-200) was used. The SiO_2 (NM-200) was supplied by JRC Nanomaterials Repository (IHCP, JRC, Ispra) and used as detailed in Section 6.2.1.

A6.1.2 Mandatory Protocol: Ultrasonic Probe Calibration

The ultrasonic probe calibration protocol is based on measuring liquid temperature increase over time, as a result of immersing a sonication probe in a liquid. This allows the direct measurement of effective acoustic energy delivered to the sonicated liquid.

The calibration protocol can be divided into three parts. The first part involved drawing 3 l of DI water into a flask; the flask was then left in a fume hood for 2 h. A 500 ml aliquot of DI water from this flask was then transferred into a separate 600-ml glass beaker. The mass and temperature of the water in the 600-ml beaker was measured before a sonicator probe was inserted at a depth of 2.5 cm below the water surface. A temperature probe was also inserted, at a depth of 2.5 cm below the water surface but positioned 1 cm away from the sonicator probe. The temperature of the water was recorded before the sonicator was turned on, at an amplitude of 5%, under a continuous mode for 5 min. During sonication, the temperature of the water was recorded, at least every 30 s. The entire step of drawing 500 ml of water, probe sonication and measuring temperature of the water was repeated twice; thus measurements were performed in triplicate. The second part of the calibration protocol involved repeating the first part of the protocol, this time with 20% amplitude. The third part involved repeating the first part with 30% amplitude.

The final part of the protocol involved the analysis of the recorded data. This involved plotting a temperature versus time curve and fitting a best linear fit to estimate the delivered power of the probe (P_{ac}, Watt), according to

$$P_{ac} = \frac{\Delta T}{\Delta t} M C_p \tag{6.1}$$

where $\frac{\Delta T}{\Delta t}$ is the slope of the regression curve, T is the temperature (K), t is sonciation time period (s), C_p is the specific heat of the liquid (4.18 J/g K for water), and M is the mass of liquid (g).

Once the delivered power of the probe (in Watts) is known, the relationship between acoustic energy (E, Joule – J) delivered as a function of sonication time (t_s, s) can be established:

$$P = E / t_s \tag{6.2}$$

According to the NanoGenotox dispersion protocol, it is necessary for the probe to deliver a total acoustic energy of 7056 ± 103 J. Equation 6.2 will allow an estimation of sonication time, t_s, at a given % amplitude.

A6.1.3 Mandatory Protocol: Benchmarking of SiO_2 (NM 200)

Once an estimation of sonication time and the % amplitude have been established, a benchmarking exercise was carried out on the materials SiO_2 (NM-200) to "fine–tune" the sonication time. In order to do this, three different SiO_2 (NM 200) vials were individually assessed. For each vial, the nanomaterial dispersion was made in accordance to the NanoGenotox protocol and then subsequently characterized using dynamic light scattering (DLS) (see below for DLS protocol). The corresponding DLS particle size distribution was acquired ×10 times without pause. The procedure was repeated with the other two SiO_2 (NM-200) vials. Results were analyzed, from which the DLS Z-average diameter and polydispersity index (PDI) values were obtained and compared against a benchmark figure. If the result did not coincide with the benchmark figure provided, then the sonication time was changed in order to hit this target. Our findings show that at 5% amplitude, a sonication time of 8 min was needed (according to instructions detailed in Section 6.2).

A6.1.4 Mandatory Protocol: Preliminary Characterization of ZnO (NM 110)

The ZnO (NM 110) test material used in the study was dispersed and characterized in accordance to the NanoGenotox protocol, using sonicator setting of 5% amplitude for 8 min. As with the benchmarking of SiO_2 (NM 200), ×10 DLS measurements were conducted (without pause) to obtain the corresponding Z-average diameter and PDI values. Results were then compared with an indicated benchmark figure.

A6.1.5 Mandatory Protocol: Dynamic Light Scattering (DLS)

Particle size distributions were acquired using a Zetasizer Nano ZS (Malvern Instruments, UK) with 633-nm red laser. Malvern Instruments Dispersion Technology software (Version 7.1) was used to control and analyze all data from the instrument.

The instrument was allowed to warm up *about* 30 min prior to its use. For the analysis of nanomaterial dispersion samples, disposable 1 ml cuvettes (DTS0012, Malvern Instruments, US) were used throughout. Prior to use, cuvettes were inspected to ensure that no dust, defects or scratches were present in the measurement area. After inspection, 1 ml of the sample was introduced into the cuvette via pipette, ensuring that no air bubbles were formed in the cell. The cuvette was placed into the Zetasizer, and measurements were carried out at 25 °C. The cells were equilibrated for 120 s at the selected temperature prior to data acquisition.

A6.2 RESULTS AND INTERPRETATION

A6.2.1 Probe Sonication

Figure A6.1 shows the average power and corresponding SD of the three replicate measurements acquired at three different amplitude setting of 5, 20, and 30%. When a line of best fit is drawn, the scatter plot shows a good linear relationship. From Figure A6.1, for a 5% amplitude setting, the delivered power is estimated to be 12.9 W. If a total delivered energy of 7056 J is needed, then this can be achieved by having a 5% amplitude with an estimated sonication time of 9 min and 8 s.

A6.2.2 Benchmarking with SiO_2 (NM 200)

The final sonication time was fine-tuned with respect to the benchmark figure given for SiO_2 (NM 200). In order to reach this target, the sonication time was changed

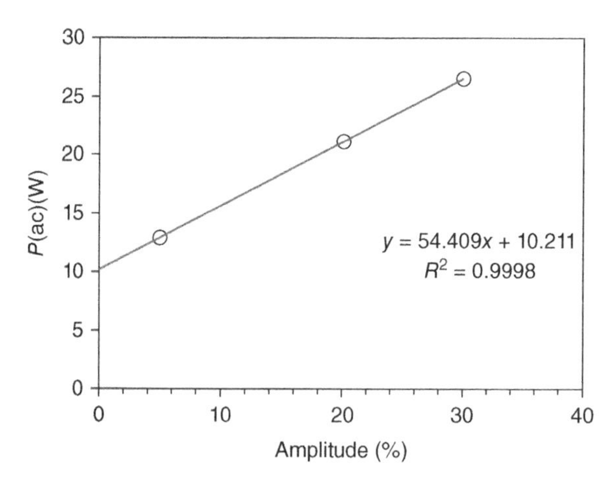

Figure A6.1 Plot of amplitude versus acoustic power to establish the performance of the ultrasonic probe.

TABLE A6.1 SiO_2(NM 200) Benchmark Data: Our Values versus the Expected Benchmark Values

	Our Values	Benchmark Target Values
Mean Z-average (nm)	244 ± 10 nm	210–270 nm
Mean PDI	0.39 ± 0.03	<0.46

TABLE A6.2 ZnO (NM 110) Data: NPL Values Versus Indicated Benchmark Value

	Mean Z-average (nm)	Mean PDI
Indicated benchmark values, as provided by one other laboratory	225 ± 3	0.09 ± 0.03
NPL: vial no 030012	217 ± 3	0.114 ± 0.015
NPL: vial no 030015	216.3 ± 1.6	0.13 ± 0.02
NPL: vial no 030054	220 ± 2	0.10 ± 0.04
NPL: vial no 030008	214.0 ± 1.8	0.101 ± 0.012
NPL: vial no 020061	220.0 ± 1.7	0.111 ± 0.012

to 8 min. The mean Z-average and PDI values were compared with the benchmark figures provided. As shown in Table A6.1, our data is within the expected benchmark values.

A6.2.3 NM 110: Characterizing Batch Dispersions ZnO (NM 110)

The resultant mean Z-average and PDI values were obtained and compared with the indicated benchmark (from one other laboratory). During the course of the study, dispersions from different ZnO (NM 110) vials have been employed and the DLS particle sizes for these were also acquired. The overall results compared to the indicated benchmark are summarized in Table A6.2. Our findings show similarities in the results, indicating no real issues associated with repeatability of dispersion protocol with respect to ZnO (NM 110) dispersions.

REFERENCES

1. Information on NanoReg project. 2013. Available at http://www.nanowerk.com/nanotechnology-labs.php?url2=NANoREG.php. Accessed 2015 Oct 16.
2. Dimeski G, Badrick T, John AS. Ion selective electrodes (ISEs) and interferences: A review. Clin Chim Acta 2010;411(5-6):309–317.
3. Walczak AP, Fokkink R, Peters R, Tromp P, Rivera ZEH, Rietjens IMCM, Hendriksen PJ, Bouwmeester H. Behaviour of silver nanoparticles and silver ions in an in vitro human gastrointestinal digestion model. Nanotoxicology 2013;7(7):1198–1210.

4. Jacob JDC, Rubianes FHC, Johnson-Beebout SE, Buresh RJ. Zinc fertilizer test kit for semi-quantitative verification of fertilizer quality. J Plant Nutr 2014;37(8):1237–1254.
5. Seregin IV, Kozhevnikova AD. Histochemical methods for detection of heavy metals and strontium in the tissues of higher plants. Russ J Plant Physiol 2011;58(4):721–727.
6. Seregin IV, Kozhevnikova AD, Gracheva VV, Bystrova EI, Ivanov VB. Tissue zinc distribution in maize seedling roots and its action on growth. Russ J Plant Physiol 2011;58(1):109–117.
7. Saebel CE, Neureuther JM, Siemann S. A spectrophotometric method for the determination of zinc, copper, and cobalt ions in metalloproteins using Zincon. Anal Biochem 2010;397(2):218–226.
8. Peixoto RRA, Macarovscha GT, Cadore S. On-line preconcentration and determination of zinc using zincon and flame atomic absorption spectrometry. Food Anal Methods 2012;5(4):814–820.
9. Platte JA, Marcy VM. Photometric determination of zinc with zincon Application to water containing heavy metals. Anal Chem 1959;31(7):1226–1228.
10. Ghasemi J, Seifi S. Beta-correction as a preprocessing method for partial least squares in simultaneous determination of zinc and lead. Talanta 2004;63(3):751–756.
11. Ghavami R, Abdollahi H, Shamsipur M. Spectrophotometric study of cobalt, nickel, copper, zinc and lead complexes with methylthymol blue in binary water-methanol mixtures. Iran J Chem Chem Eng 1997;16(1):22–28.
12. Homsher R, Zak B. Spectrophotometric investigation of sensitive complexing agents for the determination of zinc in serum. Clin Chem 1985;31(8):1310–1313.
13. Makino T. A sensitive, direct colorimetric assay of serum zinc using nitro-PAPS and microwell plates. Clin Chim Acta 1991;197(3):209–20.
14. Shum-Cheong-Sing J, Arnaud J, Favier A. Automatization of colorimetric serum zinc determination using the Bayer RA 1000 autoanalyzer. Ann Biol Clin 1994;52(11):765–8.
15. Hayashibe Y, Takeya M, Sayama Y. Direct determination of zinc in human serum by flow-injection spectrophotometric analysis. Anal Sci 1994;10(5):795–799.
16. Johnsen Ø, Eliasson R. Evaluation of a commercially available kit for the colorimetric determination of zinc in human seminal plasma. Int J Androl 1987;10(2):435–440.
17. Shibukawa M, Shibukawa M, Shirota D, Saito S, Nagasawa S, Saitoh K, Minamisawa H. Simple spectrophotometric determination of trace amounts of zinc in environmental water samples using aqueous biphasic extraction. Bunseki Kagaku 2010;59(10):847–854.
18. Versantvoort CHM, Oomen AG, Van de Kamp E, Rompelberg CJM, Sips A. Applicability of an in vitro digestion model in assessing the bioaccessibility of mycotoxins from food. Food Chem Toxicol 2005;43(1):31–40.
19. Zhang Y, Muhammed M. Critical evaluation of thermodynamics of complex formation of metal ions in aqueous solutions VI. Hydrolysis and hydroxo-complexes of Zn^{2+} at 298.15 K. Hydrometallurgy 2001;60(3):215–236.
20. Reichle RA, McCurdy KG, Hepler LG. Zinc hydroxide: Solubility product and hydroxy-complex stability constants from 12.5–75°C. Can J Chem 1975;53(24):3841–3845.
21. Tantra R, Tompkins J, Quincey P. Characterisation of the de-agglomeration effects of bovine serum albumin on nanoparticles in aqueous suspension. Colloids Surf B 2010;75(1):275–281.

7

SURFACE AREA

R. TANTRA, K. N. ROBINSON, AND J. C JARMAN

Quantitative Surface Chemical Spectroscopy Group, Analytical Science, National Physical Laboratory, Teddington TW11 0LW, UK

7.1 INTRODUCTION

One nanomaterial characteristic that has become increasingly recognized as important is surface area; this is because of the huge surface area exhibited by particles on the nanoscale. Surface area has been recently defined by ISO [1] as the "the quantity of accessible surface of a sample when exposed to either gaseous or liquid adsorbate phase. Surface area is conventionally expressed as a mass specific surface area or as volume specific area where the total quantity of area has been normalised either to the sample's mass or volume". In addition "specific surface area is defined as the surface area of a substance divided by its mass, unit [m^2/g]; or the surface area of a substance divided by its volume, unit [m^2/cm^3]. The research should also consider reporting results in both m^2/g and m^2/cm^3." As stated in Chapter 1, this ISO definition has been published as a result of its relevance in the science field of nanotoxicology but can be adopted in other science fields.

The importance of surface area property stems from the fact that nanomaterials present a much larger surface area (compared to corresponding bulk) and as such nanomaterial interactions are governed mainly by the surface, as oppose to bulk mass interactions. As a result, it is not surprising that surface area is thought to be the more relevant dose metric in nanotoxicology, as opposed to mass. Certainly, this is in contrast to conventional chemical toxicology, in which researchers generally use mass as the metric to describe the dose [2–4].

Nanomaterial Characterization: An Introduction, First Edition. Edited by Ratna Tantra.

Besides nanotoxicology, the measurement of surface area is important in other science fields. For example, in material science, surface area is important in the characterization of catalysts, as nanomaterials can have excellent catalytic properties [5]. In cars, for example, catalytic converters consist of alumina coating impregnated with platinum and rhodium nanomaterials, among others. These nanomaterials act as active sites for catalysis. For example, platinum nanomaterials can oxidize hydrocarbons and carbon monoxide, while rhodium functions to reduce nitrogen oxides [6]. Another example of a nanomaterial-based catalyst involves the use of gold nanoparticles attached to a titania support; this results in a catalyst that promotes the chemical oxidation of carbon monoxide [7–9]. As the function of a catalyst is to promote reactions at surfaces, a high specific surface area is required in order to increase the number of active sites, resulting in a highly active catalyst [6]. Overall, the need to measure the specific surface area is important to evaluate activity and adsorption capacity of materials.

Lastly, specific surface area is an important property in relation to nanoregulation as was highlighted by the European Commission (EC). The EC has referred to this property in its definition of what constitutes a nanomaterial. It has been noted that in certain cases, specific surface area can be used as a proxy to identify a potential nanomaterial, but having said this, the regulation clearly stated that results from particle number size distributions should prevail.

In this chapter, an overview of the various methods to measure the surface area of nanomaterials is given. The background of the different methods will be given. Furthermore, the methods will be assessed relative to each other against a certain set of analytical criteria requirements. As an example, the analytical criteria will be chosen in reference to what is needed ideally in nanotoxicology research.

The second half of the chapter presents a case study that evaluates two different methods for surface area measurements: nuclear magnetic resonance (NMR) and Brunauer–Emmett–Teller (BET). The case study will compare their performance when used to assess nanomaterial powder homogeneity in different vials. The powders in the vials have been subsampled using a spinning riffler, and it is important to know the suitability of the spinning riffler as a subsampling tool. The topic of the spinning riffler has been discussed in Chapter 1 and will not be further covered here.

7.2 MEASUREMENT METHODS: OVERVIEW

The measurement of surface area is not trivial as it is important to consider surface morphology. As illustrated in Figure 7.1, the total surface area of a particle is the sum of external and internal surface areas. Hence, the reporting of surface area values is associated with the extent to which the method can access the internal and external surface features. It is not surprising therefore that surface area values reported from different techniques may not be comparable, as this depends on the degree in which a given technique can accurately measure both inner and outer surface areas. The reported value for surface area can thus be associated with the following [10]:

a) *Total or geometric surface area*: The measurement of this type of surface area employs adsorbate molecules to take into account all surface features of the particle, that is, both inner and outer surface areas. However, how much of that area that is available to interact with the environment will depend on the size of the adsorbate molecules. For example, the commonly used adsorbate in BET studies is N_2; this molecule is quite small and thus will be able to access more of the surface features than if a larger molecule, such as an organic dye, is used.

b) *Superficial surface area*: The measurement of this type of surface area is often unreliable as it is obtained from the particle size information and the need to estimate what we shall refer to as "larger scale geometry of a particle." For example, in the case of particle sizing techniques such as dynamic light scattering, estimation of particle size assumes that every particle is a sphere. Yet, when the surface area is estimated from the particle size measurements, no correction for shape factors has been taken into account. Hence, inaccurate information can arise if a technique assumes or makes a general assumption with regards to the shape of the particles. Furthermore, the reporting a superficial surface area will mean the fine detailed structures such as micro/nano pores, fissures, and capillaries (as illustrated in Fig. 7.1) will not be taken into account.

c) *Active surface area* or *Fuchs surface area*: The measurement of this type of surface area is specific to aerosolized sample and thus associated with specific techniques such as epiphaniometer and diffusion charger [11]. These methods involve measuring the interactions between a particle and ions (in the carrier gas) and making adjustments for electrical effects [12, 13]. Hence, the measurement is governed by the surface of the (charged) particle that is active in exchanging energy and momentum with the carrier gas. Surface area is thus proportional to the aerosol particle charge.

Table 7.1 gives a list of different analytical techniques that can potentially be used to measure the surface area of nanomaterials. In the table, the principle of each

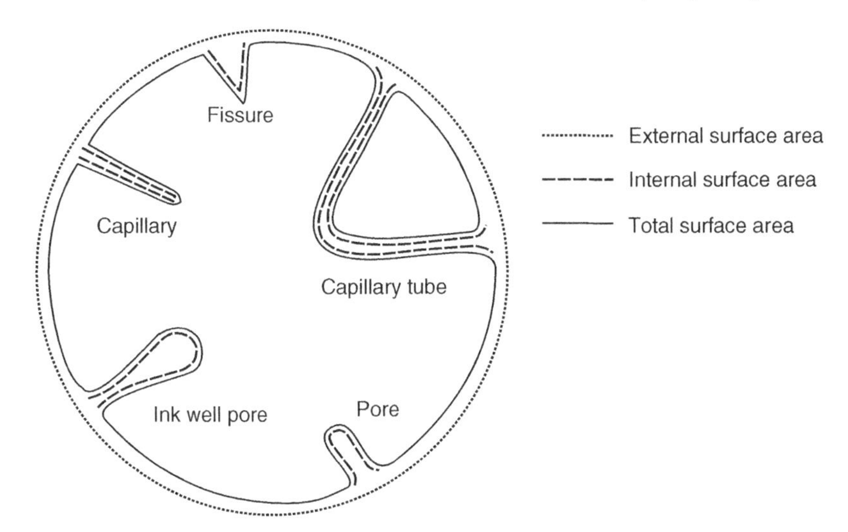

Figure 7.1 Illustration of common surface features (internal, external, and total surface area) of a particle.

TABLE 7.1 Surface Area Measurement Methods, Evaluated Against Some of the Demanding Analytical Requirements, Found in Nanotoxicology Research

Technique	Principle	Measurand	Value Reported	Units of Measurement	Can Measure When Particles Are in Liquid Dispersion?	Selectivity: Able to Differentiate from Other Particles in the Sample?	Sensitivity: Able to Detect Single Particles?
Scanning Electron Microscopy	Electron beam is fired at sample and resultant emissions such as secondary and backscattered electrons and X-rays are used to form an image of the sample. Surface area can be estimated from the 2D images produced [14].	Particle size and shape	Superficial surface area	(m^2/cm^3)	No, due to vacuum requirements	Yes, but if a chemical identity is possible, for example, SEM–EDX	Yes
Transmission Electron Microscopy	Electron beam is fired through a very thin sample; a CCD detector picks up the electrons that make it through allowing a projection image of the sample to be formed. Surface area can be estimated from these 2D projections, or directly calculated from 3D models generated from multiple images [15].	Particle size and shape	Superficial surface area	(m^2/cm^3)	No, due to vacuum requirements	Yes, but if a chemical identity is possible, for example, TEM–EDX	Yes
Atomic Force Spectroscopy	Measures inter- and intramolecular forces between an ultra-sharp tip and surface of the sample. The tip is mounted onto a cantilever and scanned across a surface, measuring the deflections caused by the surface allow a topographic image to be built [16].	Particle size and shape	Superficial surface area	(m^2/cm^3)	No, but can measure in liquid environment if particles are fixed to a substrate	No, unless contaminant particles look inherently different from the particles of interest	Yes
Small Angle X-ray Scattering	Measurement of the elastic scattering of X-rays at small angles from the beam line provides information about the size and shape of particles as well as the structure of aggregates/agglomerates [17].	Particle size, shape, and spatial arrangement	Superficial surface area	(m^2/g), (m^2/cm^3)	Yes	No	No

Dynamic Light Scattering/Nanoparticle Tracking Analysis	Particles in a suspension are illuminated with a coherent light source such as a laser; this is used to track their Brownian motion in the fluid, which in turn allows their size to be determined. Surface area can be estimated from this value using simple mathematics and assumptions about particle shape [18, 19].	Hydrodynamic diameter	Superficial surface area	(m^2/cm^3)	Yes	No	No
Laser Diffraction	Laser light is passed through a dispersion of particles; the scattering light patterns produced allow the size of the particles to be ascertained. Surface area can be estimated from this value using simple mathematics and assumptions about particle shape [20].	Particle size and shape	Superficial surface area	(m^2/cm^3)	Yes	No	No
BET Gas Adsorption	Uses the adsorption of gas molecules onto the surface of particles to determine surface area. Adsorption is measured by gravimetric or volumetric methods [21].	Mass or volume of gas adsorbed	Geometric surface area	(m^2/g)	No	No	No
Flow Micro-calorimetry (Liquid Flow Calorimetry)	First, a pure solvent is flowed through a powder bed of sample. Then, a solution containing an adsorbate in the solvent is introduced to the flow, in which the adsorbate adsorbs onto the sample, producing a measureable change in enthalpy; this can be used to calculate the surface area of the sample [22].	Enthalpy of adsorption	Geometric surface area	(m^2/g)	No, but can measure in liquid environment	No	No
Static Calorimetry	This is when a dry solid is immersed in a liquid and the subsequent enthalpy change measured. This enthalpy change occurs as a result of the sample's surface interface changing from a solid/gas to a solid/liquid, i.e. wetting, can be used to calculate its surface area [22].	Enthalpy of immersion	Geometric surface area	(m^2/g)	No, but can measure in liquid environment	No	No

(continued)

TABLE 7.1 (*Continued*)

Technique	Principle	Measurand	Value Reported	Units of Measurement	Can Measure When Particles Are in Liquid Dispersion?	Selectivity: Able to Differentiate from Other Particles in the Sample?	Sensitivity: Able to Detect Single Particles?
Adsorption from Solution	Particles are introduced to a solution containing an adsorbate, the drop in concentration of the adsorbate in the free solution allows the surface area of the sample to be determined [10].	Reduction in concentra- tion of adsorbate	Geometric Surface Area	(m^2/g)	Yes	No	No
Nuclear Magnetic Resonance	Adsorbed and free molecules of a substance exhibit different NMR relaxation times. The ratio between the two signals is dependent on the area of the adsorbing surface, allowing surface area to be calculated [23].	Spin relaxation time [23]	Geometric Surface Area	(m^2/g)	Yes	No	No
Epiphaniometry	Aerosolized sample is exposed to the source of ^{211}Pb atoms; these atoms adhere to the surface of the particles. The ^{211}Pb atoms decay, emitting α-radiation, which is proportional to the number of adsorbed atoms on a particles surface, allowing the Fuchs surface area to be calculated [24].	α-Decay per unit area	Fuchs Surface Area	(m^2/cm^3)	No	No	No
Diffusion Charger	Aerosolized sample is charged by passing through a coronal discharge. Particles then travel on to an electrometer, where upon impacting the meter's surface they release their charge. The current measured can be related to the charge carried by the particle, which is in turn related to its active surface area [25].	Charge per unit area	Active Surface Area	(m^2/cm^3)	No	No	No

technique is presented and the table indicates whether the method reports superficial surface area, geometric surface area or Fuchs surface area (sometimes referred to as active surface area in the literature).

The ability to measure the surface area reliably will be dependent on whether the measurement method chosen is suitable for the specific sample under analysis and the nano-specific application. In the field of nanotoxicology, for example, it is necessary to take into account several analytical requirements. Ideally, analytical requirements for nanotoxicology may include the need (i) to measure particles when suspended in liquid dispersion, (ii) to differentiate particles of interest, in the presence of other particles in the sample, and (iii) to measure in the dilute concentration regime, with the ideal being able to measure to single particle level. These are some examples of analytical requirements that may be considered, which have been used in Table 7.1 to compare the different methods; the findings are presented in Table 7.1. It is apparent from Table 7.1 that there is currently no single technique that can fulfil all three analytical requirements.

Of all of the methods listed in Table 7.1, the least reliable are those associated with the measurement of superficial surface area. These include imaging based (TEM, SEM, and AFM) and non-imaging-based techniques (such as DLS, NTA, SAX, and laser diffraction). Having said this, the extent of unreliability depends on the sample itself as the method can be reliable if the sample is nonporous and monodisperse. If the sample is nonporous, then the surface area estimate due to porosity will not be taken into account during the measurement. If the material is monodisperse, then this will minimize the variations in shape and size of the particles arising from the sample population.

One major issue in having a highly polydisperse sample is the lack of analytical techniques that can measure particle size distribution information accurately. Anderson et al. [26] shows that techniques such as NTA and DLS are good at detecting "single population" of particles, corresponding to either the largest or smallest particle in a multimodal sample. By this we mean that if a particle size distribution in a sample is bi-modal (i.e., having two different populations), then techniques such as NTA and DLS may give the appearance of a monomodal distribution, thus indicating only a "single" population. If the reporting of particle size is unreliable, then the corresponding estimated surface area value will also be unreliable.

The rest of the techniques listed in Table 7.1 do not make any assumptions with regard to particle shape and are thus considered more accurate for polydisperse samples. These techniques can be further differentiated on the basis as to whether the nanomaterial analyzed is in powder, liquid dispersion or aerosol form. For powder samples, by far the most commonly used is the BET gas adsorption method. Calorimetry is similar to BET in that it relies on the sample being dried during some stage in the analysis but is less commonly used.

When the nanomaterial is in colloidal suspension, then two methods should be considered: adsorption from solution and NMR. In some ways, these two methods share some similarity to the BET method, in that they use molecular probes to access external and interior surface of particles to determine the surface area. However, compared to BET, these methods are not so well developed. Out of the two, "adsorption from solution" is more difficult to implement as it relies on a wide and varying range of adsorbents that can be potentially used [10]. Furthermore, the need to add adsorbents is also not ideal as they can interact with species already present in the sample and can even result in agglomeration, thus affecting reliability of the measurement. Lastly, the measurement is complicated by the following effects, which will pose several limitations on the accuracy of the surface area estimate [27]:

a) Both solute and solvent molecules compete in the adsorption process.
b) Total coverage is difficult to compute, as only the effect of solute molecules being removed from solution is determined.
c) Large and complex shape of some solute molecules result in uncertainty in determining their orientation at the interface. Giles and coworkers [28] have examined a huge number of published isotherms and concluded that only a limited number of adsorbates can be used for surface area determination.
d) Unknown state of the adsorbent when on the surface.

The NMR method, however, does not suffer from the above-listed limitations [23]. The theory behind NMR measurement will not be further discussed in this section, as this is covered in the case study section. A major advantage of NMR is that the analysis is relatively quick to perform. However, NMR has several limitations that should be highlighted [27]. First, a similar material of a known surface area must be used as a reference for the spin–lattice relaxation time to estimate the surface area of the nanomaterial. Secondly, the liquid in which the particles are suspended must contain protons and materials having a high paramagnetic content cannot be examined. Thirdly, the accuracy is inferior to that of other methods when the specific surface area is low. Finally, the NMR method is only ideal for highly concentrated dispersions, being mainly suitable for the analysis of nanomaterial slurries. The NMR measurement precision improves with increasing concentration because the analysis is based on a difference in relaxation time between the liquid without particles and particle dispersion. The greater the particle concentration, the greater the relaxation time signal difference. Large particles have a correspondingly small wetted surface area, and as a result, the relaxation time difference between the liquid and the dispersion is smaller. The power of the NMR technique is in measuring the relative surface area, rather than absolute specific surface area.

For aerosolized samples, only two techniques are currently available, diffusion charger and epipahniometer. The former is much simpler in construction and, unlike epipahniometer, it does not require the use of radioactive lead from an actinium source being deposited on the particle [29].

7.3 CASE STUDY: EVALUATING POWDER HOMOGENEITY USING NMR VERSUS BET

As discussed in Chapter 1, the process of subsampling nanomaterial powders is not trivial. There is a need to ensure that the method used should result in subsampled powders that are homogeneous and representative of the entire sample. The degree of homogeneity between the subsamples can potentially be assessed by measuring their corresponding specific surface area and assessing data variability between the different subsamples. Out of all the techniques listed in Table 7.1, the BET technique seems to be the ideal choice as it is a mature technology and suitable for powder analysis. However, one potential downside of BET is that experiments are often laborious to perform, for example, extensive sample conditioning prior to analysis, sometimes involving an overnight degassing step. In homogeneity testing, where there are likely to be multiple samples required for analysis, a less laborious method is needed. In this case study, NMR-based measurement is explored. Its performance will be assessed and compared to that of BET. In order to assess the degree of data variability, replicate measurements were taken. For each method, six separate subsamples were taken from six vials for analysis. For each subsampled material analyzed, three replicate measurements were acquired.

7.3.1 Background: NMR for Surface Area Measurements

The NMR technique relies on the shorter relaxation time of liquid that is bound to a particle surface compared to unbound liquid. Using the assumption that all bound molecules have the same relaxation time, knowledge of the unbound relaxation time permits the average relaxation time to be analyzed [23]. Two relaxation rate constants can be correlated to the surface area (S) in the following equation:

$$R_{\text{av}} = \psi_p S L \rho_b (R_s - R_b) + R_b \tag{7.1}$$

where:

R_{av} = Average spin relaxation rate constant
ψ_p = Particle volume to solvent volume ratio
S = Total surface area per unit weight
L = Thickness of liquid surface layer
ρ_b = Bulk density of particles
R_s = Relaxation rate constant of bound solvent
R_b = Relaxation rate constant of bulk solvent

This can be reduced to

$$R_{\text{av}} = K_a S \psi_p + R_b \tag{7.2}$$

where

$$K_a = L\rho_b(R_s - R_b) \tag{7.3}$$

Further rearrangement of the equation yields surface area (S):

$$S = \frac{R_{\text{sp}} R_b}{K_a \psi_p} \tag{7.4}$$

where,

$$R_{sp} = \frac{R_{av}}{R_b} - 1 \tag{7.5}$$

One of the main issues with the NMR method is the need to estimate the K_a value in Equation 7.3. The K_a value depends on the combination of the particle and liquid, so K_a for silica in water is different from K_a for silica in ethanol. To determine K_a for a particular particle–liquid combination, there is a need to have an estimate of the specific surface area (from the particles in their wet state) from some other means, for example, the use of DLS, centrifugal sedimentation, laser diffraction, and microscopy. Each of these methods has its limitations and potentially can result in high uncertainty associated with the K_a value.

7.3.2 Method

7.3.2.1 Materials DI water (Elga Purelab; resistivity = 18.2 M cm) with a typical pH value of just less than pH 7 was used throughout the study. The nanomaterial ZnO (NM-110) was supplied by the JRC (Joint Research Centre) through an EU-funded Framework Programme MARINA project [30]. The nanomaterial has been subsampled using a spinning riffler, resulting in the production of vials identified by their respective vial numbers. Each NM-110 vial was packed under argon. Once the vials were opened, the contents were used immediately and the remainder discarded.

The ZnO nanomaterial used here had been previously characterized under PROSPEcT [31]; the PROSPEcT project represents the UK's contribution to an OECD WPMN (Organisation for Economic Co-operation and Development, Working Party on Manufactured Nanomaterials) sponsorship programme. The physicochemical data of ZnO has been previously reported [31] and a summary of relevant properties is given in Appendix A7 below.

7.3.2.2 Sample Preparation for NMR One gram of NM-110 was weighed out into a clean vial after which a few drops of deionized water (18.2 MΩ) were added. The powder was then mixed into a thick paste using a clean spatula before more water was added to make up a dispersion of 10 wt%. The resultant dispersion was then sonicated in accordance to guidelines under PROSPEcT protocol. In short, a 130-W Ultrasonic Processors (Cole-Palmer, UK) with a Ti probe (6.0-mm-diameter tip) running at 20 kHz was used; the sonicator probe was lowered halfway into the dispersion. The sample was sonicated constantly for 20 s at 90% amplitude. After sonication, a glass pipette was used to transfer some of the NM 110 dispersion into a clean NMR tube.

7.3.2.3 Protocol: NMR Analysis NMR measurements were taken on newly prepared dispersions. Before acquiring the data, the contents of the NMR tube were inverted several times. The tube was then inserted into the NMR instrument for analysis. NMR investigations were performed with an Acorn Area particle analyzer (Xigo Nanotools, USA) at room temperature; this is a commercially available bench-top NMR instrument. Experimental data was analyzed using the instrument's AreaQuant software (Version 0.9.2). The instrument was left on for at least 2 h before starting the experiment. With the NM-110 data collection, the instrument was left on for >24 h. Data was acquired using the following settings: resonant frequency (13,077,853 Hz), pulse lengths (90° = 6.78 μs, 180° = 13.55 μs), pre-amp tuning value (220), and $R(x)$ gain of 13 dB. A copper sulfate calibration standard solution (Xigo Nanotools, USA) was used to verify the performance of the instrument before use. The T_2 value of DI water was measured; a value of 2289.4 ms was obtained and this was used as the bulk relaxation value for the remainder of the test. Surface area was estimated using the "Area by T_2 CPMG" sequence in the software. A specific surface relaxivity $K_a = 0.002239$ was used, with a particle to liquid volume ratio of 0.02 and an anticipated T_2 of 1100 ms.

The specific surface relaxivity (K_a) value was determined using a calculator in the AreaQuant software, which required the surface area estimated from nominal particle size, particle liquid volume ratio (0.02), and relaxation time T_2 (1129.84 ms) information for NM 110. For nominal surface area estimation, the D_{90} oversize percentile from DCS (or analytical centrifugation) data was used; the D_{90} value was 107 nm. Note that D_{90} is defined as the size value corresponding to over 90% of particles having a mean size of 107 nm or larger.

7.3.2.4 BET Protocol BET measurements were outsourced to MCA Services, UK. A Micromeritics TriStar II (3020) was used for the collection of nitrogen adsorption/desorption isotherm data up to a saturation pressure of approximately 0.995 P/Po. The analysis was typically conducted to measure 45 adsorption relative pressure points and 23 desorption relative pressure points. Samples were outgassed overnight in vacuum at 300 °C using a Micromeritics VacPrep apparatus prior to analysis. The sample tube dead space was measured for each analysis using helium (CP grade), thus providing warm and cold freespace values. BET surface area was calculated using partial pressures in the nominal range of 0.07–0.25.

7.3.3 Results and Interpretation

Figure 7.2 shows that data obtained from the two measurement methods are comparable, with reported SSA value ranging from 9.5 to 12.0 m^2/g. Overall, our findings indicate the suitability of using the spinning riffler to subsample the NM-110, as data variation between the different subsamples is minimal. As discussed in Chapter 1, the spinning riffler is only suitable for subsampling powders that are considered to be "free flowing." In our case, the suitability of spinning riffler is somewhat expected as the NM-110 material can be considered as "sufficiently free flowing" (indicated by its corresponding Carr index of 20; see Appendix, Table A7.1).

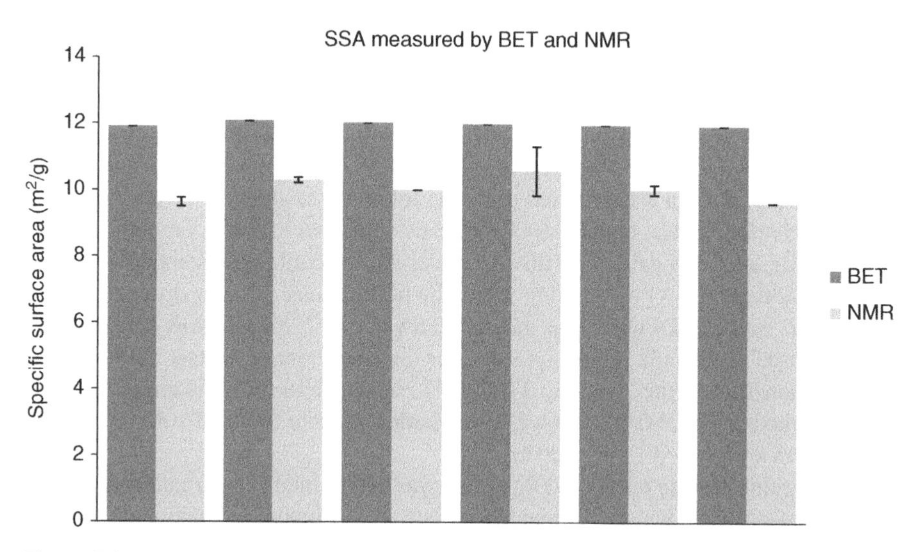

Figure 7.2 Plot of specific surface area values of NM 110. A comparison between BET and NMR; the values plotted are the mean of three replicates (±1 SD).

The results show that although NMR gave comparable measurement relative to the BET, the estimate is consistently lower than that of the BET. There are several possible explanations for this. First, the BET uses N_2, which is nonpolar and therefore is free to adsorb on all surfaces, whereas water (in the case of NMR) has better affinity for some surfaces over others (hence will not wet all parts of the surface equally). According to the XPS data (see Appendix, Table A7.1), the composition at the surface is not only made up of Zn and O but also C (22.7%) and to a smaller extent Si (0.2%). The presence of carbon surface contamination (potentially from the air) may mean an uneven coverage of carbon on the surface of the nanomaterial. Having said this, a more likely explanation is that the possibility that the data is governed by an error associated with the estimation of the K_a value.

Another interesting observation to highlight from the result is the extent of data variability in Figure 7.2, in which more data variability is associated with NMR compared to BET. This may stem from the fact that sample preparation with NMR requires a dispersion step, which is not only sample dependent but also operator dependent. From Table A7.1 (see Appendix section), a TEM primary particle size is reported to be ~78.9 ± 50.2 nm. As the SD here represents the broadness of the particle size distribution, this indicates a highly polydisperse powder. When the powder is dispersed in DI water, reported DCS average particle size becomes 193 nm (see Table A7.1, Appendix section). This indicates that particles are not dispersed as primary particles but as agglomerated particles. Recently, Tantra et al. [32] have discussed the difficulties in producing reproducible dispersions from highly polydisperse nanomaterial powder, which further confirms a much larger data variability associated with NMR. This is complicated by the fact that when ZnO is dispersed in DI water, the nanomaterial stability comes into question. ZnO nanomaterial is known to undergo dissolution through time, as indicated by the solubility data in Table A7.1 (see Appendix). Furthermore, zeta-potential value for these particles has been reported to be ~24 mV (Table A7.1, Appendix), which is lower than ±30 mV required to ensure stability. Overall, the inability to produce stable and reproducible dispersions may contribute to a greater data variability associated with NMR-based measurements.

7.3.4 Conclusion

The study compares the two methods, that is, BET versus NMR to assess the homogeneity of subsampled powders. Although initially NMR may be considered to be less laborious, there is a great deal of uncertainty associated with measurement, compared to the BET. In particular, errors that can arise from the need to produce reproducible nanomaterial dispersions and the uncertainty associated with the estimation of the K_a value should be taken into account when using liquid NMR.

7.4 SUMMARY

This chapter gives a review of different technologies that can potentially measure the surface area of nanomaterials. Table 7.1 is a useful starting point as it serves as an initial guidance for researchers to help identify potential methods for different scenarios. To assess the different methods, the analytical criteria associated with nanotoxicology were chosen as an example. Undoubtedly, the final choice of methods will be dependent on the nano-specific applications, for example, type of nanomaterial, state of the nanomaterial (if the sample is in powder form, aerosol or dispersed in a liquid suspension) and type of experimental investigation. The chapter also presents a case study, which assessed the performance of the NMR versus BET for specific surface area measurement of subsampled powders. This was potentially useful in the case of homogeneity testing of subsampled powders. The limitations of the NMR method compared to the BET have been highlighted, indicating that BET is very much a gold standard for measurement of specific surface area.

ACKNOWLEDGMENTS

The authors would like to thank Framework 7 Project (MARINA) for providing funding and to Alex Shard (NPL) for providing useful comments during the preparation of this manuscript. We gratefully acknowledge feedback/comments from Kirstin Rasmussen and other JRC colleagues.

APPENDIX A7

Physicochemical data on ZnO nanomaterial powder. Adapted from PROSPEcT report [31] (Tables A7.1 and A7.2).

TABLE A7.1 Physicochemical Properties of the ZnO Nanomaterial

Property	Analytical Technique	Reported Values	Notes
Particle size of powder	Transmission electron microscope (TEM)	78.9 ± 50.2 nm	The size of primary particles, as defined by their corresponding Feret's diameter. Mean diameter (±1 SD) of ~200 particles measured in the TEM images; the SD here represents the broadness of the size distribution. Data analysis of the images was processed with the aid of a Tablet PC and a digital pen
Particle size (when dispersed in DI water)	Differential centrifugal sedimentation (DCS)	193 ± 3 nm (DI water)	Measured equivalent spherical particle diameter; the mean and ± SD of three replicates are shown
Zeta-potential (when dispersed in DI water)	Micro-Doppler electrophoresis	24.3 ± 0.4 mV (DI water) 20.8 ± 0.8 (5 mM NaCl in DI*)	The mean values of zeta-potential Values are the mean and ±1 SD of six replicates * DI water +5 mM NaCl; this medium was used to compare with the DI results in the presence of inert background electrolyte
Solubility	Supernatant extraction using centrifugation, followed by inductively coupled plasma mass spectrometry (ICP-MS) measurements	Zinc concentration of the supernatant extracted on Day 2 = 2536 (ng/g) Day 6 = 3360 (ng/g) Day 9 = 3130 (ng/g) Day 14 = 3772 (ng/g) Day 22 = 5030 (ng/g)	The dissolved species in the extracted supernatant solutions were measured using ICP-MS. Solubility of the nanomaterial was measured as a function of time
Surface composition	X-ray photoelectron spectroscopy	C(1s) = 22.7% O(1s) = 40.7% Si(2s) = 0.2% Zn(2p3/2) = 36.3%	The powders were made into pellets prior to XPS analysis
Pour density, Tapped density, and Carr index	N/A see protocol below (Table A7.2)	Pour (bulk) Density (g/cm^3) 0.415 Tapped density (g/cm^3) 0.519 Carr index of 20 = Fair flow	Pour density and tapped density measurement results (and the calculated Carr index) obtained Flow properties are indicated by the Carr index.
Specific Surface Area and Porosity	BET method for specific surface area	11.95 ± 0.01 (m^2/g) ±2SD	This BET analysis gives a measurement of the area of exposed particle surface, as determined by the adsorption of nitrogen gas

Also refer to Table A7.2 for detailed protocols.

TABLE A7.2 Corresponding Analytical Protocols Used to Acquire Data Presented in Table A7.1

Analytical Technique	Protocol
TEM	Sample preparation was carried out by putting a small amount of the nanomaterial (10 mg) into a clean glass container and then dispersing in 3 ml of ethanol. The particles were deposited on TEM grids and images acquired using a Hitachi 2300A instrument operated at 200 kV. An adequate magnification (in which the shape and limits of the primary particles should become apparent) was chosen for image acquisition, and images were subsequently analyzed for the estimation of particle size
Image analysis from SEM and TEM images	TEM images were analyzed with a Tablet PC and a digital pen to outline the contours of primary particles. Image analysis was carried out using ImageJ software, which automatically calculates particle diameter dimensions
Differential Centrifugal Sedimentation (DCS)	Particle size distribution by centrifugal sedimentation was acquired with CPS Disc Centrifuge Model DC 20000 instrument (Analytik Ltd, UK). At the start of the method, the centrifuge was brought up to speed by partially filling the disc with a sucrose gradient fluid and dodecane cap fluid. The purpose of the gradient fluid was to stabilize the sedimentation; the purpose of the cap fluid was to maintain the gradient inside the disc. The disc centrifuge was then allowed to equilibrate at 6000 rpm for 1 h; this stable gradient was used within the following 6 h. 0.2 ml of the nanoparticle sample (50 mg/l) was injected into the disc; a calibration standard was injected after every three samples. Analysis was run against a calibration standard, NIST traceable standard, PVC 0.377 μm. The Disc Centrifuge Control System software (CPS Instruments Inc.) was used to acquire and process the data
Zeta-potential	Electrophoretic measurements were obtained using a Zetasizer Nano ZS (Malvern Instruments, UK) equipped with a 633-nm wavelength laser. The reference standard (DTS1230, zeta-potential standard from Malvern) was used to qualify the performance of the instrument. Sample preparation involved the filling of a disposable capillary cell (DTS1060, Malvern). Prior to their use, these cells were thoroughly cleaned with ethanol and deionized water, as recommended by the instrument vendor. For analysis, the individual cell was filled with the appropriate sample and flushed before refilling; measurements were carried out on the second filling. Malvern Instrument's Dispersion Technology software (Version 4.0) was used for data analysis, and zeta-potential values were estimated from the measured electrophoretic mobility data using the Smoluchowski equation
Extraction of supernatant in a colloidal suspension (part of protocol for solubility measurement)	This is the removal of solid nanomaterial in the dispersion prior to sample analysis using ICP-MS (so that the sample analyzed contains only the dissolved fraction) The particle extraction involved three main steps, to ensure complete removal. First, extraction of particles was carried out using filtration, using Millipore Express PES membrane, 0.1 μm pore size filter (Fisher, UK) under vacuum. The second step involved collecting the resultant filtrant and transferring to an appropriate centrifugation vial. The vial containing the sample was centrifuged (Centrifuge 5430, Eppendorf, UK) (7500 rpm for 1 h) resulting in the formation of a pellet at the bottom of the vial. Lastly, the resultant clear supernatant was extracted using a Peri-Star Pro peristaltic pump (World Precision Instruments, UK); this was done carefully, so as not to disturb the pellet. Only half of the supernatant was collected, stored in the freezer for further analysis using ICP-MS

(*continued*)

TABLE A7.2 (*Continued*)

Analytical Technique	Protocol
ICP-MS for measurement of the extracted supernatant	The analysis of the supernatant was subcontracted and performed by personnel at LGC (Laboratory Government Chemist, UK). The ICP-MS analysis was carried out using an Agilent 7500ce ICP-MS Octopole Reaction System, operating in standard (no collision cell gas) mode for Cerium (Ce) and Helium mode for Zinc (Zn). The instrument is UKAS accredited and was set up following standard operating procedure (SOP) INS/A1-0013. The samples were equilibrated at room temperature and agitated to ensure homogeneity. An aliquot of 0.2–0.23 g was taken from each sample and digested in a CEM Discover microwave, SOP INS/A1-0014, using a mixture of HNO_3/H_2O_2. All samples were digested and analyzed over a period of 5 days. Validation was carried out following SOP INS/A1-0015, which includes spiked recoveries and replicate analyses. The limit of detection (LoD) and limit of quantitation (LoQ) are given. The estimated uncertainty at 95% confidence ($k = 2$) is 12% for Zn and 20% for Ce. Results below the LoQ are likely to have a higher error. The limit LoD for Zn concentration is 5 ng/g and the corresponding LoQ is 15 ng/g
XPS	XPS measurements were obtained in ultra-high vacuum using a Kratos AXIS Ultra DLD (Kratos Analytical, UK) instrument fitted with a monochromated Al $K\alpha$source, which was operated at 15 kV and 5 mA emission. Photoelectrons from the top few nanometers of the surface were detected in the normal emission direction over an analysis area of approximately 700 × 300 μm. Spectra in the range 1400–10 eV binding energy and a step size of 1 eV, using a pass energy of 160 eV were acquired from selected areas of each sample. The peak areas were measured after removal of a Tougaard background. The manufacturer's intensity calibration and commonly employed sensitivity factors were used to determine the concentration of the elements present. High-resolution narrow scans of some peaks of interest were acquired with a step size of 0.1 and 20 eV pass energy; the manufacturer calibrated the intensity calibration over the energy range. The energy scale was calibrated according to ISO 15472 Surface chemical analysis – X-ray photoelectron spectrometers – Calibration of energy scales. However, the charge neutralizer was used when acquiring the spectra, which shifted the peaks by several eV. The C 1s hydrocarbon peak (285 eV binding energy) was used to determine the shift for identifying the peaks *Sample preparation for XPS analysis* Pellets were made from the nanomaterial powder. Pellets from the sample powders were produced using the KBr Quick Press pellet presser. The powder was loaded into the presser (½ to ¾ full) and then pressed to produce the pellets. The pellet presser was sufficiently cleaned in between sample preparations in order to avoid cross contaminations; this was done by washing the presser sufficiently with DI water, detergent, and isopropanol.

(*continued*)

TABLE A7.2 ***(Continued)***

Analytical Technique	Protocol
Pour (and tapped density measurements)	This was outsourced to Escubed, Leeds The pour density (sometimes referred to as bulk density) was calculated by measuring a known weight of the solid material, and placing it in a glass measuring cylinder to obtain the volume. In order to measure its tapped density, the cylinder was then tapped mechanically (using a Copley JV2000) by raising and lowering by a set distance until a consistent volume was reached. The volume was measured in order to determine the tapped density; this corresponds to the maximum packing density of the material
BET SSA and Porosity	This was outsourced to MCA Services Micromeritics TriStar II (3020) was used for the collection of nitrogen adsorption/desorption isotherm data up to a saturation pressure of approximately 0.995 P/Po. The analysis was typically conducted to measure 45 adsorption relative pressure points and 23 desorption relative pressure points. Samples were outgassed overnight in vacuum at 300 °C using a Micromeritics VacPrep apparatus prior to analysis. In order to indicate any possible microporous nature of the materials, additional relative pressure data were also collected at pressures lower than the usual starting point for analysis using this instrument. These were in the approximate range of 0.005–0.01 P/Po. The sample tube dead space was measured for each analysis using helium (CP grade), thus providing warm and cold freespace values. The same equipment, with the application of the same freespace measurement technique, was used with samples requiring only BET surface area analysis (i.e., the JRC homogeneity measurements). BET surface area was calculated using partial pressures in the nominal range of 0.07–0.25

REFERENCES

1. ISO. *ISO/TR 13014:2012: Nanotechnologies – Guidance on Physico-Chemical Characterization of Engineered Nanoscale Materials for Toxicologic Assessment*. ISO; 2012.
2. Donaldson K, Schinwald A, Murphy F, Cho WS, Duffin R, Tran L, Poland C. The Biologically Effective Dose in Inhalation Nanotoxicology. Acc Chem Res 2013;46(3):723–732.
3. Ju-Nam Y, Lead JR. Manufactured nanoparticles: An overview of their chemistry, interactions and potential environmental implications. Sci Total Environ 2008;400(1–3):396–414.
4. Handy RD, Owen R, Valsami-Jones E. The ecotoxicology of nanoparticles and nanomaterials: current status, knowledge gaps, challenges, and future needs. Ecotoxicology 2008;17(5):315–325.
5. Li YM, Somorjai GA. Nanoscale advances in catalysis and energy applications. Nano Lett 2010;10(7):2289–2295.
6. Bell AT. The impact of nanoscience on heterogeneous catalysis. Science 2003;299(5613):1688–1691.
7. Bond GC, Louis C, Thompson DT, editors. *Catalysis by Gold*. Catalytic Science Series - Vol 6. Imperial College Press; 2006.
8. Haruta M. Size- and support-dependency in the catalysis of gold. Catal Today 1997;36(1):153–166.

9. Haruta M, Tsubota S, Kobayashi T, Kageyama H, Genet MJ, Delmon B. Low-temperature oxidation of Co over gold supported on TiO_2, alpha-Fe_2O_3, and Co_3O_4. J Catal 1993;144(1):175–192.
10. Allen T. *Particle Size Measurement: Volume 2: Surface Area and Pore Size Determination*. Vol. 2. Springer; 1996.
11. Gini MI, Helmis CG, Eleftheriadis K. Cascade epiphaniometer: An instrument for aerosol "Fuchs" surface area size distribution measurements. J Aerosol Sci 2013;63(0):87–102.
12. Heitbrink WA, Evans DE, Ku BK, Maynard AD, Slavin TJ, Peters TM. Relationships among particle number, surface area, and respirable mass concentrations in automotive engine manufacturing. J Occup Environ Hyg 2008;6(1):19–31.
13. Pérez ANK. *Electron Spectroscopy on Nanosized Particles in a Carrier Gas*. ETH Zurich; 2001.
14. Echlin P. *Handbook of sample preparation for scanning electron microscopy and X-ray microanalysis*. Springer; 2009.
15. Van Doren EAF, De Temmerman P-JRH, Francisco MAD, Mast J. Determination of the volume-specific surface area by using transmission electron tomography for characterization and definition of nanomaterials. J Nanobiotechnol 2011;9:17–17.
16. West PEP. *Atomic Force Microscopy*. Oxford; 2010.
17. Marega C, Causin V, Saini R, Marigo A, Meera AP, Thomas S, Devi KSU. A direct SAXS approach for the determination of specific surface area of clay in polymer-layered silicate nanocomposites. J Phys Chem B 2012;116(25):7596–7602.
18. Berne BJ, Pecora R. *Dynamic Light Scattering: With Applications to Chemistry, Biology, and Physics*. Dover Publications; 2000. ; Unabridged edition.
19. Filipe V, Hawe A, Jiskoot W. Critical evaluation of nanoparticle tracking analysis (NTA) by nanosight for the measurement of nanoparticles and protein aggregates. Pharm Res 2010;27(5):796–810.
20. Xu R. *Particle Characterization: Light Scattering Methods*. Springer; 2001.
21. Lowell S, Shields JE, Thomas MA, Thommes M. *Characterization of Porous Solids and Powders: Surface Area, Pore Size and Density (Particle Technology Series)*. 1st ed. Springer; 2006. 2004. Corr. 2nd printing edition.
22. Auroux A. *Calorimetry and Thermal Methods in Catalysis*. Vol. 154. Springer; 2013.
23. Davis PJ, Gallegos DP, Smith DM. Rapid surface area determination via NMR spin-lattice relaxation measurements. Powder Technol 1987;53(1):39–47.
24. Gäggeler HW, Baltensperger U, Emmenegger M, Jost DT, Schmidt-Ott A, Haller P, Hofmann M. The epiphaniometer, a new device for continuous aerosol monitoring. J Aerosol Sci 1989;20(5):557–564.
25. Ntziachristos L, Polidori A, Phuleria H, Geller MD, Sioutas C. Application of a diffusion charger for the measurement of particle surface concentration in different environments. Aerosol science and technology 2007;41(6):571–580.
26. Anderson W, Kozak D, Coleman VA, Jämting ÅK, Trau M. A comparative study of submicron particle sizing platforms: Accuracy, precision and resolution analysis of polydisperse particle size distributions. J Colloid Interface Sci 2013;405:322–330.
27. Kissa EE. *Dispersions: Characterization, Testing, and Measurement*. CRC Press; 1999.
28. Giles CH, Giles CH, MacEwan TH, Nakhwa SN, Smith D. 786. Studies in adsorption. Part XI. A system of classification of solution adsorption isotherms, and its use in diagnosis of adsorption mechanisms and in measurement of specific surface areas of solids. J Chem Soc 1960;(0):3973–3993.

29. Ku BK, Kulkarni P. Comparison of diffusion charging and mobility-based methods for measurement of aerosol agglomerate surface area. J Aerosol Sci 2012;47:100–110.
30. Institute of Occupational Medicine. MARINA: Managing Risks of Nanomaterials. 2010 Available at http://www.marina-fp7.eu/. Accessed 2015 Oct 16.
31. Tantra R, Boyd R, Cackett A, Fry AT, Gohil DD, Goldberg S, Lee JLS, Minelli C, Peck R, Quincey P, Smith S, Snowden J, Spencer S, Tompkins J, Wang J, Yang L. *Final Report on the Physico-Chemical Characterisation of PROSPEcT Engineered Nanomaterials*. National Physical Laboratory; 2012.
32. Tantra R, Sikora A, Hartmann N, Sintes JR, Robinson KN. Comparison of different protocols on the particle size distribution of TiO_2 dispersions. Particuology 2014;19:35–44.

8

SURFACE CHEMISTRY

N. A. Belsey, A.G. Shard, and C. Minelli

Analytical Science Division, National Physical Laboratory, Teddington TW11 0LW, UK

8.1 INTRODUCTION

Surface chemical science deals with the study of materials' surfaces and interfaces, their chemical properties, and interactions. Because a large percentage of the atoms in nanomaterials are located at a surface or interface, the behavior of nanomaterials in their working environments is largely dictated by their interfacial surface chemistry; therefore, accurate description and understanding of the materials' properties, performance, and safety require adequate characterization of their surfaces [1, 2].

Nanomaterials are currently of major industrial interest owing to their unique properties, which afford many advantages in a wide range of technologies such as energy, catalysis, and medical sectors such as drug delivery and imaging [3]. Their efficient synthesis, translation into safe and effective products, incorporation into advanced engineered devices, and blending into multifunctional composites requires thoughtful and extensive characterization, including that of their sometimes overlooked surface and interface. This is why the roles of surface chemical analysis methods for nanomaterial characterization has been and currently is the subject of an intense debate at international level, for example, for the formulation of normative documents such as the standard ISO/TR 14187 [4]. However, it is somewhat surprising to note that surface chemical study of nanomaterials is significantly under-reported with respect to other properties such as particle size, zeta-potential, and solubility. Considering that much of the international effort in the study of nanomaterials over the past years has concentrated on their toxicology aspects (due to the need to develop regulation for

Nanomaterial Characterization: An Introduction, First Edition. Edited by Ratna Tantra.

nanotechnologies), this is even more surprising. As well as potentially being intrinsically toxic, nanomaterials, in fact, can act as a vehicle for toxic compounds adsorbed on their surface, whose presence can only be revealed with surface analysis [2]. It seems then natural that surface chemistry of nanomaterials should be carefully studied when assessing their potential toxicological impact.

Nanomaterials exist in an ever-growing variety of geometries, for example, nanoparticles (NPs) [3], rods [5], tubes [6], ribbons [7], stars [8], bubbles [9], or biological structures including micelles and liposomes [10], and nanoscale-layered structures, for example, graphene [11], or materials with features such as holes on the nanoscale [12]. These nanomaterials may be intentionally coated with a wide variety of molecules, such as capping agents, specific chemical functional groups, polymers [13], drugs [3], fluorophores [14], peptides [15], proteins [15], antibodies [16], DNA [17], or alternatively they may acquire unintentional adsorbates as a result of their environment, for example, contaminants, or a protein "corona" in biological media [18]. Given this wide variety of systems, a universal protocol for their preparation and characterization does not exist. More often, nanomaterials require the development of *ad hoc* methods to allow quantitative measurements, which are extremely important for manufacturers of nanoparticle products in areas such as diagnostics or drug delivery [19], where the performance is dictated by the number of probes attached to the nanoparticle surface [20].

Characterization of the surface chemistry of nanomaterials is not limited to the surface elemental composition, but may also include coating thickness, identification of additives, contaminants, defects, phase (separation), crystalline state, oxidation state, photocatalytic properties, and hydrophilicity/lipophilicity, charge, and topology/surface roughness. It should also be noted that the size of nanomaterials is inherently linked with their surface chemistry [21], since the composition of the molecular exterior is often affected by this property. For example, the surface adsorption and thickness of protein coatings is sometimes reported to be strongly linked to nanoparticle diameter [22]. Surface curvature is also known to affect the extent of protein denaturation when bound to nanoparticle surfaces [23].

Extending the use of traditional surface analysis tools to the characterization of nanomaterials is not straightforward as these present new challenges to even the most established surface characterization techniques. For example, although it is preferable to analyze materials in the "natural" environment [2, 24], many nanomaterials are synthesized and stored in liquid environments, which is not compatible with many surface analysis techniques that operate in high vacuum. Rigorous protocols for the removal of nanomaterials from the natural or working environments and exposure to ambient and/or vacuum environments need development [4].

Within a nanoparticle population, there is dispersion in particle size, shape, and other parameters. This is also true for surface chemistry; thus, monitoring the population spread of the nanomaterials' surface chemical properties is also desirable. How much deviation is there within the sample compared with the mean value? Ideal nanoparticle characterization techniques would be able to take measurements of individual particles; however, currently this is not a widespread capability. Due to the broad range of nanomaterial structures and compositions, not all measurement techniques are well suited for their study, and careful consideration is essential to choose the most relevant and insightful. As is often the case in many disciplines, cross-validation of results by a number of techniques, where possible, is a valuable strategy for accurate characterization, since each technique may offer complementary information [15]. Although desirable, access to a wide range of characterization techniques is not always possible; for example, many surface analysis techniques require ultra-high vacuum to operate and are consequently expensive and not usually available in nonspecialist laboratories.

Nanomaterials are also associated with unique challenges with regard to their preparation, often suffering from reproducibility issues, and thus the measurement feedback loop is very important throughout the synthetic development process. The dynamic nature of nanoparticles (NPs) and their coatings also complicate matters. Such materials often undergo changes over time; they can be sensitive to temperature changes and light; and introduction to new environments often leads to structural transformation, agglomeration, dissolution, or acquisition of surface coatings. The sensitivity is such that even the choice of storage vessel material can be vital, for example, glass versus plastic [25], as nanoparticles can potentially interact with the surface of a storage container.

This chapter discusses some of the challenges associated with characterizing the surface chemistry of nanomaterials and reviews the various surface science techniques that have been previously used. Two case studies to illustrate the application of X-ray photoelectron spectroscopy (XPS) and time-of-flight secondary ion mass spectrometer (TOF-SIMS) for surface chemistry measurements are given.

8.2 MEASUREMENT CHALLENGES

Preparation of nanomaterial samples for analysis can be a very lengthy process, often following a tortuous path of trial and error. There are very few generic approaches, since each nanomaterial behaves differently, and may need to be prepared on a variety of substrates depending on the technique and the information needed, with different requirements for coating thickness and uniformity. For example, a thick layer of nanomaterial on a substrate is preferred for XPS, whereas for medium-energy ion scattering (MEIS) the most desirable sample is a single layer of nanomaterial on a substrate [2].

Placing particles in conditions very different from their working environments can lead to problems with characterization; for example, placing them in a vacuum can distort shape. In addition, the act of performing measurements can inherently cause chemical and physical changes to the sample; for example, some polymer nanoparticles shrink under electron beam irradiation, and biomolecular coatings may be damaged by a beam of ionizing radiation. It is important to monitor and be aware of the extent of such changes over time to minimize sample damage and subsequent misconceptions. Commonly reported examples include X-ray damage [26], melting/damage of samples by electron beams [27–30], and changes in oxidation states [31].

Depositing solution-based nanomaterial onto substrates in an optimal configuration for analysis can be nontrivial. Issues to be considered in extracting particles from solution for analysis and further testing include removing residual ions and molecular species present in solution but not at the surface of the nanomaterial; removing solvent in a manner that minimizes aggregation and interference with the measurement process; eliminating nonstructural water or other solvent without significantly altering particle phases; minimizing erosion of original surface coatings; and avoiding reactions with the medium or its contaminants that may occur upon exposure to oxygen and other potentially reactive species that will alter the samples either immediately or as a function of time [32]. For these samples, there are a number of useful approaches. For example, a centrifugation-based method has been used to concentrate the nanomaterial into a pellet form [15] and subsequently "washing" the nanomaterial by re-suspending the pellet in water (to remove solutes such as buffer salts, or other solution species). Care should be taken to use as gentle conditions as possible, since this process can trigger sample aggregation. Other methods include spin coating [33, 34], dialysis [35, 36], and filtration-based procedures [32, 37]. The production of uniform deposits on the desired substrate can also be a challenge; for example, undesirable "coffee-ring" drying effects are a common feature formed when nanomaterial dispersions are allowed to dry on a substrate in an uncontrolled manner.

Nanomaterial powder samples also present challenges. First, precautions must be taken during handling to ensure that the powders are safely contained and not accidently inhaled. Such fine powders are often affected by static and can easily become airborne during transfer between storage vials or substrates; therefore, working within a containment or glove box is recommended. Second, once deposited on a substrate it must be ensured that the particles are properly immobilized, otherwise the sample may be displaced during mounting or analysis. Carbon tape or indium foil, both of which have the advantage of being conductive, has been used for this purpose with some success, as described in the case study in Section 8.4 Part II. In the case of the indium foil, the powder sample is pressed into the malleable surface. However, these approaches would not generally produce a sample sufficiently flat for analysis by techniques such as TOF-SIMS and XPS, whose spectra would be affected by artifacts as a result of how samples are prepared. TOF-SIMS and XPS analysis of powder nanomaterial samples have been performed with success by preparing the sample in a form of a compressed pellet [38]. The size and shape of the mould for the pellet can be customized to the requirement of the analysis instrument.

Consideration must also be given to the choice of substrate; this must be compatible both with the measurement technique and the sample. It is generally advisable to select a substrate with a distinct chemical signature that will not interfere with that of the sample. Silicon wafers (sometimes gold-coated) are a common choice, since they are conductive and sufficiently flat and uniform in structure for most techniques, but cleaning procedures (before their use) such as solvent wash cycles and UV–ozone treatment are important to remove surface contamination [39]. Polytetrafluoroethylene (PTFE) is another useful material when substrate conductivity is not required since it is less affected by contaminants and exhibits a clear chemical signature for analysis by, for example, XPS [15]. Furthermore, PTFE can be sourced in sheets that can be neatly folded around a flat silicon wafer support of any required size or shape. Due to its high hydrophobicity, aqueous dispersions tend to "bead" heavily on the surface resulting in small but thick sample deposits, which can be convenient for precious samples where only a small analysis area is required.

Sample handling should always be undertaken with care, for example, using clean tweezers and powderless polyethylene gloves. Vinyl gloves, often used in clean rooms, are usually coated with residues from the molding process, and should be avoided. Sometimes a short storage period is inevitable between sample preparation and measurement. In such instances, storing samples in an air-tight bag filled with inert gas such as argon can offer enhanced stability [40]. Shielding from light by the use of aluminum foil is also recommended to avoid the possible occurrence of photo-oxidation reactions, which may alter surface properties.

Finally, there are also challenges associated with dispersion of the powders into a liquid suspension. As discussed in Chapter 1, huge data variability in dispersion quality can arise from varying the dispersion protocol.

8.3 ANALYTICAL TECHNIQUES

A wide range of tools are available for the characterization of materials' surfaces and interfaces. These include those based on electron spectroscopy, such as Auger electron spectroscopy (AES) and XPS, those involving incident ion beams, such as secondary ion mass spectrometry (SIMS) and low-energy ion scattering (LEIS), and those based on scanning probe microscopy (SPM) including atomic force microscopy (AFM) and scanning tunneling microscopy (STM). There are, in addition, many other techniques useful for surface chemical analysis, some of which are introduced in this chapter. However, new techniques are constantly emerging, and those discussed here should not be considered exhaustive. For example, for a number of techniques that traditionally operate in high vacuum, a version that operates in "ambient" or "near ambient" conditions now exists such as ambient pressure XPS [41–43]. These useful techniques could potentially provide new information on the biomolecular coatings of nanomaterials or allow characterization of materials under reaction conditions.

Different surface analysis techniques provide different types of information, as illustrated in Figure 8.1. The diagram shows that the various techniques typically have different spatial resolutions, with only a few capable of resolving individual nanoparticles.

In general, it is quite common in nanomaterial research to use various analytical techniques for nanomaterial characterization, and this is certainly the case for the measurement of surface chemistry. It is desirable to utilize as many different techniques as possible, since combining information from different sources often yields some further insight. This concept is further explored within the case study (Section 8.4, Part I).

The following section is dedicated to the most widespread techniques to characterize nanomaterial surface chemistry. They are presented as an overview, and readers

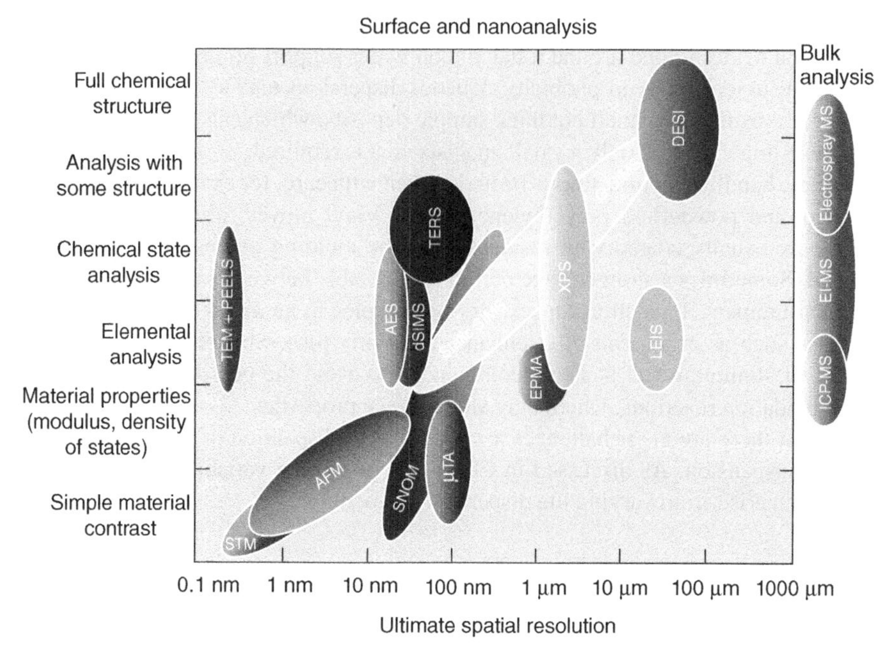

Figure 8.1 Illustrative overview of spatial resolution and types of information that can be obtained by a range of tools important for nanoanalysis, including AES, AFM, desorption electrospray ionization (DESI), dynamic SIMS (dSIMS), electron probe microanalysis (EPMA), gentle SIMS (G-SIMS), low-energy ion scattering (LEIS), micro thermal analysis (μTA), scanning near-field optical microscopy (SNOM), STM, static SIMS (sSIMS), tip-enhanced Raman spectroscopy (TERS), transmission electron microscopy (TEM), parallel electron energy-loss spectrometry (PEELS), and XPS. The diagram also shows the techniques for bulk analysis of materials of electron ionization (EI), electrospray and inductively coupled plasma (ICP) mass spectrometry (MS) [44].

are encouraged to seek further information from the references included in the text and ISO standards such as ISO 20903:2011 [45] and ISO 22048:2004 [46].

8.3.1 Electron Spectroscopies

8.3.1.1 X-ray Photoelectron Spectroscopy (XPS) XPS uses monochromatic X-ray radiation to probe a sample in a vacuum. X-ray photons collide with electrons in the sample, causing emission of photoelectrons from the sample. Photoelectrons near the sample surface are able to escape without losing energy through collisions and are subsequently detected; their energies are characteristic of the atomic and molecular orbitals from which they originated. The kinetic energy of photoelectrons is simply related to the difference between the energy of the X-ray photon and the binding energy of the electron. Since the binding energies of core electrons (i.e., those electrons not involved in chemical bonding) are well known, identification of elements present in the sample surface is straightforward. XPS can identify all elements except hydrogen and helium, which possess no core electrons. Although core electrons are not directly involved in chemical bonding, their binding energies are affected by the chemical environment of the atom from which they originate and typically this is related to the charge, or oxidation state, of the atom. Thus, it is possible to distinguish, for example, silicon from silicon oxides. An example of this type of "chemical shift" for carbon, where carbon in PTFE is distinguishable from carbon in protein, is provided in Part I of the case study (Section 8.4).

The number of photoelectrons is related to the concentration of the chemical species within the sample; thus, direct, quantitative, chemically specific information is obtained. Typically, detection limits for XPS are in the region of 0.1 at%, but this depends upon the element and the matrix in which it is present. A comprehensive table of elemental detection limits in elemental matrices has been produced by Shard [47]. The lateral resolution of commercial XPS instruments is approximately 10 μm, so it is not usually possible to distinguish individual nanostructures; therefore, XPS can only be used to provide population averaged data.

XPS is particularly sensitive to surface chemistry because of the relatively short attenuation depths through which photoelectrons can travel without energy loss through inelastic scattering (in the region of 10 nm, depending on the material). Photoelectrons emitted from deeper in the sample will on average undergo a greater proportion of collisions during their path to the sample surface, so the inelastic scattering tail will consequently have a greater intensity compared with the photoemission peak for subsurface elements. Thus, some additional information can be inferred from the relative changes in the inelastic scattering background, for example, whether the species is present in a superficial over-layer or the subsurface.

Depth profiling is possible on flat samples either by changing the angle between the sample and the detector, changing the energy of the X-ray excitation, or by sputtering through the sample with an ion beam. The latter approach provides information over depths much larger than ~10 nm, but analysis is not trivial due to the possible effects of ion beam damage and preferential elemental sputtering, which changes the composition of the material. For samples with topographic features, including nanostructured materials, the interpretation of depth profiling data is far from trivial.

Methods for determining the thickness of an over-layer on a substrate have been reported, including measuring the shell thickness of nanoparticle coatings [36, 48]. An example is presented in Part I of the case study (Section 8.4).

8.3.1.2 Auger Electron Spectroscopy (AES) AES utilizes an electron beam to probe conductive samples, resulting in the excitation of "Auger electrons." These electrons have energies characteristic of their emitting elements, akin to XPS, thus many experimental considerations are shared. The experimental setup is similar to SEM, in which the electrons are used for imaging the sample; but with the added advantage of reporting chemical information. This technique is particularly valuable due to its spatial resolution, which is sometimes sufficient to report on individual nanoparticles [49].

8.3.2 Incident Ion Techniques

8.3.2.1 Secondary Ion Mass Spectrometry (SIMS) SIMS involves bombardment of the sample with an energetic ion beam, typically with several thousand electron volt energy. Among the many processes that occur upon ion beam irradiation is the sputtering, or erosion, of the material being irradiated. A small fraction of the sputtered material is, itself, ionized and these "secondary ions" (the primary ions are the energetic incoming ions) can be analyzed with a mass spectrometer to provide chemical information about the sample. TOF-SIMS involves the acceleration of secondary ions to a fixed potential, before drifting through a field-free zone to the detector. Heavier ions travel more slowly; thus, mass can be determined from the measured flight time.

SIMS may be used as a surface analytical technique, by restricting the dose received by the sample to a sufficiently small value so that the primary ions do not impact areas of the surface damaged by previous impacts. This is called "static" SIMS. Alternatively, the dose of primary ions can be increased so that significant erosion occurs during the experiment and a depth profile of the material is obtained. The latter approach, termed "dynamic" SIMS, is used extensively in the semiconductor industry. For nanoparticle analysis, the latter approach provides many challenges due to the initial topography of the sample, enhanced sputtering yields from small particles, and the possible melting of small particles due to energy confinement after a primary ion impact [27, 30].

The SIMS technique is highly surface sensitive because it is limited to the sputtering depth, which is typically only a few nanometers. Although the process is very energetic (far in excess of chemical bond energies), the observation of atomic secondary ions demonstrates that it is still possible to observe complete molecular species. The mechanisms that permit the retention of molecular structure in SIMS, yet provide sufficient energy for ionization, are the subject of speculation and debate. Nevertheless, SIMS has become a highly utilized technique for the analysis of surface chemical structure, and the growing use of primary ions that contain more than a single atom has advanced the field enormously in the past few decades. For example, advances in depth profiling of organic layers have been possible with argon cluster ion beams, enabling the retention of molecular information [50].

SIMS has excellent sensitivity compared with many other surface analytical techniques and very good lateral resolution (currently $\sim$100 nm). It has proven itself a valuable tool in characterization of surface coatings and functionalization of nanomaterials [2, 4]. Although the use of SIMS as an identification method for chemical species on the surface of nanomaterials is recommended, it remains a significant challenge to employ the method to measure the surface concentration of these species. Currently, this is only possible where significant efforts have been made to generate calibration standards [51].

8.3.2.2 Low- and Medium-Energy Ion Scattering (LEIS and MEIS) In LEIS, sometimes known as ion scattering spectroscopy (ISS), samples are exposed to low-energy noble gas ions (typically from 1 to 10 keV); the scattering of which reports on the outermost atomic layers of the sample surface. By scrutinizing the positions, velocities, and energies of the backscattered ions that have interacted with the surface, information can be deduced regarding the elemental composition and relative positions of atoms in a surface lattice. LEIS is uniquely sensitive to both structure and composition and is one of the few surface-sensitive techniques capable of directly detecting hydrogen atoms. LEIS can be used to perform both static and dynamic depth profiling [52]. Recent advancements have significantly reduced sample damage, leading to less degradation and an associated increase in sensitivity, allowing accurate thickness measurement of thin films. LEIS has successfully been employed to determine shell thicknesses of functionalized gold nanoparticles [53].

This technique has been shown to be particularly effective at composition analysis of surface adsorbates, in addition to catalysts [54], alloys [55], metals (oxides), and semiconductor surfaces [52]. However, because this technique is so surface sensitive, it is sometimes desirable to perform a very brief sputter clean to remove hydrocarbon contamination from the surface.

LEIS is closely related to medium-energy ion scattering (MEIS) and high-energy ion scattering (HEIS). The techniques differ primarily by the energy of the ion beam used to probe the surface; MEIS typically operates between 20 and 200 keV, whereas HEIS operates between 200 and 2000 keV.

LEIS is sometimes used in conjunction with AES, which is less surface-sensitive than LEIS, but can offer better lateral resolution; therefore, the two techniques can be used in a complementary manner [56].

8.3.3 Scanning Probe Microscopies

Scanning probe techniques such as scanning STM and AFM are widespread techniques employed to generate 3D structural "mapping" of nanomaterials deposited upon flat surfaces [57]. It can be very useful for revealing information on the nanoscale, such as size, distribution patterns of nanomaterials on a flat surface, roughness, and topological information. Although AFM can relay a huge amount of very useful information, it lacks the ability to report direct information on chemical composition. To provide sensitivity toward target chemical elements or compounds, AFM tips have been functionalized with engineered molecules with affinity for such materials, producing what is known as chemical force microscopy [58]. This method has been applied to chemically map surfaces and measure adhesion and frictional forces between molecules [59], with important applications in biotechnology [60].

Tip-enhanced Raman spectroscopy (TERS) combines the 3D mapping capability of AFM with the chemical specificity offered by Raman spectroscopy. Although the technique requires refinement to improve reproducibility [61], it has been successfully used to generate chemical imaging of nanomaterials, for example, graphene [62, 63], and the mapping of catalytic activity on surfaces [64].

8.3.4 Optical Techniques

In addition to TERS, mentioned earlier, there exist a number of Raman-based techniques; one of the most widely used is surface-enhanced Raman scattering (SERS), a surface-sensitive technique, in which Raman scattering is enhanced by using substrates featuring metallic nanostructures [65]. For the analysis of some nanomaterials (e.g., silver), the nano-objects themselves form these structures, and SERS is utilized to directly probe their interfaces. It is thought that localized plasmon resonances are responsible for the enhancement in the electric field, the magnitude of which peaks when the plasmon frequency resonates with that of the incident light. This magnifies the intensity of the light beam and further stimulates the Raman-active modes of the molecule. The effect is twofold, since the emitted light from the Raman-active molecules then experiences the same boost via its return path to the detector. SERS can be performed within a colloidal solution, although the more common approach involves application of the sample to a flat substrate modified with noble metal nanostructures. Spectra often show some differences from conventional Raman, due to changes relating to selection rules; for example, molecules might lose a centre of symmetry when bound to a surface. In such cases, the mutual exclusion rule is no longer discriminatory, and modes that are usually only infrared (IR) active can appear in SERS spectra. SERS has been utilized in a wide range of nanomaterial studies, including interfacial chemistry [66], catalysis [67], and ligand coordination studies [68].

Another spectroscopic technique worthy of note is sum frequency generation (SFG). This is a nonlinear optical technique, which is performed by focusing two laser beams at an interface generating a new frequency equal to the sum of the two incident beams. Due to the ability to tune the wavelength of one of the incident beams, it is possible to obtain vibrational spectra for nanostructures [69, 70], essentially creating a surface-sensitive version of infrared spectroscopy. It is an excellent tool for probing heterogeneous catalysis, since surface intermediates can be observed during catalytic reactions [71]. Determination of the molecular conformation of ligands bound to nanoparticles has also been demonstrated [70]. Second harmonic generation (SHG) is a special case of SFG, in which the incident beams are of the same wavelength (i.e., only one beam is required). New photons are generated with twice the frequency and half the wavelength. SFG and SHG are only generated from non-centrosymmetric structures; typically no signal is generated from bulk materials, which gives rise to the surface sensitivity. SHG and SFG can occur from a sample containing spherical nanoparticles [72], since at the surface of a small sphere, inversion symmetry is broken.

8.3.5 Other Techniques

This section gives a short overview (by no means exhaustive) of other techniques available to characterize surface chemistry of nanomaterials. Techniques capable of quantification of surface functional groups include the use of titration methods [73] and electrochemical analysis [74]. Techniques such as infrared (IR) spectroscopy and thermogravimetric analysis (TGA) can be utilized for surface analysis if the analyte is known to be confined to the surface [75]. In addition, one- and two-dimensional nuclear magnetic resonance (NMR) has been previously employed to determine the ligand shell structure of nanoparticles covered with varying ligand compositions, to distinguish between "Janus," patchy, or striped nanoparticle ligand arrangement patterns [76]. X-ray diffraction has been used to determine the crystalline phase of nanomaterials [38]. Electron spin resonance (ESR) has also proved useful for the study of multicompound monolayers at the surface of nanoparticles [77, 78].

There exist a number of techniques that may not be capable of reporting chemically specific information but can be considered to be useful. Contact angle analysis (to measure wettability of a surface) is one such example. For instance, the presence of surface contaminants on a flat substrate, such as a silicon wafer, would be expected to result in a change in contact angle. Ellipsometry and quartz crystal microbalance (QCM) have been used for some time to characterize thin film deposition especially in conjunction with other techniques [39]. Metallic nanomaterials exhibit localized surface plasmon resonances (LSPR), which are affected by changes in surface chemistry. LSPR shifts due to changes in shell thickness have been monitored by ultraviolent (UV)–visible spectroscopy [15, 16].

8.4 CASE STUDIES

8.4.1 Part I: Surface Characterization of Biomolecule-Coated Nanoparticles

This case study demonstrates how valuable XPS can be used to gather a range of information on the surface molecular state of nanoparticles. In particular, XPS is used to extract chemical information from the nanoparticles' surface, in which the surface elemental composition of nanoparticles is measured. In addition, an estimate of the thickness and number of protein molecules contained within nanoparticle coatings [15, 48] is given. These results are cross-validated by a solution-based approach to measure the number of biomolecules bound to nanoparticles involving the complementary use of ultraviolet (UV)–visible spectroscopy and particle sizing techniques such as differential centrifugal sedimentation (DCS) and dynamic light scattering (DLS) [15, 16].

In order to demonstrate that XPS gives consistent results, a set of model nanoparticles were used; gold nanoparticles ranging in core diameter, with coatings made of two proteins of different shape and size, namely immunoglobulin G (IgG) and bovine serum albumin (BSA). The preparation of the protein-coated gold nanoparticles involved incubation of gold colloids with an equal volume of protein solution for 1 h, and then centrifuged to remove unbound material. The resulting nanoparticle pellet was resuspended in ultrapure water; the step of centrifugation and resuspension of the pellet in water was performed three times. In relation to the substrate material used to fix the nanoparticles, PTFE-wrapped silicon wafer was employed. The choice was attributed to some noteworthy advantages previously discussed. Other advantages also include PTFE being relatively free of contaminants, the signal emanating from the substrate could be easily separated from that of the sample due to the large chemical shift of the C 1s peak when bound to fluorine (refer to the C 1s narrow scans in Figure 8.2). In addition, the hydrophobic nature of the surface results in the production of relatively thick sample deposits from a minimal amount of sample. For detailed information on the protocol, the reader should refer to the literature [15].

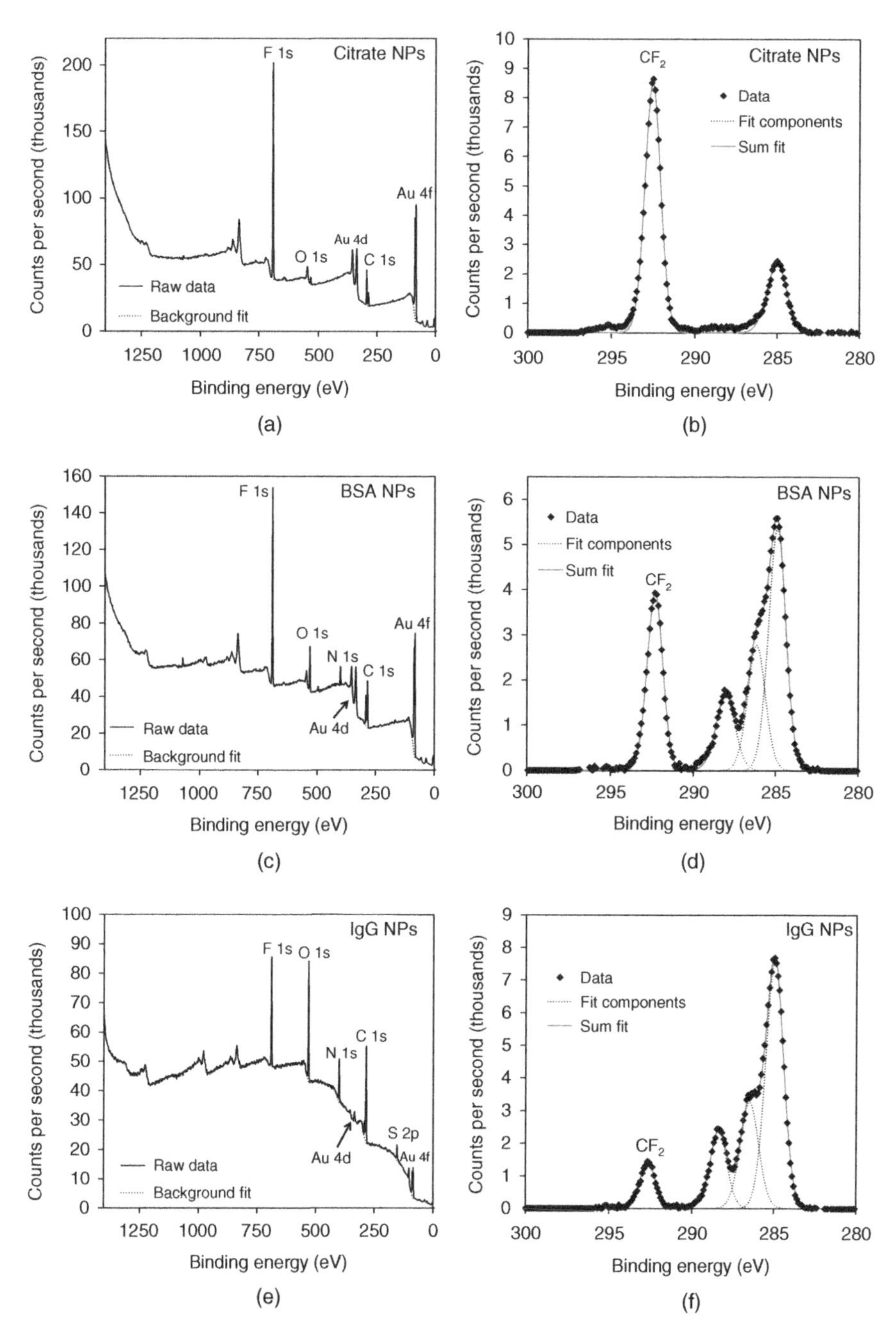

Figure 8.2 XPS Survey (a, c, and e) and C 1s narrow spectra (b, d, and f) of 40 nm gold nanoparticles coated with citrate (a and b), BSA (c and d) and IgG molecules (e and f), deposited on PTFE-wrapped silicon wafer.

Figure 8.2 shows the XPS survey and C 1s narrow spectra of 40 nm gold nanoparticles coated with citrate, BSA, and IgG molecules, deposited on PTFE-wrapped silicon wafer. Note that the chemical composition of the nanoparticle coatings cannot be directly derived from these spectra as is the case for flat surfaces. Nanoparticle core sizes and shell thicknesses are, in fact, on the same length scale of the attenuation lengths of the ejected photoelectrons; therefore, one would incur significant errors by approximating the system to a flat surface. A number of approaches have been developed to overcome this problem [36, 48, 79].

One of these, described by Shard [48], enables the measurement of the average chemical composition and thickness of the dry protein shell of a nanoparticle sample. A basic requirement of this method is that the dry shell is on the same length-scale of the attenuation length of the photoelectrons, which is in the region of 10 nm, depending on the type of material. Furthermore, it is necessary to separate the core XPS signals from that of the shell, which can be challenging when the core and the shell contain the same elements. In this example, the core (gold) and shell (organic) have clearly distinctive chemical features.

From the work of Shard [48], we obtain the following equations:

$$A = \frac{I_1 I_2^{\infty}}{I_2 I_1^{\infty}} \tag{8.1}$$

where I_i is the measured XPS intensity and I_i^{∞} is the measured or calculated intensity for the pure material of unique photoelectrons from the shell (overlayer), $i = 1$, and the core (substrate), $i = 2$, respectively.

$$B = \frac{L_{1,a}}{L_{2,a}} \tag{8.2}$$

$$C = \frac{L_{1,a}}{L_{1,b}} \tag{8.3}$$

where $L_{i,j}$ is the attenuation length of photoelectrons arising from material i traveling through material j, where $j = a$ represents the shell and $j = b$ represents the core. The core radius of the particle, R, and shell thickness, T, are expressed in units of $L_{1,a}$. The following equation and ancillaries may be used to obtain T_{NP} to a precision of better than 4%.

$$T_{\mathrm{NP}} = \frac{T_{R\sim 1} + \beta T_0}{1 + \beta} \tag{8.4}$$

$$T_0 = R\left[(ABC + 1)^{1/3} - 1\right] \tag{8.4a}$$

$$T_{R\sim 1} = \frac{T_{R\to\infty} R}{R + \alpha} \tag{8.4b}$$

$$\alpha = \frac{1.8}{A^{0.1}B^{0.5}C^{0.4}} \tag{8.4c}$$

$$\beta = \frac{0.13\alpha^{2.5}}{R^{1.5}} \tag{8.4d}$$

$$T_{R\to\infty} = \frac{0.74A^{3.6}\ln(A)B^{-0.9} + 4.2AB^{-0.41}}{A^{3.6} + 8.9} \tag{8.4e}$$

Equations 8.4a–e apply to data where the core and shell intensities are measured and normalized to the intensities from the respective pure materials, as encapsulated in A [48]. This requires knowledge of or the ability to estimate the pure material intensities using the same XPS instrument and settings.

In this case, the pure material intensities cannot be obtained and, therefore, an estimate must be made. We assume that XPS relative sensitivity factors may be employed to describe the intensity arising from different elements in the same material and, therefore, the intensity of element "k" in the pure organic overlayer material can be expressed as

$$I_k^{\infty} \propto X_k S_k \tag{8.5}$$

where X_k is the mole fraction of the element in the pure organic overlayer material and S_k is the sensitivity factor of the XPS signal from that element used in the analysis. One may be tempted to use an expression such as that shown in Equation 8.6a to obtain A, but this cannot be recommended because the sensitivity factors only relate to signals arising from the same material and, because attenuation lengths, densities, and intrinsic loss processes in the two materials may differ significantly, serious errors can result.

$$A_k = \frac{I_k I_{\mathrm{Au}}^{\infty}}{I_{\mathrm{Au}} I_k^{\infty}} \neq \frac{I_k S_{\mathrm{Au}}}{I_{\mathrm{Au}} S_k X_k} \tag{8.6a}$$

To simplify and generalize the analysis of organic materials on gold substrates and nanoparticles, an additional factor, f, should be used to compensate for the different attenuation length, density, and intrinsic loss processes between the two materials: the gold core and organic shell (Au and Org). Thus, for each element in the shell, we have a distinct value of A:

$$A_k = \frac{I_k I_{\mathrm{Au}}^{\infty}}{I_{\mathrm{Au}} I_k^{\infty}} = f\frac{I_k S_{\mathrm{Au}}}{I_{\mathrm{Au}} S_k X_k} = f\frac{[k]}{[\mathrm{Au}]X_k} \tag{8.6b}$$

Here, $[k]$ and $[\mathrm{Au}]$ are the mole fractions or atomic% of those elements determined by XPS analysis in the standard manner.

To determine the value of f, one could measure the XPS intensities from a pure organic material and from sputter-cleaned gold under the same experimental conditions on the same day. The value of f would be simply given by the ratio of the normalized and summed intensities for each material. Our measurements yielded a

TABLE 8.1 Photoelectron Attenuation Lengths and Calculated Values for Terms B and C

XPS Line	*L* (Org), nm	*L* (Au), nm	*B*	*C*
O 1s	2.74	0.99	0.72	2.81 ± 0.04
C 1s	3.32	1.17	0.88	
N 1s	3.05	1.09	0.80	
S 2p	3.61	1.27	0.95	
Au 4f	3.79	1.33	NA	

value of $f = 0.56$, which appears both useful and consistent for two different types of organic material tested [15], and should be generally applicable. This value is somewhat different from the ~0.4 expected from a consideration of material densities and electron attenuation lengths, but accounts for the systematic errors in background subtraction during data analysis and, therefore, is used here.

The values of $L_{i,j}$ may be obtained from the equation S4 in the paper by Seah [80]. Table 8.1 lists the relevant values.

If the elements in the shell are homogenously distributed, then application of Equation 8.4 should provide the same shell thickness (the product $L_{k,\mathrm{Org}} \cdot T_{\mathrm{NP},k}$) for all elements, provided the values of X_k are correct in Equation 8.6b. We use the constraint that the sum of all X_k is equal to 1, along with an iterative calculation using the "Solver" function in Excel, where X_k are varied iteratively from an initial set of trial values to obtain the same shell thickness for all elements from the data. The result provides the nanoparticle shell thickness, T_{NP}, and the composition of the shell, X_k.

The empirical elemental composition can be compared with the expected structural composition to determine the proportion of other species within the shell. For example, the abundance of nitrogen can be used to quantify the relative amount of protein.

The number of protein molecules, N_{XPS}, can thus be calculated as

$$N_{\mathrm{XPS}} = \frac{V_{s,\mathrm{XPS}} \rho_p \mathrm{N_A}}{M} \tag{8.7}$$

where $V_{s,\mathrm{XPS}}$ is the volume of the protein shell as measured by XPS, ρ_p is the density of the dry protein, $\mathrm{N_A}$ is the Avogadro number, and M is the molecular mass of the protein in the shell.

The average number of molecules attached to each nanoparticle calculated by XPS was validated against a solution-based *in situ* approach: The average number of protein molecules bound to nanoparticles (N_{opt}) can be calculated by using a modified De Feijter equation [81] which takes into account the nonplanar geometry of nanoparticles:

$$N_{\mathrm{opt}} = \frac{\mathrm{N_A} V_{s,\mathrm{DLS}} (n_s - n_w)}{M \left(\frac{dn}{dc} \right)} \tag{8.8}$$

TABLE 8.2 Shell Thickness Determined by DLS versus XPS

Protein	Mass (kDa)	Average DLS Thickness (nm)	Average XPS Thickness (nm)
IgG	160	13.9 ± 1.1	7.71 ± 0.11
BSA	66.5	8.9 ± 0.7	3.5 ± 0.2

Here, $V_{s,\, DLS}$ is the volume of the shell (which can be calculated from the knowledge of the nanoparticle core size and the shell thickness) and *dn/dc* is the refractive index increment of the proteins with the concentration, *c*, of protein in the shell, for which a typical value for proteins is $0.182\,cm^3/g$ [82].

In order to calculate N_{opt}, measurements in solution were performed to determine the thickness and refractive index of the protein shell enveloping the nanoparticle core. DLS was used to measure the shell thickness, by comparing the diameters of protein-coated versus uncoated nanoparticles (DCS was used to verify that the nanoparticles were not aggregated). UV visible spectroscopy was used to monitor the LSPR to calculate the refractive index. More detailed information on this approach can be obtained from the original publications [15, 16].

Data were collected and analyzed for each nanoparticle sample by both the XPS method and the *in situ* solution-based approach. The average shell thicknesses determined by each method are displayed in Table 8.2.

Note that the shell thicknesses are much lower when measured by XPS compared with DLS. This is because XPS measurements were performed in ultra-high vacuum and, therefore, the water molecules have been extracted from the molecular shell of the nanoparticles before measurement. In fact, using the two thickness measurements, one can elucidate an estimation of the water content trapped in the protein layer. This can also be calculated from the measurement of the refractive index of the shell. We demonstrate the close agreement of these two estimation methods [15].

Calculations of the number of protein molecules (using Eqs. (8.7) and (8.8) for XPS and *in situ* analysis respectively) demonstrate excellent agreement between the two methods, as shown in Figure 8.3.

Power law fits of the number of molecules determined from XPS data result in indices of ~2, suggesting that the number of molecules bound to the surface of nanoparticles is proportional to the surface area. It should be noted that the solution-based approach relies on a number of assumptions and is heavily dependent upon the modeling of the LSPR shifts as a function of refractive index (requires a metallic core) and shell thickness. In addition, the sample must be completely free from aggregation, since this would affect both the DLS and LSPR results. The XPS-based approach is not laden by so many assumptions and produced very dependable results across the sample range investigated. However, it should be noted that for XPS the shell thickness must be within the same length scale as the attenuation of photoelectrons, with the shell elementally distinct from the core. Finally, both liquid-based and XPS approaches rely on the nanoparticles being spherical; currently, no straightforward solution to the same problem exists for more complex

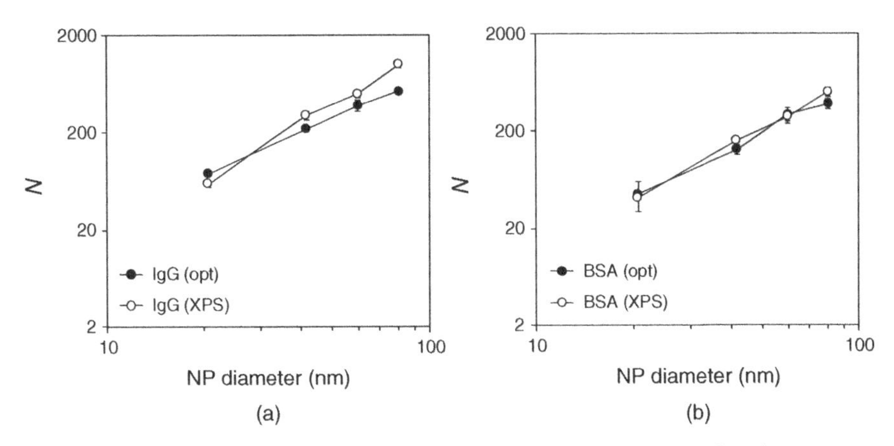

Figure 8.3 Average number of (a) IgG and (b) BSA molecules, N, plotted against nanoparticle diameter (calculated from Eq. (8.7) (XPS) and Eq. (8.8) (opt)). Data show excellent agreement of N values independently estimated from a combination of DLS and LSPR shift measurements and XPS measurements.

nanomaterial shapes. However, provided that the shape is known, it is possible to calculate this from the geometry using numerical or simulation methods [36, 79].

In conclusion, this study demonstrates that XPS is a robust approach to determine the elemental composition and shell thickness of nanoparticle coatings. This is of notable significance since not all nanoparticle samples are suitable for analysis by solution-based techniques.

8.4.2 Part II: Surface Characterization of Commercial Metal-Oxide Nanomaterials by TOF-SIMS

This case study describes the development of methodology for sample preparation and surface chemical analysis by TOF-SIMS of commercially sourced ZnO and CeO_2 nanomaterial samples. TOF-SIMS probes the first few nanometers and is, therefore, an excellent tool to assess purity and to identify surface contaminants. Nanomaterial samples and relative sample codes are described in Table 8.3.

TABLE 8.3 Specification of Nanopowders Analyzed by TOF-SIMS

Sample Code	Sample Name	Supplier
A	Nanograin CeO_2	Umicore, Belgium
B	Nanosun ZnO	Micronisers, Australia
C	Z-COTE HP 1 ZnO	BASF, Germany
D	Micron ZnO	Sigma Aldrich, UK
E	Z-COTE ZnO	BASF, Germany
F	Micron CeO_2	Sigma Aldrich, UK
G	Ceria dry CeO_2	Antaria, Australia

Several strategies were explored to optimize sample preparation, in order to achieve three main objectives. First, a high surface coverage is desired; although 10% coverage is the minimum acceptable, 100% coverage is preferred (since the technique does not have the spatial resolution to resolve individual particles and thus signal obtained is the summation of that emanating from the particles and the exposed substrate). Second, there is a need to use flat, conductive substrates to avoid topography and sample charging effects during analysis. Third, the robust attachment of the nanomaterial to the substrates; this is important as TOF-SIMS utilizes a strong extraction field (2000 V over 1.5 mm), any loosely bound particles are likely to be extracted into the mass analyzer and may cause damage to the instrument.

One of the approaches explored for sample preparation was the incorporation of PTFE beads into nanopowder samples to make a pellet. Here, PTFE serves to reduce the brittleness of the pellets. Furthermore, this material has a clear chemical signature that can be distinguished from that of the nanopowder. Although this approach was successfully used for XPS analysis, the composite pellets were still too brittle for TOF-SIMS measurements [38]. When sputtered with the primary ion gun and subjected to the strong extraction field, there was a significant risk of introducing loosely bound particles into the mass analyzer. For this reason, this sample preparation approach was excluded for TOF-SIMS measurements.

A second approach involved the deposition of nanopowders onto a thin indium foil. When pressed against the foil, the particles became embedded in it, as indium is sufficiently malleable. Indium also affords the advantage of providing a distinct chemical signature. However, concerns over achievable surface coverage and looseness of the nanomaterial on this surface led us to rule out this approach.

A third approach that was considered was the deposition of the nanopowders from solution onto a flat conductive substrate such as silicon wafer. However, dispersing dried nanopowder into solution is likely to modify the nanoparticle surface chemistry. Moreover, nanomaterial such as ZnO is known to dissolve in solution [38]. While this approach will be useful for the study of the effects of dispersion on nanomaterial surface chemistry, it is not suitable for characterizing the surface chemistry of the as-received nanopowders.

The most successful sample preparation method utilized conductive carbon tape. Si wafers were cleaned by rinsing with ethanol and acetone, and dried in a stream of nitrogen. Double-sided conductive carbon tape was cut to size and attached to the Si wafers, and excess nanomaterial powder was sprinkled (inside a glove box) onto the surface of the carbon tape to maximize surface coverage (typically about 170 mg of nanopowder was used per square centimeter). The sample with nanomaterial was then covered with aluminum foil, which was pressed firmly to ensure strong adhesion of the particles to the tape surface. Loosely surface bound material was removed by application of a nitrogen gas stream. Control samples, consisting of carbon tape on Si wafer without nanomaterial, were also prepared. TOF-SIMS analysis was performed with a Bi^+ cluster primary beam (incident at 45° from the sample normal) to obtain high-resolution mass spectra with an imaging resolution of approximately 5 μm.

8.4.2.1 Effect of Sample Topography The quality of the sample preparation protocol was assessed using sample E (ZnO nanomaterials). Figure 8.4a shows the TOF-SIMS spectrum of this sample.

The region around the Zn^+ peak is enlarged and shown in Figure 8.4b. As expected, a $^{69}Zn^+$ peak is observed at 69.93 u in the mass spectra. However, an unassigned satellite peak with an apparent mass 0.045 u higher than the $^{69}Zn^+$ peak was also observed. A detailed data analysis to elucidate the origin of the two peaks was performed. The total ion image, shown in Figure 8.4c, reveals significant spatial inhomogeneity on the sample. It was found that the two peaks originate from different regions of the sample. As shown in Figure 8.4d, the higher intensity peak (gray) originates from the flat central area, while the shifted peak (black) originates from a bright area that is surrounded by a dark circular region on the left of the image. Similar effects were observed for other peaks associated with the nanomaterial, both on this and other samples.

This effect was interpreted as an instrumental artifact due to sample topography [83]. Peaks such as the one marked in gray are associated with regions of the sample where the powder forms a relatively flat film on the adhesive tape. The bright areas surrounded by circular dark regions are interpreted as a large aggregation of nanomaterial (>100 μm) on the surface of the sample. Due to topography effects, secondary ions from the top of the aggregate are detected with a delayed TOF (higher apparent mass) compared with secondary ions from the flat areas of the sample. These peaks are also broadened and have poorer mass resolution compared with peaks from the flat area, due to the larger spread in ions' TOF. In the dark areas, the topography prevents the secondary ions from reaching the mass spectrometer. Only signals originating from the flat areas of the samples should, therefore. be considered in the analysis of the powders.

8.4.2.2 Chemical Analysis of Nanopowders The spectrum of each nanomaterial sample was compared against that of the carbon tape. In addition to the peaks characteristic for the carbon tape, the nanopowder samples exhibited peaks that were related to the elemental composition of the nanomaterial. Figure 8.5a,b shows these peaks for the CeO_2, while Figure 8.5c,d shows the ZnO nanomaterial samples. The secondary ion emission was normalized to the ions Ce^+ and Zn^+, respectively.

The CeO_2 nanopowder samples appeared relatively similar to each other and free from inorganic contamination, as no peak was detected except for those associated with Ce and CeO and their isotopes with the small addition of the elements C and H, which may arise from trace levels of hydrocarbon contamination from either sample handling and storage or from the adhesive on the carbon tape. Unfortunately, due to the presence of the carbon tape signal (resulting in many strong organic peaks containing the elements C, H, and Si), it is not possible to draw a conclusion regarding the identity and quantity of the contamination on the nanomaterial surface.

ZnO nanopowder samples exhibited mainly peaks associated with Zn and O and their isotopes (Figure 8.5c). In addition to H and C contaminants, N was also observed

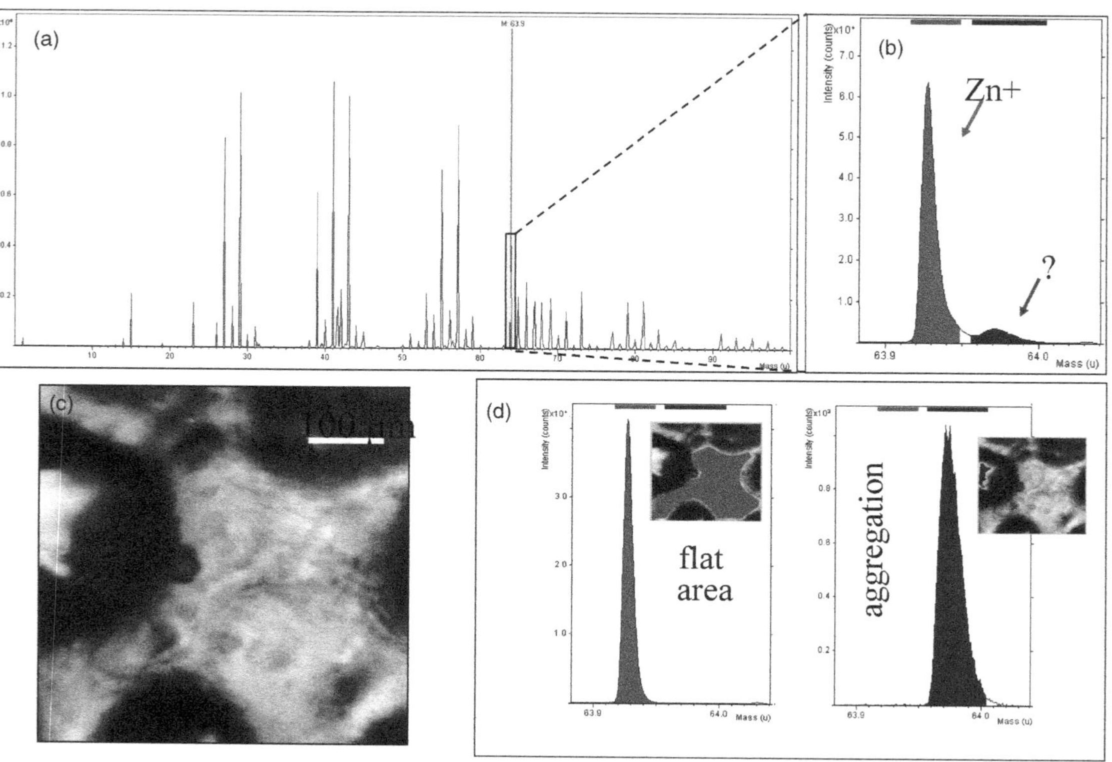

Figure 8.4 Assessment of sample preparation protocol for TOF-SIMS analysis performed on sample E (ZnO nanomaterial). (a) TOF-SIMS spectrum of the sample.(b) Enlargement of the spectrum showing the $^{69}Zn^{+}$ peak. The spectrum exhibits a satellite peak shifted 0.045 mass units from the Zn^{+} peak. (c) Total ion image of the analyzed sample surface. The dark areas refer to regions of the sample where little signal was detected. (d) Region-of-interest TOF-SIMS spectra regenerated from two areas of the image that are shown in the inset. This shows that the higher intensity peak originates from the central area of the sample, and the shifted peak originates from the bright area surrounded by dark circular regions on the left. Similar features are observed in other samples, and it is concluded that this is a typical artifact due to sample topography.

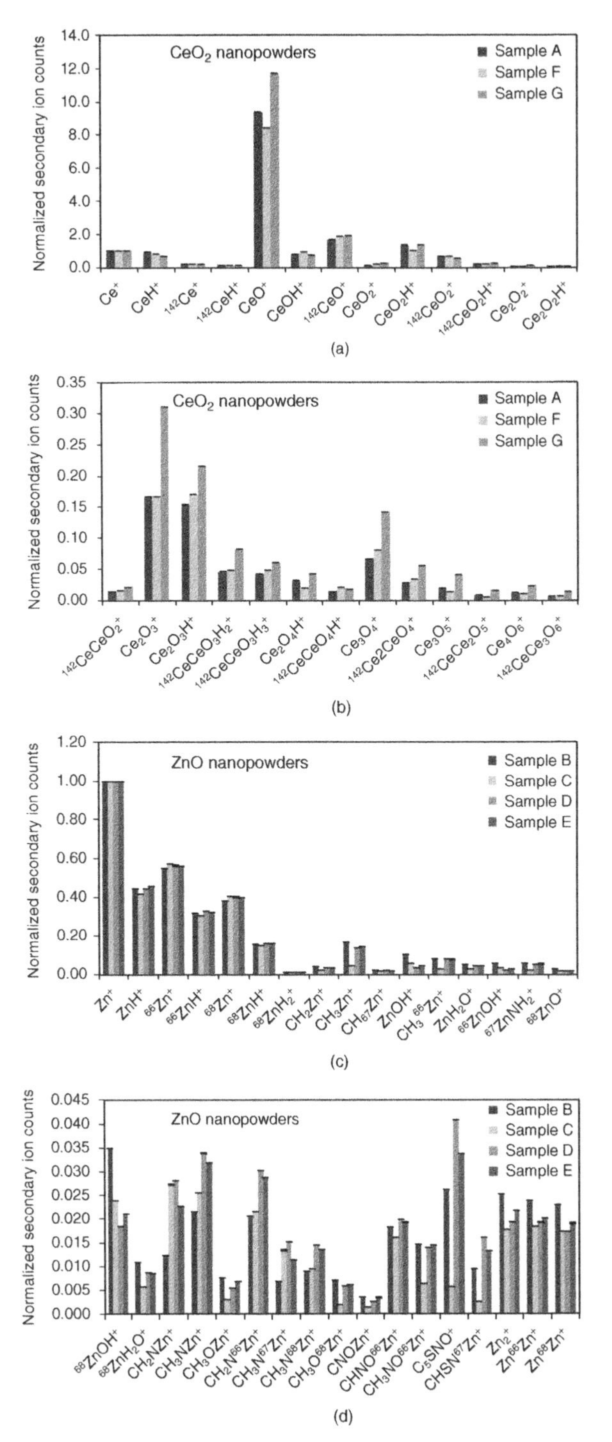

Figure 8.5 Normalized secondary ion emission of (a and b) CeO_2 and (c and d) ZnO nanopowders. The emission is normalized to the Ce^+ and Zn^+ peaks, respectively (Bi^+ beam was operated at 25 kV).

(Figure 8.5d). It is interesting to note that some of the peaks associated with the presence of CH_3 groups are low for Sample C (Figure 8.5d). This sample is the Z-cote HP1 nanomaterial, which according to the manufacturer (BASF) has a triethoxycaprylylsilane coating at the surface. It is possible that CH_3 groups are characteristic of the surface of uncoated ZnO nanomaterial. However, the silicone shell could not be detected due to the presence of strong signals from the carbon tape. The strong Zn signal from Sample C is interesting. Since TOF-SIMS is sensitive only to the outermost atomic layers of a surface, comprising a thickness of few nanometers, the detection of Zn implies that the silicone shell is either thinner than a few nanometers or it does not cover the surface homogeneously.

To conclude, seven commercial ZnO and CeO_2 nanomaterial samples were analyzed by TOF-SIMS. Sample preparation and analysis were investigated, and an example of artifacts arising from inhomogeneous sample topography was provided. Mass spectral peaks characteristic of the nanopowders were identified. However, the evaluation of purity and classification of contaminants by TOF-SIMS was challenging, due to the restrictions imposed by sample mounting requirements.

8.5 SUMMARY

Surfaces and interfaces define the behavior of nanomaterials in their working environment, so greater attention should be paid to this critical parameter. Surface analysis methods tend to be underused, meaning that too often nanomaterials are inadequately characterized, limiting their use and reliability. Cross-validation using a number of different techniques should be more routinely employed.

New advances in measurement techniques are constantly driving forward the improvements in nanomaterial characterization, yet there is much to do to meet the growing demand for high-precision measurements of such materials. In addition, the ability to perform nanomaterial characterization within complex matrices such as food and cosmetics is much sought after. Despite their associated challenges, nanomaterials offer great opportunity for furthering a huge range of technologies from the medical and healthcare sectors to catalysis and energy. For this reason, the development of reliable, versatile, and accurate measurement tools and protocols is paramount.

REFERENCES

1. Grainger DW, Castner DG. Nanobiomaterials and nanoanalysis: opportunities for improving the science to benefit biomedical technologies. Adv Mater 2008;20(5):867–877.
2. Baer DR et al. Surface characterization of nanomaterials and nanoparticles: Important needs and challenging opportunities. J Vac Sci Technol A 2013;31(5):50820.
3. Bao G, Mitragotri S, Tong S. Multifunctional nanoparticles for drug delivery and molecular imaging. Annu Rev Biomed Eng 2013;15(1):253–282.
4. ISO. *ISO/TR 14187 Surface Chemical Analysis – Characterization of Nanostructured Materials*. Switzerland: ISO; 2011.

5. Lohse SE, Murphy CJ. The quest for shape control: A history of gold nanorod synthesis. Chem Mater 2013;25(8):1250–1261.
6. De Volder MFL et al. Carbon nanotubes: Present and future commercial applications. Science 2013;339(6119):535–539.
7. Si KJ et al. Giant plasmene nanosheets, nanoribbons, and origami. ACS Nano 2014;8(11):11086–11093.
8. Cao X et al. Synthesis of Au nanostars and their application as surface enhanced Raman scattering-activity tags inside living cells. J Nanosci Nanotechnol 2015;15(7):4829–4836.
9. Wang Y et al. Preparation of nanobubbles for ultrasound imaging and intracelluar drug delivery. Int J Pharm 2010;384(1–2):148–153.
10. Mozafari MR. *Nanoliposomes: Preparation and Analysis*. Humana Press; 2009. p 29–50.
11. Pollard AJ et al. Nanoscale optical spectroscopy: An emerging tool for the characterisation of 2D materials. J Mater NanoSci 2014;1(1):11.
12. Zhang A et al. Polymeric nanoporous materials fabricated with supercritical CO2 and CO2-expanded liquids. Chem Soc Rev 2014;43(20):6938–6953.
13. Marcelo G, Martinho JMG, Farinha JPS. Polymer-coated nanoparticles by adsorption of hydrophobically modified poly(*N*,*N*-dimethylacrylamide). J Phys Chem B 2013; 117(12):3416–3427.
14. Grazon C et al. Ultrabright BODIPY-tagged polystyrene nanoparticles: Study of concentration effect on photophysical properties. J Phys Chem C 2014;118(25):13945–13952.
15. Belsey NA, Shard AG, Minelli C. Analysis of protein coatings on gold nanoparticles by XPS and liquid-based particle sizing techniques. Biointerphases 2015;10(1):019012.
16. Bell NC, Minelli C, Shard AG. Quantitation of IgG protein adsorption to gold nanoparticles using particle size measurement. Anal Methods 2013;5:4591–4601.
17. Falabella JB et al. Characterization of gold nanoparticles modified with single-stranded DNA using analytical ultracentrifugation and dynamic light scattering. Langmuir 2010;26(15):12740–12747.
18. Lynch I, Dawson KA. Protein-nanoparticle interactions. Nano Today 2008;3(1–2):40–47.
19. Minelli C, Lowe SB, Stevens MM. Engineering nanocomposite materials for cancer therapy. Small 2010;6(21):2336–2357.
20. Albanese A, Tang PS, Chan WC. The effect of nanoparticle size, shape, and surface chemistry on biological systems. Annu Rev Biomed Eng 2012;14:1–16.
21. Kaur K, Forrest JA. Influence of particle size on the binding activity of proteins adsorbed onto gold nanoparticles. Langmuir 2011;28(5):2736–2744.
22. Walkey CD et al. Nanoparticle size and surface chemistry determine serum protein adsorption and macrophage uptake. J Am Chem Soc 2011;134(4):2139–2147.
23. Goy-López S et al. Physicochemical characteristics of protein–np bioconjugates: The role of particle curvature and solution conditions on human serum albumin conformation and fibrillogenesis inhibition. Langmuir 2012;28(24):9113–9126.
24. Grassian VH. When size really matters: Size-dependent properties and surface chemistry of metal and metal oxide nanoparticles in gas and liquid phase environments†. J Phys Chem C 2008;112(47):18303–18313.
25. Karakoti AS et al. Preparation and characterization challenges to understanding environmental and biological impacts of ceria nanoparticles. Surf Interface Anal 2012; 44(8):882–889.

26. Ramallo-López JM et al. XANES study of the radiation damage on alkanethiolates-capped Au nanoparticles. J Phys 2013;430(1):012034.
27. Yang L et al. Depth profiling and melting of nanoparticles in secondary ion mass spectrometry (SIMS). J Phys Chem C 2013;117(31):16042–16052.
28. Havelund R et al. Electron flood gun damage effects in 3D secondary ion mass spectrometry imaging of organics. J Am Soc Mass Spectrom 2014;25(9):1565–1571.
29. Tanuma S et al. Quantitative evaluation of surface damage on SiO(2)/Si specimen caused by electron beam irradiation. Appl Surf Sci 2005;241(1–2):122–126.
30. Yang L et al. Sputtering yields of gold nanoparticles by C60 ions. J Phys Chem C 2012;116(16):9311–9318.
31. Baer DR et al. Challenges in applying surface analysis methods to nanoparticles and nanostructured materials. J Surf Anal 2005;12(2):101–108.
32. Nurmi J et al. Recovery of iron/iron oxide nanoparticles from solution: comparison of methods and their effects. J Nanopart Res 2011;13(5):1937–1952.
33. Kumar N, Rae A, Roy D. Accurate measurement of enhancement factor in tip-enhanced Raman spectroscopy through elimination of far-field artefacts. Appl Phys Lett 2014;104(12):123106.
34. Rafati A et al. Quantitative XPS depth profiling of codeine loaded poly(L-lactic acid) films using a coronene ion sputter source. J Control Release 2009;138(1):40–44.
35. Lee SH et al. Amine-functionalized gold nanoparticles as non-cytotoxic and efficient intracellular siRNA delivery carriers. Int J Pharm 2008;364(1):94–101.
36. Techane S, Baer DR, Castner DG. Simulation and modeling of self-assembled monolayers of carboxylic acid thiols on flat and nanoparticle gold surfaces. Anal Chem 2011;83(17):6704–6712.
37. Torelli MD et al. Quantitative determination of ligand densities on nanomaterials by X-ray photoelectron spectroscopy. ACS Appl Mater Interfaces 2015;7(3):1720–1725.
38. Tantra R et al. *Final Report on the Physico-Chemical Characterisation of PROSPEcT Engineered Nanomaterials*. National Physical Laboratory; 2012. p 62.
39. Ray S, Shard AG. Quantitative analysis of adsorbed proteins by X-ray photoelectron spectroscopy. Anal Chem 2011;83(22):8659–8666.
40. ISO. *ISO 18117 Surface Chemical Analysis – Handling of Specimens Prior to Analysis*. Switzerland: ISO; 2009.
41. Tao F. Design of an in-house ambient pressure AP-XPS using a bench-top X-ray source and the surface chemistry of ceria under reaction conditions. Chem Commun 2012;48(32): 3812–3814.
42. Toyoshima R et al. In situ ambient pressure XPS study of CO Oxidation reaction on Pd(111) surfaces. J Phys Chem C 2012;116(35):18691–18697.
43. Klyushin AY et al. A near ambient pressure XPS study of Au oxidation. Phys Chem Chem Phys 2014;16(17):7881–7886.
44. Gilmore IS, Seah MP, Johnstone JE. Quantification issues in ToF-SSIMS and AFM co-analysis in two-phase systems, exampled by a polymer blend. Surf Interface Anal 2003;35(11):888–896.
45. ISO. *ISO 20903:2011 Surface Chemical Analysis – Auger Electron Spectroscopy and X-ray Photoelectron Spectroscopy – Methods Used to Determine Peak Intensities and Information Required When Reporting Results*. Switzerland: ISO; 2011.

46. ISO. *ISO 22048:2004 Surface Chemical Analysis – Information Format for Static Secondary-Ion Mass Spectrometry*. Switzerland: ISO; 2004.
47. Shard AG. Detection limits in XPS for more than 6000 binary systems using Al and Mg Kα X-rays. Surf Interface Anal 2014;46(3):175–185.
48. Shard AG. A straightforward method for interpreting XPS data from core-shell nanoparticles. J Phys Chem C 2012;116(31):16806–16813.
49. Rades S, Wirth T, Unger W. Investigation of silica nanoparticles by Auger electron spectroscopy (AES). Surf Interface Anal 2014;46(10–11):952–956.
50. Shard AG et al. Argon cluster ion beams for organic depth profiling: Results from a VAMAS interlaboratory study. Anal Chem 2012;84(18):7865–7873.
51. Vickerman J, Gilmore I. *Surface Analysis the Principal Techniques*. 2nd ed. Wiley; 2009.
52. ter Veen R et al. Combining dynamic and static depth profiling in low energy ion scattering. J Vac Sci Technol A 2013;31(1):3.
53. Rafati A, ter Veen R, Castner DG. Low-energy ion scattering: Determining overlayer thickness for functionalized gold nanoparticles. Surf Interface Anal 2013;45(11–12): 1737–1741.
54. Déronzier T et al. Catalysis on nanoporous gold–silver systems: Synergistic effects toward oxidation reactions and influence of the surface composition. J Catal 2014; 311(0):221–229.
55. Stamenkovic VR et al. Trends in electrocatalysis on extended and nanoscale Pt-bimetallic alloy surfaces. Nat Mater 2007;6(3):241–247.
56. Sharp JC, Yao YX, Campbell CT. Silver nanoparticles on Fe3O4(111): energetics by Ag adsorption calorimetry and structure by surface spectroscopies. J Phys Chem C 2013;117(47):24932–24936.
57. Eaton P, West P. *Atomic Force Microscopy*. Oxford: OUP; 2010.
58. Frisbie CD et al. Functional group imaging by chemical force microscopy. Science 1994;265(5181):2071–2074.
59. Moy V, Florin E, Gaub H. Intermolecular forces and energies between ligands and receptors. Science 1994;266(5183):257–259.
60. Muller DJ, Dufrene YF. Atomic force microscopy as a multifunctional molecular toolbox in nanobiotechnology. Nat Nano 2008;3(5):261–269.
61. Blum C et al. Tip-enhanced Raman spectroscopy – An interlaboratory reproducibility and comparison study. J Raman Spectr 2014;45(1):22–31.
62. Stadler J, Schmid T, Zenobi R. Nanoscale chemical imaging of single-layer graphene. ACS Nano 2011;5(10):8442–8448.
63. Su W, Roy D. Visualizing graphene edges using tip-enhanced Raman spectroscopy. J Vac Sci Technol B 2013;31(4):041808.
64. Kumar N et al. Nanoscale mapping of catalytic activity using tip-enhanced Raman spectroscopy. Nanoscale 2015;7:7133–7137.
65. Stiles PL et al. Surface-enhanced Raman spectroscopy. Annu Rev Anal Chem 2008;1(1):601–626.
66. Carim AI, Gu J, Maldonado S. Overlayer surface-enhanced Raman spectroscopy for studying the electrodeposition and interfacial chemistry of ultrathin Ge on a nanostructured support. ACS Nano 2011;5(3):1818–1830.

67. Xie W, Walkenfort B, Schlücker S. Label-free SERS monitoring of chemical reactions catalyzed by small gold nanoparticles using 3D plasmonic superstructures. J Am Chem Soc 2012;135(5):1657–1660.
68. Cure J et al. Monitoring the coordination of amine ligands on silver nanoparticles using NMR and SERS. Langmuir 2015;31(4):1362–1367.
69. Bordenyuk AN et al. Vibrational sum frequency generation spectroscopy of dodecanethiol on metal nanoparticles. J Phys Chem C 2007;111(25):8925–8933.
70. Weeraman C et al. Effect of nanoscale geometry on molecular conformation: Vibrational sum-frequency generation of alkanethiols on gold nanoparticles. J Am Chem Soc 2006;128(44):14244–14245.
71. Krier JM et al. Sum frequency generation vibrational spectroscopy of colloidal platinum nanoparticle catalysts: Disordering versus removal of organic capping. J Phys Chem C 2012;116(33):17540–17546.
72. Karam TE, Haber LH. Molecular adsorption and resonance coupling at the colloidal gold nanoparticle interface. J Phys Chem C 2013;118(1):642–649.
73. Hennig A et al. Simple colorimetric method for quantification of surface carboxy groups on polymer particles. Anal Chem 2011;83(12):4970–4974.
74. Wain AJ, Pollard AJ, Richter C. High-resolution electrochemical and topographical imaging using batch-fabricated cantilever probes. Anal Chem 2014;86(10):5143–5149.
75. Muthukumaran T, Philip J. A single pot approach for synthesis of phosphate coated iron oxide nanoparticles. J Nanosci Nanotechnol 2015;15(4):2715–2725.
76. Liu X et al. Determination of monolayer-protected gold nanoparticle ligand–shell morphology using NMR. Nat Commun 2012;3:1182.
77. Gentilini C et al. Formation of patches on 3D SAMs driven by thiols with immiscible chains observed by ESR spectroscopy. Ange Chem 2009;121(17):3106–3110.
78. Lucarini M, Pasquato L. ESR spectroscopy as a tool to investigate the properties of self-assembled monolayers protecting gold nanoparticles. Nanoscale 2010;2(5):668–676.
79. Werner WSM et al. Interpretation of nanoparticle X-ray photoelectron intensities. Appl Phys Lett 2014;104(24):243106.
80. Seah MP. Simple universal curve for the energy-dependent electron attenuation length for all materials. Surf Interface Anal 2012;44(10):1353–1359.
81. De Feijter JA, Benjamins J, Veer FA. Ellipsometry as a tool to study the adsorption behavior of synthetic and biopolymers at the air–water interface. Biopolymers 1978; 17(7):1759–1772.
82. Vörös J. The density and refractive index of adsorbing protein layers. Biophys J 2004;87(1):553–561.
83. Lee JLS et al. Topography and field effects in the quantitative analysis of conductive surfaces using ToF-SIMS. Appl Surf Sci 2008;255(4):1560–1563.

9

MECHANICAL, TRIBOLOGICAL PROPERTIES, AND SURFACE CHARACTERISTICS OF NANOTEXTURED SURFACES

C.A. Charitidis, D.A. Dragatogiannis, E.P. Koumoulos, and D. Perivoliotis

School of Chemical Engineering, Laboratory Unit of Advanced Composite, Nanomaterials and Nanotechnology, National Technical University of Athens, Athens 15780, Greece

9.1 INTRODUCTION

The term nanotextured surfaces can be defined as surfaces that are covered with nano-sized structures and thus fall under the umbrella of nanomaterials. Nanostructured surfaces have in the past attracted considerable attention because they show novel physical properties, thus leading to various applications. In surface and coating research, such materials have raised much attention as they can potentially possess numerous novel properties: high mechanical strength, chemical inertness, broad optical transparency, high refractive index, wide band gap, excellent thermal conductivity, extremely low thermal expansion, and very attractive friction (wear) properties. Such properties have resulted in the use of nanotextured surfaces in a wide range of applications: tribological applications (e.g., rolling bearings, machining, mechanical seals, biomedical implants, micro-/nanoelectromechanical systems, sensors, microfluidics), self-cleaning, anti-fogging, anti-icing, and antibacterial action.

A major goal of surface and coating research is to design and fabricate surfaces with special anti-wetting properties, which repel not only water [1–3] but also oils (or even lower surface energy liquids). Note that in the literature, amphiphobicity is

---

Nanomaterial Characterization: An Introduction, First Edition. Edited by Ratna Tantra.

the term used to describe a surface/coating that is both water and oil repellent; this is in contrast to oleophobicity (repelling only oil) and omniphobicity (repelling everything) [4]. One solution to achieve special anti-wetting properties is to employ surface nanotexturing (as well as microtexturing) with chemical modification using low surface energy polymers (with organic layers). In the past, researchers have applied such processes to achieve the above-mentioned properties, not only on open surfaces but also in the domain of microfluidics [5–7].

Till now, numerous procedures to prepare nanotextured surfaces have been reported. These procedures can be classified as either "top-down" or "bottom-up." Top-down denotes methods of inducing roughness by removing material from a bulk matrix such as patterning by lithography [8] or etching techniques [9]. Conversely, bottom-up describes roughness and functionality constructed from the substrate upward, typically as a thin film, through methods such as crystal growth, [10, 11] layer-by-layer [12] chemical vapor deposition (CVD) [13], electrospinning [14], and sol–gel [15].

The choice of method to produce nanostructured surfaces will depend on the surface properties desired. Random nanotexturing of polymeric surfaces with pillar-like micro–nanostructures can be achieved with the use of several technologies, such as plasma processing [16, 17], replication from Si molds [18], and ultra-short pulsed-laser irradiation [19]. Short plasma nanotexturing can modify the chemistry and topography of polymeric surfaces, without affecting its bulk properties. This creates nanoscale surface roughness (i.e., nanotextured surfaces) that can be used for controlling the optical, wetting/flow properties (as well biomolecule adsorption and cell adhesion on such surfaces) [20]. Such surface treatment has been used in various applications: polymer-based microdevices, flow control [21], micro-/nanofiltration, flexible electronics, and bio-MEMS [22–24]. In the case of O_2 plasma treatment [25] of a poly(methyl methacrylate) (PMMA), the surface has been shown to result in a high aspect ratio (HAR) topography with pillar-like structures [26–28] and acquires superhydrophilicity. The superhydrophilicity property was mainly due to –OH and –COOH groups found on the surface after its exposure to O_2 plasma [27, 29].

Although there is clear potential on the application of nanotextured surfaces (and coatings), there are potential drawbacks, such as low mechanical durability (particularly in the preparation of superhydrophobic films). Also, while significant effort has been devoted to the development of nanostructured surfaces (by surface micro-/nanotexturing), there is little work on the investigation of mechanical durability, mechanical stability, and wear resistivity of such materials. In addition with the difficulty of preparing mechanically robust superhydrophobic films, the determination of local mechanical characterization is not trivial (yet, the mechanical and tribological properties of nanotextured surfaces are of critical importance to the development of highly stable and durable applications).

The purpose of this chapter is twofold: to better understand how nanotextured surfaces can be created and how to characterize the resultant surface properties (i.e., mechanical and tribological). The chapter starts off by presenting several methods commonly employed to develop nanotextured surfaces (e.g., plasma processing and chemical vapor deposition). The benefits of surface-texturing for various applications and the size-dependent mechanical properties of such nanostructures are also discussed. The chapter then presents an overview of some of the methods used to characterize mechanical behavior. The methods of focus are nanoindentation and nanoscratch tests. Several case studies are presented to show how such tests can be applied to characterize the mechanical properties of nanotextured surfaces. The material presented in the case studies is based mainly on our work, surrounding mechanical (and tribological) properties of hydrophobic fluorocarbon (FC) coatings, superhydrophobic, and superamphiphobic polymeric surfaces.

9.2 FABRICATING NANOTEXTURED SURFACES

9.2.1 Plasma Treatment Processes

Gas plasma (glow discharge) treatment has been extensively used for the surface modification of polymeric materials (e.g., thermoplastic films, fibers, nonwoven, membranes, biomedical devices). The main advantage of this versatile technique is that it is confined to the surface layer of a material, without affecting its bulk properties. Moreover, it is a dry (solvent free), clean, and time-efficient process with a large variety of controllable process parameters (e.g., discharge gas, power input, pressure, treatment time) within the same experimental setup [30]. However, due to the complexity of the plasma process and the variety of chemical and physical reactions that can occur, it is difficult to predict and control the chemical (and structural) composition of a plasma-treated surface [31].

The plasma process (usually using a low pressure plasma system) can be used to perform various functions, for example, clean, etch, and sputter. In polymers, plasma process has been used to etch and sputter (with oxygen or noble-gas plasmas) to cause roughening of the polymer surface. The process of roughening a surface can consist of several steps. First, the polymer can be etched by a chemical reaction of reactive plasma species (e.g., radicals, ions); this is referred to as chemical plasma etching. Second, ion bombardment on the polymer surface can result in sputtering of the surface (thus, roughening is established via a physical process). Third, UV radiation from the plasma phase can cause dissociation of chemical bonds, which leads to formation of low-molecular-weight (LMW) material. In general, these three steps can occur simultaneously during the plasma treatment of a polymer and induce a flow of volatile LMW products from the substrate to the plasma, causing a gradual weight loss of the treated polymeric material [31].

Reactions can occur between a polymer substrate and reactive neutral species (formed in the plasma phase) to result in chemical etching. This in turn can be accelerated by ion bombardment (i.e., ion-enhanced) etching [32]. In the absence of such (energetic) ions, the oxygen plasma etching step proceeds at relatively low rates. It is only the combination of chemical etching accelerated by ion bombardment that results in anisotropical etching of the polymer surface. This combined process, which is also commonly referred to as reactive ion etching (RIE), is often used for patterning of surfaces in the semiconductor and (nano)lithographic industry [5, 17, 33–38]. Anisotropicity (or directionality) in etching can be achieved when the substrate is properly biased, that is, when the substrate serves as self-bias (voltage) of cathode electrode. This is an important feature that distinguishes RIE from other plasma etching techniques [31]. However, it must be remembered that RIE does not affect all polymers equally. For example, semicrystalline polymers can show preferential etching behavior, resulting in preferential amorphous etching.

For a given plasma condition, some polymers can exhibit a higher etching rate than others. The etching rate (Å/min) of a polymer largely depends on the plasma treatment conditions (e.g., discharge gas, pressure, discharge power, substrate temperature, and treatment time) and on the polymer's chemical and physical properties. The most susceptible polymer segments in a phase-separated polymer system (e.g., block copolymers, semicrystalline polymers, polymer blends) can, therefore, removed by preferential etching. Over the course of treatment time, plasma etching will reach an equilibrium state (i.e., constant etching rate). Any initial fluctuations of etching rate may arise due to changes of surface temperature and surface chemistry. For example, crosslinking in the surface layer of the substrate during plasma treatment could suppress the initial etching rate of a polymer [39].

Overall, plasma treatment is used mainly to increase the surface energy of a material, often polymer. For example, cold plasmas produced can generate O_2, N_2, air, or NH_3, which subsequently introduce oxygen- or nitrogen-containing groups on the surface of the polymer [40, 41]. Hence, this will introduce polar hydrophilic groups on the surface. In addition to oxygen- and nitrogen-containing discharges, plasmas generated in pure helium or argon will lead to the creation of free radicals (leading to crosslinking or grafting of oxygen-containing groups when the surface is exposed to oxygen or air after the treatment). Finally, it should be mentioned that the induced surface characteristics are not permanent; the treated surfaces will tend to partially recover to their untreated state during storage in, for example, air (so-called hydrophobic recovery) and can also undergo postplasma oxidation reactions [42].

In this section, only a brief summary of plasma treatment processes have been presented so far. For more detailed descriptions (e.g., theory), the reader is referred to the literature [36, 43–51]. In the following sections, several examples of nanotextured surfaces are presented.

9.2.2 Randomly Nanotextured Surfaces by Plasma Etching

Figure 9.1 shows the scanning electron microscope (SEM) image of a nanotextured surface typically produced on cyclo-olefin polymer (COP) (under 4 min exposure to highly anisotropic plasma). Similar findings, that is, the formation of nanotextured surface resulting in increased wettability, has been observed with other samples that were processed under the same conditions but on different days.

Figure 9.2 shows SEM images polyether ether ketone (PEEK) (Figure 9.2a) and poly(methyl methacrylate) (PMMA) (Figure 9.2b) surfaces, as a result of plasma

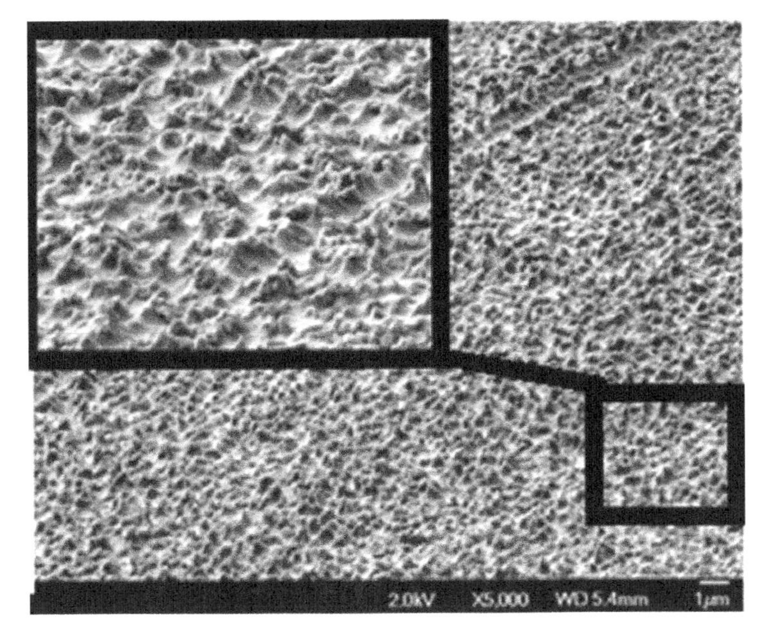

Figure 9.1 SEM image of 4 min nanotextured COP surface. A higher magnification (×20,000) image is given as inset.

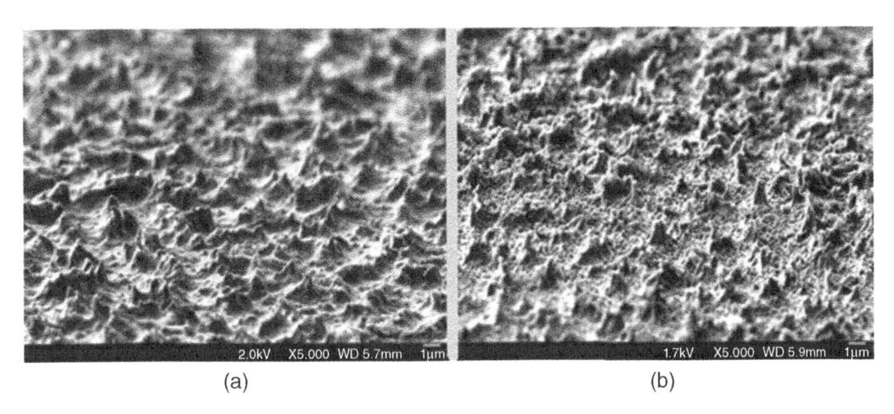

Figure 9.2 SEM images of micro–nanotextured, water-immersed and dried PEEK (a) and PMMA (b) surfaces after perfluorosilane modification in cyclohexane (70° tilted). Curved microhills (re-entrant-like structures) are produced after etching and grafting of the polymeric surfaces. Figure has been adapted from [28].

treatment (and carrying out a sequential postplasma process of being immersed in water, dried, followed by plasma-induced reactivation and grafting by perfluorodecyltrichlorosilane (FDTS)). The images clearly show the presence of curved microhills (re-entrant-like structures) as a result of such an etching process.

Figure 9.3 SEM images of 10 min SF6 plasma-nanotextured PDMS surfaces before silanization and (as inset) after silanization. Figure has been adapted from [28].

The wettability of a surface can subsequently be altered to produce superhydrophobic surfaces, in which a thin hydrophobic film can be deposited on the nanotextured surface. This can be done either by using a C_4F_8 plasma deposition (that can deposit a fluorocarbon (FC) film) or by spin coating with TEFLON® AF1600 [27, 52, 53]).

As indicated by the SEM images, nanotextured surfaces are not robust as they can easily be damaged by, for example, by impact (or simple rubbing [54]), wear-induced chemical degradation. The inherent aging of such a surface can, therefore, limit the successful application of such superhydrophobic surfaces, which has been highlighted by several researchers [55–58].

Figure 9.3 shows an SEM image of nanotextured polydimethylsiloxane (PDMS) surface. Results show that this nanotextured surface was less homogeneous than corresponding PMMA or PEEK (Fig. 9.2) surfaces. The PDMS nanotextured surface shown was produced after using 10 min plasma processing time. Results clearly showed surface topography to consist of high aspect ratio surface features. Upon further testing, the PDMS surfaces were not stable when samples were immersed in water. Interestingly, when silanized (in cyclohexane), the high aspect ratio features became more stable, as the surface exhibits grass-like, re-entrant-like structures (Fig. 9.3 inset). The image in Figure 9.3 clearly shows the differentiation in topography with vertical pillars (in Fig. 9.3) as opposed to bent pillars (Fig. 9.3 inset) being associated before and after silanization, respectively. The re-entrant-like structure profiles associated with the silanization step was needed to achieve superoleophobicity. Results also showed that fluorosilanized PDMS exhibited a slightly lower contact angle (with hexadecane) than corresponding fluorosilanized PMMA and PEEK; this observation has been attributed to differences in surface inhomogeneity [28].

Figure 9.4 SEM images of PMMA surfaces (60° tilted) displaying the hierarchical, hexagonally ordered packed pillars obtained upon plasma etching using 1 μm (left) and 3 μm polystyrene particles (right). Figure has been adapted from [28].

9.2.3 Ordered Hierarchical Nanotextured by Plasma Etching

Figure 9.4 shows a SEM image of uniform, mushroom-like micropillars produced on PMMA; this has been grown by a combination of colloidal lithography (using polystyrene (PS) particles) and plasma etching. Result shows the difference in the nanotextured surface when plasma etching was conducted using 1 μm PS particles (Fig. 9.4, left) as opposed to 3 μm (Fig. 9.4, right) [28]. The detailed procedure to produce such nanotextured surfaces is not given here, as it has been described elsewhere [53]. Interestingly, the mushroom-like re-entrant micropillars do not display any coalescence after immersion in water or silane solvent solution (for 1 h), which indicate their relative stability in comparison to the randomly nanotextured surfaces.

9.2.4 Carbon Nanotube Forests by Chemical Vapor Deposition (CVD)

Nanotexured surfaces can also be formed using CVD method. This method has been developed as a novel manufacturing process in many industrial sectors, such as in the semiconductor and ceramic industry. The CVD method has been used in the past to produce carbon nanotubes on substrates. These materials are capable of creating CNT forests (sometimes also called turfs or brushes) that prove important in a multitude of applications , for example, field emission electron sources [59], electrical interconnects [60], and thermal interface materials [61]. In addition CNT forests show possible application and use in microelectronic devices and microelectromechanical systems [62]. Except the conventional vertically aligned (VA) CNTs, which are grown on planar substrates, there are reports on large-scale CNTs that have been grown with microscale carbon fiber bundles as their substrate [63, 64]. A potential use of these materials are stress and strain sensors for modifying the fiber–matrix interfaces in composite materials [65], and as artificial hair flow sensors, that are used on micro air vehicles.

The importance and synthesis of CNTs using the CVD method (and corresponding novel properties of CNTs) have been introduced in the beginning of the book (Chapter 2) and thus are not covered here. Nonetheless, it is important to highlight that the CVD method offers potential for the production of high-purity CNTs in a controlled manner. Various experimental conditions (e.g., different substrates, precursors, and catalyst systems) in the CVD method will govern the property of the nanotextured surface produced. Carbon precursors (of solid, liquid, or gas phase) can be produced by different carbon sources, such as hydrocarbons (e.g., CH_4, C_2H_4, C_2H_2), alcohols (e.g., C_2H_5OH), and camphor ($C_{10}H_{16}O$). Camphor is considered a source of carbon to produce nanotubes, due to its hexagonal and pentagonal carbon ring structure. For the catalysts, nanometer-size metal particles can be used, in order to enable carbon source decomposition, leading to growth of CNTs. The most commonly used metals are Fe, Co, and Ni due to their high catalytic activity and relatively low reaction temperature. Ferrocene is a good catalyst precursor for the production of iron nanoparticle catalysts in the formation of nanotubes. Different CNT structures can be produced, depending on the experimental conditions, for example, temperature and other growth parameters (flow rate, pressure, catalyst concentration, etc.) [66].

Figure 9.5 shows a SEM image of multiwalled CNT carpet consisting of well-aligned carbon nanotubes. Such nanotextured surface was produced using the CVD method under the following conditions: (i) when camphor was used as a carbon source, (ii) when ferrocene was used as a catalyst compound (with a 20:1 mass ratio), (iii) where deposition temperature of 850 °C was used, and (iv) when carrier gas (nitrogen) flow was 330 ml/min. The image shows vertically aligned (VA) multiwalled CNT carpet with a diameter distribution ranging from 60 to 100 nm

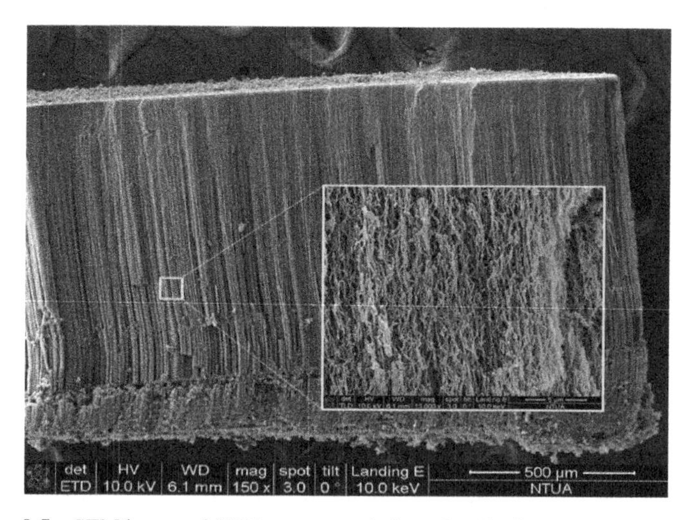

Figure 9.5 SEM image of CNT carpet consisting of well-aligned carbon nanotubes.

(grown on silicon substrate). The average thickness of the CNTs was in the range of 1–2 mm and an average distance between the tubes was approximately 100–150 nm [67]. Although the fabrication of such structures has been reported widely, relevant studies on the mechanical behavior of such structures are limited.

9.3 MECHANICAL PROPERTY CHARACTERIZATION

In almost all applications, material performance depends on its surface properties. For example, effective catheterization requires that biochemical and tribological properties of the polymer surface be preserved during operation. Undoubtedly, this places a huge emphasize on the need to characterize micro–nanotextured and/or structured surfaces. Having said this, very little is known about the mechanical properties of nanotextured surfaces. There is a need to universally adopt methodology beyond the use of basic tests (e.g., film hardness by pencil method) in order to determine relative hardness. Better methods to characterize mechanical property of nanotextured surfaces can include the use of nano-/microindentation and nano-/microscratch tests; these tests have been found to be useful to understand mechanical durability, mechanical stability, and wear resistance [68, 69].

According to Dyett et al., the mechanical durability of nanotextured surfaces is frequently overlooked [69] and yet it is important to have tests in place to characterize this. Such tests will be useful as they give an insight into the deformation mechanisms associated with mechanical damage of the surface features and loss of superhydrophobicity, which is the main drawback of many practical applications [69]. In this section, nanoindentation and nanoscratch tests (for characterization of mechanical durability) are described.

9.3.1 Nanoindentation Testing

When probing nanoscale features such as those present on nanotextured surfaces, it would be desirable to characterize the corresponding mechanical properties. One option is to include an examination of length-scale effects in indentation measurements, that is, nanoindentation [70, 71].

Nanoindentation studies can be performed using a nanoindentation platform, in which a nanoindenter is able to record continuously the displacement of a tip through a material. The corresponding curve that results will directly reflect the mechanical behavior and deformation mechanism of the material. Resultant curve (of displacement versus load) and contact area of the indent can be utilized to determine material properties, namely elastic modulus and hardness. The analysis of depth-sensing nanoindentation is unfortunately complicated due to several factors such as the presence of surface roughness. If nanoindentation is carried out at shallow penetration depths (in order to study rough films), there are additional concerns, namely effects associated with substrate influences [72], scale effects [73–76], and surface effects [77].

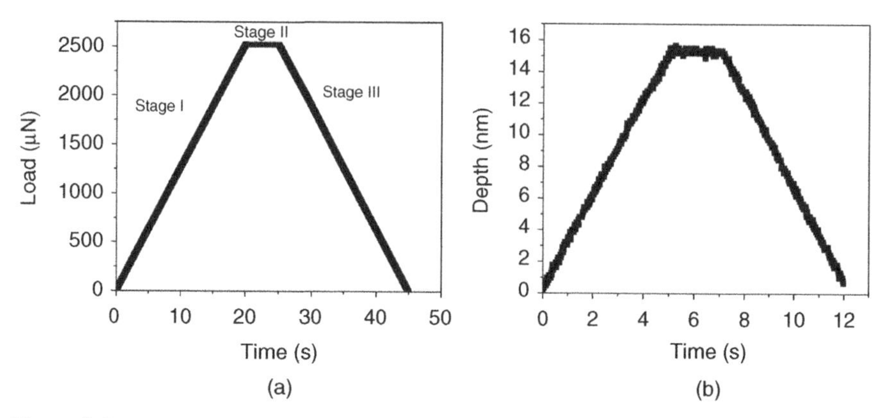

Figure 9.6 Representative (a) load versus time curve and (b) depth-displacement versus time curve during the load-control experiment.

The procedure and analysis of nanoindentation experiments were initially developed by Doerner and Nix [78], then further developed by Oliver and Pharr [79] and then Joslin and Oliver [80], in which the method proposed is more suited to deal with rough surfaces. Overall, a succinct review on nanomechanical characterization using nanoindentation has been provided by Bhushan [81] (and more recently by Palacio and Bhushan [82]). As a starting point, film penetrations are usually lower than 10% of the film thickness, so as to avoid substrate influence. At the same time, to avoid the scattering from surface roughness, indents are performed to a depth of 20 Ra [72]. It is immediately apparent that thin superhydrophobic films are not ideal for these measurements and yet mechanical parameters associated with superhydrophobic thin films are imperative in rationalizing coating performance.

A typical protocol to carry out nanoindentation testing often involved the use of a three-sided pyramidal Berkovich diamond indenter (radius of ~120 nm). Tip radius is calculated before each experimental procedure following a calibration process. Experiments are performed in a clean area environment with ~45% humidity and 23 °C ambient temperature [83]. In all depth-sensing tests, a total of five indents are averaged to determine the mean hardness (H) and elastic modulus (E) values for statistical purposes, with a spacing of 50 μm.

Nanoindentation tests are performed under two conditions, that is, load and displacement controlled. In the first case of a load-controlled experiment, load–unload curves (typically illustrated in Fig. 9.6) can be extracted at several applied loads (ranging from 100 to 4000 μN). The load-controlled experiment is useful to study the effect of the treatment processes on the mechanical properties of the substrate. The duration of loading and unloading segments of the performed indentation tests have a course of 20 s, respectively, and holding time at the maximum load is 5 s. In the second case of a displacement-controlled test, load–unload curves can also be extracted at several applied displacements (from 5 to 50 nm), to elucidate the effect of treatment process on the surface, to quantify the substrate effect in the response, and to study

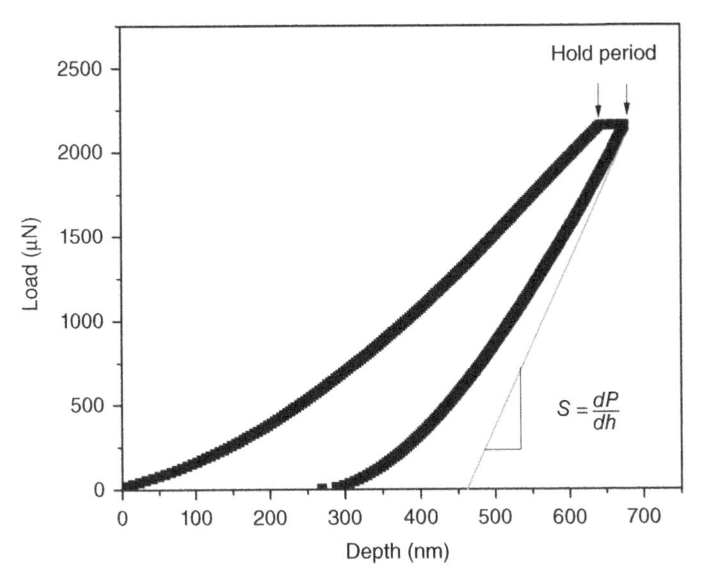

Figure 9.7 Typical load–depth displacement curve from a nanoindentation experiment.

the coating–substrate mechanical behavior. Overall, the first case requires high loads to probe the bulk region, whereas second case will only probe the top surface, that is, the coating and the nanotextured surface.

Following a nanoindentation test, Oliver–Pharr method is applied, in which load versus depth–displacement can be plotted (as depicted in Fig. 9.7), to determine three important values: hardness and modulus at maximum applied load, the penetration depth in the surface at the maximum load, and unloading contact stiffness (that corresponds to the slope at the beginning of the curve in Fig. 9.7).

According to the Oliver and Pharr method, the curve can fit to the following function [79]:

$$P = C(h - h_r)^m \tag{9.1}$$

where P is the applied load at a point in the unloading curve, C, m, and the residual depth h_r are empirically fit constants, and h is the depth into the surface corresponding to a load P. The value of m is taken to be between 1 and 2.

Based on the half-space elastic deformation theory, elastic modulus values can be extracted from the experimental data (load–displacement curves) [79]. The derived expressions for calculating the elastic modulus are based on Sneddon's elastic contact theory for unloading between a conical indenter and the contact surface [84]:

$$E_r = \frac{S\sqrt{\pi}}{2\beta\sqrt{A_c}} \tag{9.2}$$

where A_c is the projected contact area between the tip and the substrate, and β is a constant that depends on the geometry of the indenter ($\beta = 1.167$ for Berkovich tip).

This relation was originally derived for unloading between a conical indenter and the contact surface, but also holds true for the Berkovich and spherical indenters. The reduced modulus is related to the material modulus E by the relation

$$\frac{1}{E_r} = \frac{1 - v_i^2}{E_i} + \frac{1 - v_s^2}{E_s} \tag{9.3}$$

where v is the Poisson ratio of the material and the subscript i refers to those properties of the indenter. An inherent weakness in this method is that v of the tested material must be known or closely estimated beforehand. Conventional nanoindentation hardness refers to the mean contact pressure; this hardness, which is the contact hardness, H_c, is actually dependent upon the geometry of the indenter.

$$H_c = \frac{F}{A_c} \tag{9.4}$$

where

$$A_c = A(h_c) = 24.5h_c^2 + a_1 h_c + a_{1/2} h_c^{1/2} + \cdots + a_{1/16} h_c^{1/16} \tag{9.5}$$

and

$$h_c = h_m - \varepsilon \frac{P_m}{S_m} \tag{9.6}$$

where h_c is the contact depth and ε is an indenter geometry constant, equal to 0.75 for Berkovich and spherical indenter [79].

Lastly, plasticity is quantified based on the following relations:

$$\frac{W_{\text{tot}} - W_u}{W_{\text{tot}}} \tag{9.7}$$

where W_{tot} is the work of total indentation process and W_u is the work during unloading.

9.3.2 Tribological Characterization by Nanoscratching

If indentation is carried out in a lateral direction, then micro- and nano-scratching tests can be performed, to allow the interpretation of qualitative trends to be established. To an extent, this avoids some of the strenuous experiment or analysis accompanied by corresponding indentation tests.

The scratch behavior can be influenced by tip geometry and normal load [85]. These experiments can provide crucial understanding toward improving the abrasive resistance of surface micro- and nanostructures. By simultaneously performing a scratch and monitoring the friction coefficient, mechanistic information can be inferred with regard to failure of surface asperities [85–87]. During the nanoscratching process, the interface will experience diverse loads with both normal and lateral components. In this sense, scratch tests are perhaps more meaningful to determine

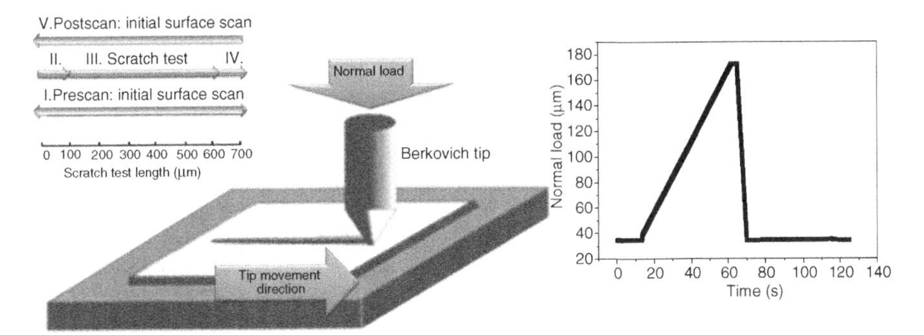

Figure 9.8 (a) Representative scheme of scratch segments and (b) scratch load protocol (with increasing load).

durability through determining resistance to shear forces and wear at multiple length scales.

The nanoscratching test consists mainly of three stages, namely prescan, scratch, and postscan (as illustrated in Fig. 9.8). A typical protocol often consists of first a prescan under a very small load (1 μN). The indenter is then used to scrape the sample under a certain force, to generate a scratch; applied loads normally used are 50 and 140 μN and length of the scratches being typically ~10 μm. A postscan under the same load as the prescan (1 μN) is then conducted to get an image of the sample after scratch. For each scratch experiment (performed at a given load), three measurements are acquired from different regions of the samples (with a spacing of 50 μm) for statistical reason.

Nanoscratch and post-nanoscratch tests can be performed in order to investigate the resistivity of a textured surface to both normal and tangential loading. By comparing nanoscratch and post-nanoscratch data, the lower coefficient of friction of the coated surface as compared with the uncoated surface can be demonstrated. Following the nanoscratch tests, SEM images can be acquired to confirm the scratch resistance of the surface.

9.4 CASE STUDY: NANOSCRATCH TESTS TO CHARACTERIZE MECHANICAL STABILITY OF PS/PMMA SURFACES

9.4.1 Method

Two nanoscratch tests were performed: a duplicate scratch (over the same path) at a steady load and another single scratch with an increasing load. Both tests comprised three-step typical nanoscratch process, that is, prescan, scratch, and postscan (see section 9.3.2).

During the steady normal load scratch experiment, the indenter follows a loading protocol presented in Figure 9.9. Initially, a normal load of 30 μN is applied for a

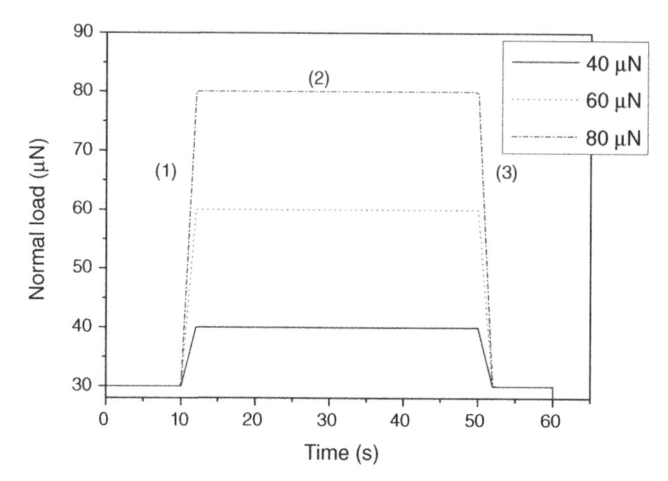

Figure 9.9 Scratch and postscratch load protocol: (1) preload, (2) constant load stage (for various maximum loads 10–100 μN), and (3) unloading.

duration of 10 s, before the load is increased up to the preferred value (for approximately 40 s), and finally decreased to 30 μN for another 5 s. The scratch protocol consists of two passes, both following the same loading protocol (Fig. 9.9), along the same motion path. Under that steady normal load protocol (duplicate scratch test), the durability of the coating and nanotextured structures in two consecutive repeated wear cycles can be evaluated. Nanoscratch tests were performed at various loading rates, with constant maximum applied loads being between 10 and 100 μN. This scratch protocol is designed to probe the scratch resistance of the coating as well as its texture [88]

The nanoscratch test part of the protocol comprised a longer motion path (40 μm with an increasing load from 0 to 100 μN, which corresponds to increased penetration. The maximum penetration depth was chosen to be less than ~850 nm, to make sure that the tip does not exceed the height of the microstructures in depth. Also, the maximum applied load should be lower than the critical yield and buckling and bending loads. Indeed, when a load is applied on a micropillar, pillar destruction, bending, or buckling may occur. For a given pillar geometry (e.g., diameter ~300 nm and height ~1000 nm, as shown in Fig. 9.4), the maximum vertical stress caused by an 80 μN load was calculated to be at approximately 1.1 GPa, well below PMMA elastic modulus Ebulk PMMA ~5.6 GPa [28]. Details about the theoretical formulas used for the calculations of bending stress and critical buckling force are not discussed here, as this has been reported elsewhere [89].

9.4.2 Results and Discussion

This section discusses scratch tests performed on the 1 μm PS/PMMA ordered pillar surface, to test the wear of the micro–nanostructures as well as the coating adhesion on them. The PS/PMMA pillar surface was chosen due to its orderly form as

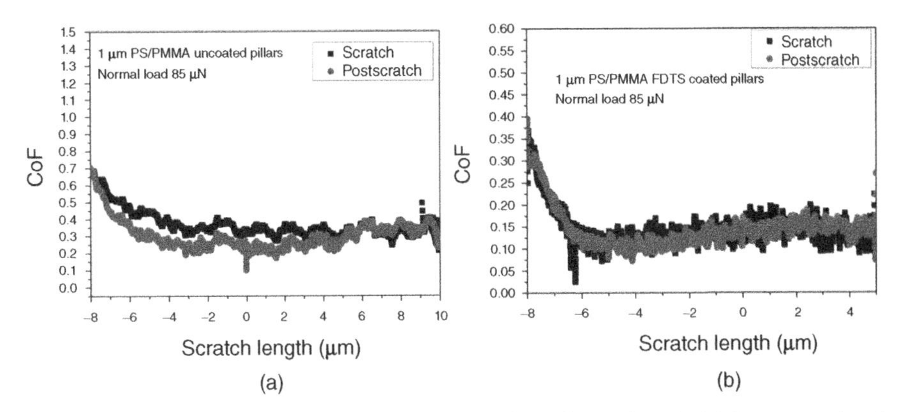

Figure 9.10 Coefficient of friction (CoF) as a function of scratch path: (a) for uncoated and (b) perfluorodecyltrichlorosilane (FDTS)-coated 1 μm PS/PMMA samples.

opposed to random nanotextured surfaces. The coefficient of friction (CoF) curves obtained from the scratch and postscratch tests on uncoated PS/PMMA samples under an 85 μN load are shown in Figure 9.9a. Results suggest that the tip starts sliding into and out of the spacing between the micropillars. A lower value for CoF is shown when it slides into the spacing, and a higher CoF is indicated when it travels out [90, 91]. Despite the size of the tip, the tip motion yields fluctuation in the CoFs that is proportional to the shape of the surface morphology (as indicated by Fig. 9.10a). Figure 9.10a also emphasizes the excellent mechanical stability of the ordered 1 μm PS/PMMA pillars during both scratch and postscratch. The pillars are not destroyed by the tip motion during the scratch experiment, and the fluctuation that follows the topography remains during the postscratch for loads up to 85 μN tested here. This behavior has also been observed by He et al., who measured the CoF of microtextured poly(dimethylsiloxane) at both micro- and macroscale [92]. This reversible behavior manifested upon scratching of textured surfaces has been attributed to the comparable size of the pillars to the interpillar spacings [90, 91]. The durability of the pillars is in agreement with the approximate calculations (referred in Section 9.4.1), which indicate that the yield strength as well as the critical buckling load of the pillars has not been exceeded.

In Figure 9.10b, the results from nanoscratch experiment for the corresponding PS/PMMA pillars coated with FDTS are presented. The CoF values for FDTS-coated PS/PMMA are found to be 50% lower than those for the uncoated PS/PMMA system. In the FDTS-coated PS/PMMA surface (Fig. 9.10b), the tip motion yields a CoF fluctuation that does not follow the shape of the pillar geometry. This implies that the FDTS coating may act as a lubricant when deposited on the surface. Recently, Zhao et al. reported that the choice of pillar diameter to height ratio should be governed by final application and a compromise between mechanical and wetting stability [93]. Although this is generally accepted, careful structure design and coating adhesion optimization can lead to surfaces, in which wetting property optimization does not compromise mechanical stability. As a matter of fact, the method developed here shows surfaces that are mechanically robust and simultaneously hydrolytically stable (for long periods). To further verify the scratch resistance of the structures, the morphology of the scratched area was studied further in which SEM images of the surface was acquired. The results have been reported previously [28], which involved producing nine 10-μm-long constant load scratches (three for every load: 40, 60, 80 μN) and two longer (~40 μm) increasing load scratches on the surface. This was performed in a premarked surface area to facilitate the observation of scratch location through SEM. SEM images presented elsewhere [28] indicates that the 80 μN scratches have only managed to slightly push the pillars aside and increase the interpillar distance as a result of the deeper penetration of the large tip, while when using smaller load (40 and 60 μN), scratches could not even be detected. It can be, therefore, concluded that these smaller loads do not affect the structures. The results indicate that minimal interpillar distance increase has minimal if no effect at all on the surface wettability (if compared to usual dislocation defects that are present on a sample surface after colloidal self-assembly). The increase in minimal interpillar distance that occurs while scratching at 85 μN, may be the cause of the slightly reduced CoF during the second scan (as shown in Fig.9.10). As the load increases, the tip penetrates deeper and interpillar distance increases, but again no collapse or destruction or buckling is observed up to 100 μN load. It is important to emphasize that the maximum load used (~100 μN) corresponds to stresses of some GPa, which are below the Young's modulus and below the critical buckling load of a pillar.

The nanomechanical tests presented here have allowed us to probe the scratch resistance of both the individual nanostructures on the polymer and their FDTS coating. The range of scratch forces was noted and shown to cause less damage (compared with the defects already existing on the surface itself). However, several limitations are noted in relation to nanoscratching tests. First, due to the limited area of the scratch (compared with the size of a water drop), effects on the contact angle cannot be measured with our test. Second, it is difficult to compare our results to macromechanical testing (such as abrasion tests) where a large much larger surface is taken into account, and the effects on the contact angle are more obvious.

9.5 CASE STUDY: STRUCTURAL INTEGRITY OF MULTIWALLED CNT FOREST

Mechanical properties of the CNT forest structures have been widely explored (mostly through nanoindentation or similar techniques), being mostly associated with static, elastic-type deformation analysis. Several findings of great interest from the literature will be mentioned and explained in the following paragraphs. Mesarovic et al. [94] and McCarter et al. [62] both reported the stress relaxation behavior of carbon nanotube forests under a spherical indenter load, along with the fact that CNTs show a time (rate)-dependent viscoelastic deformation. This specific time-dependent deformation has been connected and attributed to contact movements from neighboring nanotubes and nanotube forests. Furthermore, Pathak et al. have further shown that CNTs exhibit frequency-dependent viscoelastic deformation [95]. In addition, as reported by Misra et al., the indentation load–depth responses of the nanotube forests has been rather rate dependent, that is, the higher the indentation velocities, the stiffer the load–depth curves. This specific observation has been ascribed to local densification effects, which occur directly below the compressed area [96].

Although the mechanical behavior of carbon nanotubes has been studied extensively, experimental work on the shell buckling of nanotubes are limited, despite this being fundamentally important, that is, in governing nanotube mechanics (and thus applications). In 2005, Waters et al. [97] describes an experimental technique in which individual multiwalled carbon nanotubes were axially compressed using a nanoindenter and critical shell-buckling load reported. The results were compared with predictions of existing continuum theories, which model multiwalled carbon nanotubes as a collection of single-walled shells interacting through van der Waals forces. The finding shows that the theory gives a much lower buckling load value when compared with experimental findings [97].

Figure 9.11 shows a scanning probe microscope image of a multiwalled CNT forest, and Figure 9.12 shows the corresponding load–displacement data from nanoindentation testing.

According to the results, the loading part (shown in Fig. 9.12) sporadically includes three main stages: the first stage is described by an almost initial linear increase. Following this, a sudden drop in the slope appears and the curve becomes flat. This describes the second stage. The third and final stage comprises an increasing load. The signature of the shell buckling makes up the sudden decrease in the slope. This shell buckling further indicates a collapse process under the indenter. The first critical buckling load (according to multiples experiments) has been consistently measured to be ~4.4 μN. After buckling, neighboring nanotubes come into contact

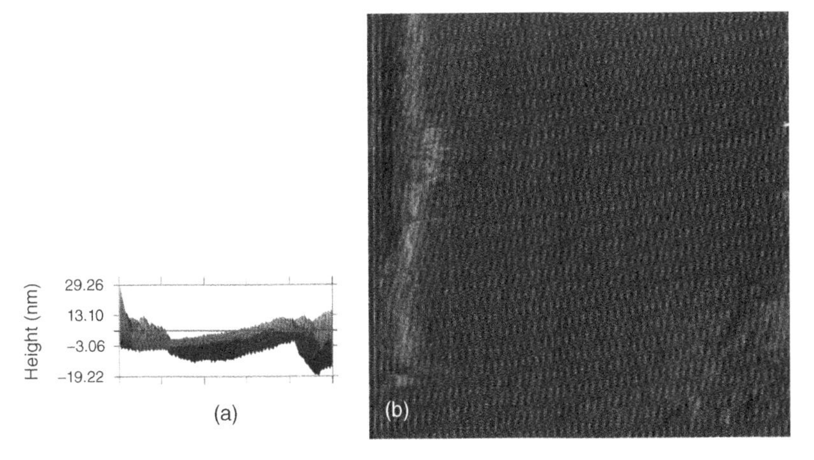

Figure 9.11 SPM images of the CNT forest (a) in 3D, and (b) from top view (5 μm × 5 μm), where the edges of CNT forest surface are observed.

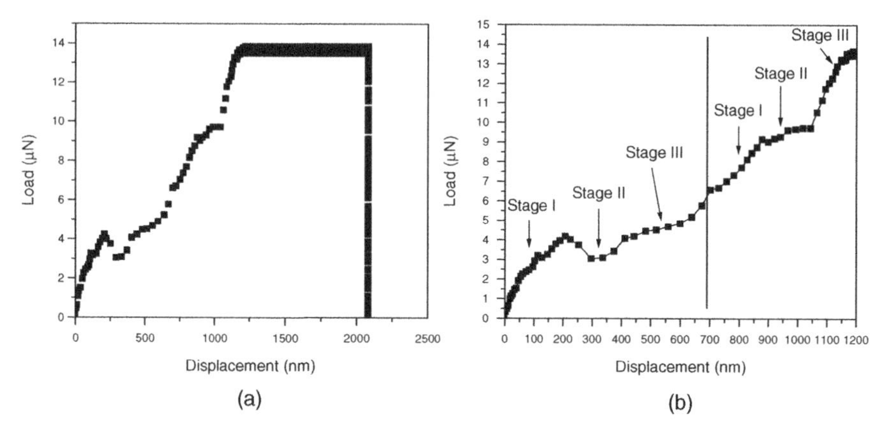

Figure 9.12 Representative load–displacement data from a loading–unloading cycle: (a) full cycle and (b) loading part).

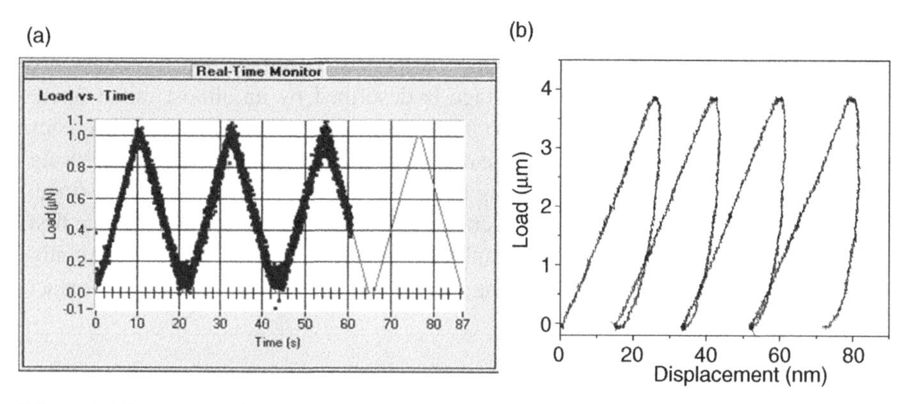

Figure 9.13 (a) Input function of CNT forest nanoindentation experiment, following repeating loading cycles and (b) representative load–unload curves.

with the indenter tip, resulting in an increase in load in the third stage [94]. The individual tubes first bend under the compression of the indenter in an elastic, time-independent deformation), which is then followed by the beginning of the indenter being in contact with neighboring tubes. In cases where the indenter load is held constant over time, the changes in contacts may still continue. Due to the contact deformation (such as sliding), energy dissipates. This has been observed by the hysteresis loops in the indenter loading–unloading cycle (shown in Fig. 9.13). It must be taken into consideration that hysteresis is a property of materials, which happens as a result of the local stress distribution (in this case the interfacial contact stresses between the inter CNTs) [97].

9.6 CASE STUDY: MECHANICAL CHARACTERIZATION OF PLASMA-TREATED POLYLACTIC ACID (PLA) FOR PACKAGING APPLICATIONS

A typical atomic force microscope (AFM) image of a plasma-treated PLA surface is shown in Figure 9.14. A nanoindentation experiment has been carried out, in which several points were carefully selected (after AFM imaging) to obtain accurate results. The corresponding (trapezoidal) load–time curve is shown in Figure 9.15.

Hardness (H) and elastic (E) modulus values versus displacement (H, E) of four different PLA samples are presented in Figures 9.16 and 9.17. Results indicate that

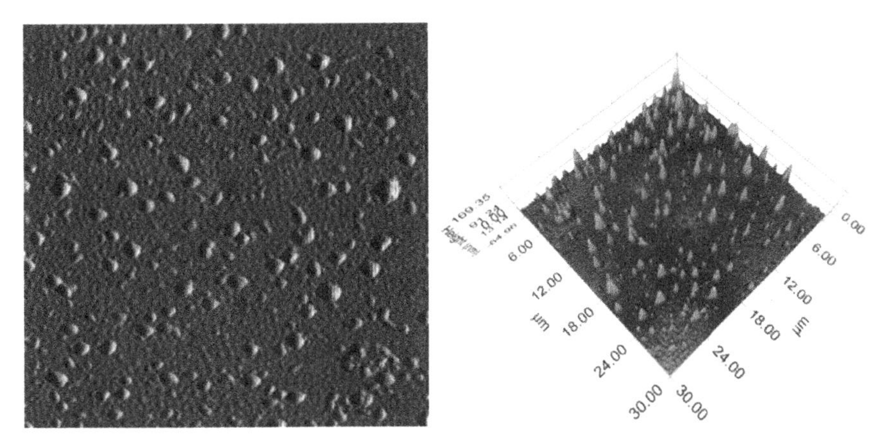

Figure 9.14 AFM imaging ($30 \times 30\,\mu m^2$) of plasma-treated PLA surface.

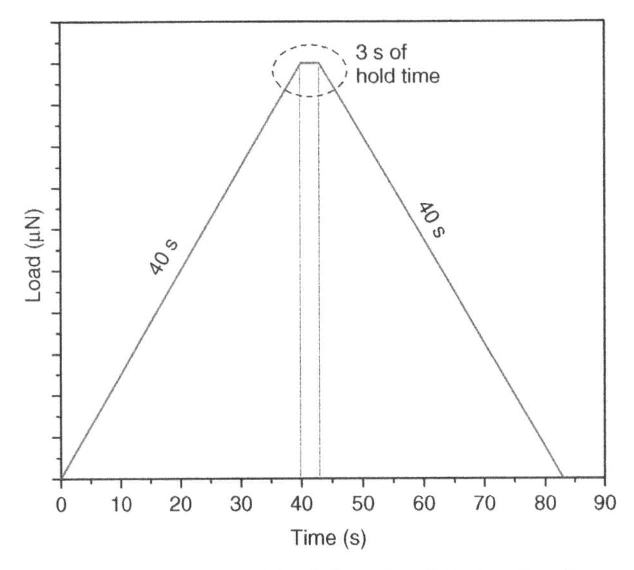

Figure 9.15 Schematic trapezoidal of load–time $P = P(t)$ function for nanoindentation experiment.

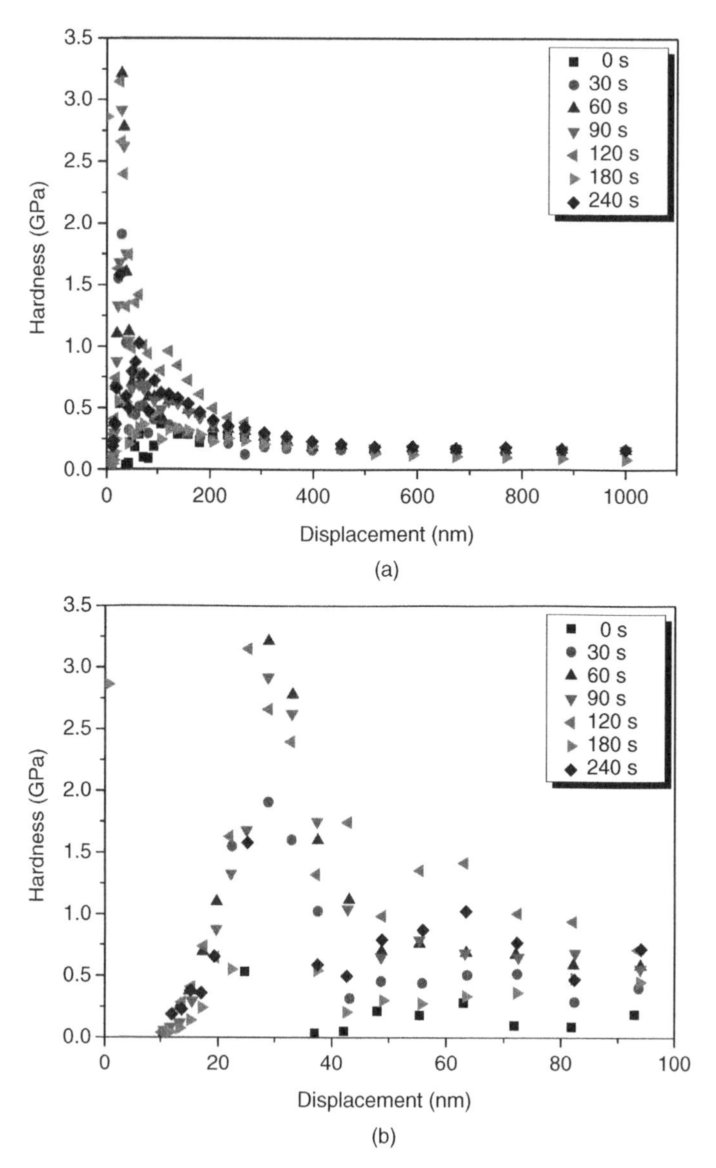

Figure 9.16 Hardness versus displacement for oxygen plasma-etched PLA, with displacement scale ranging between (a) 0 and 1000 nm (b) 0 and 100 nm.

all PLA samples exhibit an almost hard-like surface area, where enhanced H and E values were observed. As the tip penetrates further into bulk, both H and E tend to reach pristine PLA's values.

The plasma process (due to the interactions of plasma species as well as thermal effects) [98] can be used to create a topography change. As indicated by thermal

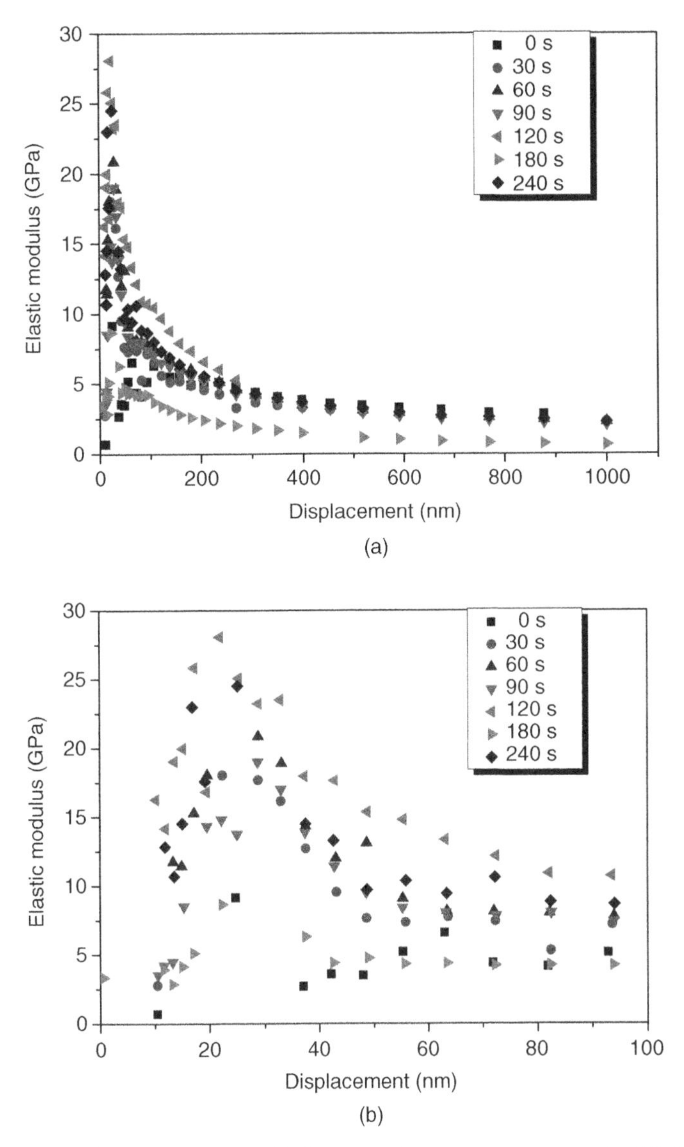

Figure 9.17 Elastic modulus versus displacement for oxygen plasma-etched PLA, with displacement scale ranging between (a) 0 and 1000 nm (b) 0 and 100 nm.

analysis and measurements of water vapor permeability measurements [98], the bulk structure of the film remains almost unchanged. The plasma can produce crosslinking, thus enhancing the performance of a surface. The activity of the plasma can create a higher crosslinking density within the material to a depth of a few thousand angstroms. This can result in an increase in hardness and chemical resistance, useful

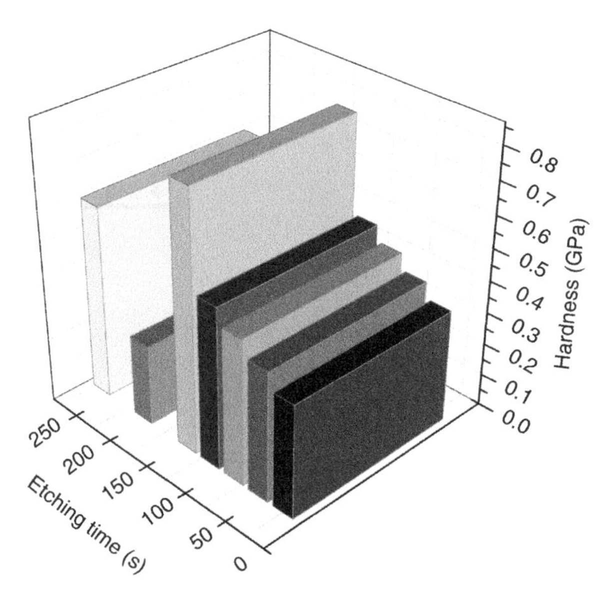

Figure 9.18 Hardness versus etching time for oxygen-plasma etched PLA, at ~100 nm of displacement.

to enhance performance of materials in many applications. For example, silicone rubber components can be modified (when treated with inert gas plasma) to form a hard "skin" on the surface. The result is a substantial decrease in surface tack and CoF. Recently, a plasma immobilization process has been developed to directly crosslink precoated molecules onto polymer surfaces. The molecules immobilized by this process can be organic compounds, surfactants, polymers, or proteins, and do not require unsaturated double bonds in the molecules [99]. Air plasma mainly adds oxygen atoms to the PLA surfaces. In the case of plasma treatment in air, researchers reported an increase in the concentration of C–O and O–C=O groups, while a decrease in C–C and C–H functional groups [42].

Figure 9.18 shows a bar graph of how the (oxygen plasma) etching time affects the hardness of PLA; here, hardness was determined using load- and displacement-sensing indentation experiments (a displacement of ~100 nm was used). The hardness values presented indicate a periodic phenomenon, possibly due to the various crystalline/amorphous states formed during the plasma treatment, in which cycles of surface etching is followed by clustering of sputtered species.

It is important to highlight at this point that the ratio of hardness/elastic (H/E) modulus is of significant interest in tribology. Higher stresses are expected in high H/E, hard materials (with high stress concentrations developed toward the indenter). In the case of low H/E, soft materials, the stresses are lower and are distributed more evenly across the cross-section of the material [99–101]. The high ratio of hardness to elastic modulus (H/E) is indicative of the good wear resistance in a disparate range of materials [101, 102], for example, ceramic, metallic, and polymeric (e.g., c-BN,

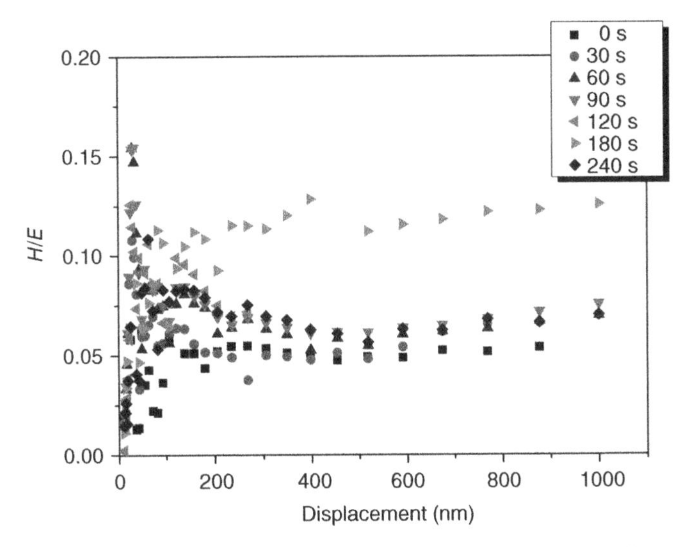

Figure 9.19 Hardness to modulus ratio for oxygen plasma-etched PLA.

tool steel and nylon, respectively), which are equally effective in resisting attrition for their particular intended application.

The goal of the majority of nanoindentation tests is to extract elastic modulus and hardness, that is, H/E ratio from load–displacement measurements. Figure 9.19, shows H/E values with respect to displacement. Results show an apparent change in the H/E slope, which indicates the strengthening of the PLA material when oxygen plasma etching time was180 s.

In summary, all PLA samples exhibited an almost hard-like surface area where enhanced H and E was observed. The activity of the plasma creates a higher crosslinking density at the surface. For higher displacements, both H and E tend to reach pristine PLA's values. Hardness values revealed a periodic phenomenon, attributed to the formation of various crystalline/amorphous states (possibly attributed to irregularly periodic events that occur during etching).

9.7 CONCLUSIONS

Nanotextured surfaces have gained significant attention due to their unique properties and their promising potential applications. However, the practical application of nanotextured surfaces and coatings remains limited by low mechanical stability, durability, and scratch resistivity. The chapter highlights some of the difficulties to characterize mechanical properties and the need to have more sensitive techniques in order to deal with issues arising in the nanoscale. Several methods that can potentially characterize mechanical properties of nanotextured surfaces have been presented here, namely nanoindentation and nanoscratch tests. The analysis of nanoindentation is complex, usually leading to scatter of calculated results, due to several factors such as the presence of surface roughness, substrate influences, and scale effects. Although making quantitative measurements remains challenging in this respect, nanoindentation test is useful to identify asperity failures (through manipulation of single asperities, which is a crucial missing element of current characterization methods).

ACKNOWLEDGMENTS

The authors thank M. Valentin, Igor Krupa, Igor Novak for providing the untreated and plasma-treated PLA sample materials and A. Tserepi, E. Goggolides, and K. Ellinas for providing the untreated and plasma-treated PS and PMMA sample materials. The research regarding mechanical stability of the untreated and plasma-treated PS and PMMA sample materials was supported by the Project "THALIS-DESIgn and fabrication of Robust superhydrophobic/hydrophilic surfaces and their application in the realization of 'smart' microfluidic valves" funded by the Hellenic and European Regional Development Funds (ERDF) under the Hellenic National Strategic Reference Framework (NSRF) 2007–2013. The work regarding mechanical integrity of CNTs carpets was supported by FP7 Collaborative project "FIBRALSPEC." The abbreviation "FIBRALSPEC" stands for "Fuctionalised Innovative Carbon Fibres Developed from Novel Precursors With Cost Efficiency and Tailored Properties" (Grant agreement no.: 604248). Financial support for author D.A. Dragatogiannis through a PhD scholarship granted by Research Committee of the National Technical University of Athens (NTUA) is also gratefully acknowledged.

REFERENCES

1. Li X-M, Reinhoudt D, Crego-Calama M. What do we need for a superhydrophobic surface? A review on the recent progress in the preparation of superhydrophobic surfaces. Chem Soc Rev 2007;36(8):1350–1368.
2. Wong T-S, Sun T, Feng L, Aizenberg J. Interfacial materials with special wettability. MRS Bull 2013;38(05):366–371.
3. Bellanger H, Darmanin T, Taffin de Givenchy E, Guittard F. Chemical and physical pathways for the preparation of super-oleophobic surfaces and related wetting theories. Chem Rev 2014;114(5):2694–2716.
4. Liu K, Tian Y, Jiang L. Bio-inspired superoleophobic and smart materials: Design, fabrication, and application. Prog Polym Sci 2012;58(4):503–564.
5. Tsougeni K, Papageorgiou D, Tserepi A, Gogolides E. "Smart" polymeric microfluidics fabricated by plasma processing: Controlled wetting, capillary filling and hydrophobic valving. Lab Chip 2010;10(4):462–469.
6. Tanzi S, Østergaard PF, Matteucci M, Christiansen TL, Cech J, Marie R, Taboryski R. Fabrication of combined-scale nano- and microfluidic polymer systems using a multilevel dry etching, electroplating and molding process. J Micromech Microeng 2012;22(11): 115008.

7. Vourdas N, Tserepi A, Boudouvis AG, Gogolides E. Plasma processing for polymeric microfluidics fabrication and surface modification: effect of super-hydrophobic walls on electroosmotic flow. Microelectron Eng 2008;85(5–6):1124–1127.
8. Öner D, McCarthy T. Ultrahydrophobic surfaces. Effects of topography length scales on wettability. Langmuir 2000;16:7777–7782.
9. Qi D, Lu N, Xu H, Yang B, Huang C, Xu M, Gao L, Wang Z, Chi L. Simple approach to wafer-scale self-cleaning antireflective silicon surfaces. Langmuir 2009;25:7769–7772.
10. Hosono E, Fujihara S, Honma I, Zhou H. Superhydrophobic perpendicular nanopin film by the bottom-up process. J Am Chem Soc 2005;127:13458–13459.
11. Xu B, Cai Z. Fabrication of a superhydrophobic ZnO nanorod array film on cotton fabrics via a wet chemical route and hydrophobic modification. Appl Surf Sci 2008;254:5899–5904.
12. Li Y, Liu F, Sun J. A facile layer-by-layer deposition process for the fabrication of highly transparent superhydrophobic coatings. Chem Commun 2009;19:2730–2732.
13. Zimmermann J, Reifler FA, Fortunato G, Gerhardt L-C, Seeger S. A simple, one-step approach to durable and robust superhydrophobic textiles. Adv Funct Mater 2008;18:3662–3669.
14. Sas I, Gorga RE, Joines JA, Thoney KA. Literature review on superhydrophobic self-cleaning surfaces produced by electrospinning. J Polym Sci B 2012;50:824–845.
15. Kartsonakis IA, Koumoulos EP, Balaskas AC, Pappas GS, Charitidis CA, Kordas GC. Hybrid organic–inorganic multilayer coatings including nanocontainers for corrosion protection of metal alloys. Corros Sci 2012;57:56–66.
16. Palumbo F, Di Mundo R, Cappelluti D, d'Agostino R. Superhydrophobic and superhydrophilic polycarbonate by tailoring chemistry and nano-texture with plasma processing. Plasma Processes Polym 2011;8:118–126.
17. Tserepi AD, Vlachopoulou ME, Gogolides E. Nanotexturing of poly (dimethylsiloxane) in plasmas for creating robust super-hydrophobic surfaces. Nanotechnology 2006;17(15):3977.
18. Sainiemi L, Jokinen V, Shah A, Shpak M, Aura S, Suvanto P, Franssila S. Non-reflecting silicon and polymer surfaces by plasma etching and replication. Adv Mater 2011;23:122–126.
19. Pazokian H, Selimis A, Barzin J, Jelvani S, Mollabashi M, Fotakis C, Stratakis E. Tailoring the wetting properties of polymers from highly hydrophilic to superhydrophobic using UV laser pulses. J Micromech Microeng 2012;22:035001.
20. Gogolides E, Vlachopoulou M, Tsougeni K, Vourdas N, Tserepi A. Micro and nano structuring and texturing of polymers using plasma processes: Potential manufacturing applications. Int J Nanomanuf 2010;6:152–163.
21. Ellinas K, Tserepi A, Gogolides E. Superhydrophobic, passive microvalves with controllable opening threshold: Exploiting plasma nanotextured microfluidics for a programmable flow switchboard. Microfluid Nanofluid 2014;17:1–10.
22. Bhushan B. Adhesion and stiction: Mechanisms, measurement techniques, and methods for reduction. J Vac Sci Technol B 2003;21:2262–2296.
23. Park JH, Lee SH, Choi KH, Noh HS, Lee JW, Pearton SJ. Comparison of dry etching of PMMA and polycarbonate in diffusion pump-based O_2 capacitively coupled plasma and inductively coupled plasma. Thin Solid Films 2010;518:6465–6468.

24. Palacio M, Bhushan B, Ferrell N, Hansford D. Nanomechanical characterization of polymer beam structures for BioMEMS applications. Sens Actuators A 2007;135:637–650.
25. Cvelbar U, Krstulovic N, Milosevic S, Mozetic M. Inductively coupled RF oxygen plasma characterization by optical emission spectroscopy. Vacuum 2007;82:224–227.
26. Vourdas N, Tserepi A, Gogolides E. Nanotextured super-hydrophobic transparent poly (methyl methacrylate) surfaces using high-density plasma processing. Nanotechnology 2007;18:125304.
27. Tsougeni K, Vourdas N, Tserepi A, Gogolides E. Mechanisms of oxygen plasma nanotexturing of organic polymer surfaces: From stable super hydrophilic to super hydrophobic surfaces. Langmuir 2009;25:11748–11759.
28. Ellinas K, Pujari SP, Dragatogiannis DA, Charitidis CA, Tserepi A, Zuilhof H, Gogolides E. Plasma micro-nano-textured, scratch, water and hexadecane resistant, superhydrophobic and superamphiphobic polymeric surfaces with perfluorinated monolayers. Appl Mater Interfaces 2014;6:6510–6524.
29. Vesel A, Mozetic M. Surface modification and ageing of PMMA polymer by oxygen plasma treatment. Vacuum 2012;86:634–637.
30. Kumar V, Sharma N. A study on hydrophobicity of silicon and a few dielectric materials. In: Vinoy KJ, Ananthasuresh GK, Pratap R, Krupanidhi SB, editors. *Micro and Smart Devices and Systems*. New Delhi, India: Springer; 2014. p 265–280.
31. Olde Riekerink MB. *Structural and Chemical Modification of Polymer Surfaces by Gas Plasma Etching*. Universiteit Twente; 2001.
32. Flamm DL. Introduction to plasma chemistry. In: Manos DM, Flamm DL, editors. *Plasma Etching Technology – An Overview*. Boston: Academic Press, Inc.; 1989. p 91–183.
33. Jansen HV. *Plasma Etching in Microtechnology*. Enschede, The Netherlands: University of Twente; 1996.
34. Haring RA, Nunes SL, McGouey RP, Galligan EA, Volksen W, Hedrick JL, Labadie JJ. Adhesion properties of a structural etch stop material for use in multilayer electronic wiring structures. J J Mater Res 1995;10:1028–1037.
35. Baggerman JAG, Visser RJ, Collart EJH. Ion-induced etching of organic polymers in argon and oxygen radio-frequency plasmas. J Appl Phys 1994;75:758–769.
36. Samukawa S, Mieno T. Pulse-time modulated plasma discharge for highly selective, highly anisotropic and charge-free etching. Plasma Sources Sci Technol 1996;5:132–138.
37. Taylor GN, Wolf TM, Moran JM. Organosilicon monomers for plasma-developed x-ray resists. J Vac Sci Technol A 1981;19:872–880.
38. Richter K, Orfert M, Drescher K. Anisotropic patterning of copper-laminated polyimide foils by plasma etching. Surf Coat Technol 1997;97:481–487.
39. Borcea V. *Radioactive Ion Implantation of Thermoplastic Elastomers*. Université catholique de Louvain; 2008.
40. Morent R, De Geyter N, Gengembre L, Leys C, Payen E, Van Vlierberghe S, Schacht E. Surface treatment of a polypropylene film with a nitrogen DBD at medium pressure. Euro Phys J 2008a;43:289–294.
41. Morent R, De Geyter N, Leys C. Effects of operating parameters on plasmainduced PET surface treatment. Nucl Instrum Methods Phys Res B 2008b;266:3081–3085.
42. De Geyter N, Morent R. Non-thermal plasma surface modification of biodegradable polymers. In: Ghista DN, editor. *Biomedical Science, Engineering and Technology*. INTECH Open Access Publisher; 2012., ISBN: 978-953-307-471-9.

43. Terlingen JGA. *Introduction of Functional Groups at Polymer Surfaces by Glow Discharge Techniques*. Enschede, The Netherlands: University of Twente; 1993.
44. Gombotz WR, Hoffman AS. Gas-discharge techniques for biomaterials modification. CRC Crit Rev Biocompat 1987;4:1–42.
45. Biederman H, Osada Y. Plasma chemistry of polymers. Adv Polym Sci 1990;95:59–109.
46. Suhr H. Application of nonequilibrium plasmas in organic chemistry. Plasma Chem Plasma Process 1983;3(1):1–61.
47. Yasuda HJ. Plasma for modification of polymers. J Macromol Sci 1976;A10:383–420.
48. Bell AT. Fundamentals of plasma chemistry. In: Hollahan J, Bell AT, editors. *Techniques & Applications of Plasma Chemistry*. New York: Wiley; 1974. p 1–56.
49. Liston EM, Martinu L, Wertheimer MR. Plasma surface modification of polymers for improved adhesion: A critical review. In: Strobel M, Lyons C, Mittal KL, editors. *Plasma Surface Modification of Polymers: Relevance to Adhesion*. Utrecht: VSP; 1994. p 3–39.
50. Yasuda H. *Plasma Polymerization*. Orlando: Academic Press Inc.; 1985.
51. Chapman B. *Glow Discharge Processes – Sputtering and Plasma Etching*. New York: John Wiley & Sons; 1980.
52. Tsougeni K, Tserepi A, Boulousis G, Constantoudis V, Gogolides E. Tunable PDMS topography in O2 or Ar plasmas for controlling surface wetting properties and their ageing Japan. J Appl Phys 2007;46:744–750.
53. Ellinas K, Tserepi A, Gogolides E. From superamphiphobic to amphiphilic polymeric surfaces with ordered hierarchical roughness fabricated with colloidal lithography and plasma nanotexturing. Langmuir 2011;27(7):3960–3969.
54. Callies M, Quéré D. On water repellency. Soft Mater 2005;1(1):55–61.
55. Chandra D, Yang S. Stability of high-aspect-ratio micropillar arrays against adhesive and capillary forces. Acc Chem Res 2010;43:1080–1091.
56. Su CH, Xu YQ, Gong F, Wang FS, Li CF. The abrasion resistance of a superhydrophobic surface comprised of polyurethane elastomer. Soft Mater 2010;6:6068–6071.
57. Zimmermann J, Reifler FA, Schrade U, Artus GRJ, Seeger S. Long term environmental durability of a superhydrophobic silicone nanofilament coating. Colloids Surf A 2007;302:234–240.
58. Xiu Y, Hess DW, Wong CP. UV and thermally stable superhydrophobic coatings from sol–gel processing. J Colloid Interface Sci 2008;326:465–470.
59. Fan S, Chapline MG, Franklin NR, Tombler TW, Cassell AM, Dai H. Self-oriented regular arrays of carbon nanotubes and their field emission properties. Science 1999;283:512.
60. Kreupl F, Graham AP, Duesberg GS, Steinhögl W, Liebau M, Unger E, Hönlein W. Carbon nanotubes in interconnect applications. Microelectron Eng 2002;64:399–408.
61. Xu J, Fisher TS. Enhancement of thermal interface materials with carbon nanotube arrays. Int J Heat Mass Transf 2006;49:1658–1666.
62. McCarter CM, Richards RF, Mesarovic S, Dj RCD, Bahr DF, McClain D, Jiao J. Mechanical compliance of photolithographically defined vertically aligned carbon nanotube turf. J Mater Sci 2006;41:7872–7878.
63. Bajpai V, Dai L, Ohashi T. Large-scale synthesis of perpendicularly aligned helical carbon nanotubes. J Am Chem Soc 2004;126:5070–5071.
64. Zhang Q, Liu J, Sager R, Dai L, Baur J. Hierarchical composites of carbon nanotubes on carbon fiber: influence of growth condition on fiber tensile properties Compos. Sci Technol 2009;69:594.

65. Patton ST, Zhang Q, Qu L, Dai L, Voevodin AA, Baur J. Electromechanical characterization of carbon nanotubes grown on carbon fiber. J Appl Phys 2009;106(10):104313.
66. (a) Porro S, Musso S, Vinante M, Vanzetti L, Anderle M, Trotta F, Tagliaferro A. Purification of carbon nanotubes grown by thermal. CVD Phys 2007;E37:58–61; (b) Kumar M, Ando Y. Chem Phys Lett 2003;374:521; (c) Musso S, Fanchini G, Tagliaferro A. Diamond Relat Mater 2005;14:784; (d) Porro S, Musso S, Giorcelli M, Tagliaferro A, Dalal SH, Teo KBK, Jefferson DA, Milne WI. J Non Cryst Solids 2006;352:1310.
67. Pavese M, Musso S, Bianco S, Giorcelli M, Pugno N. An analysis of carbon nanotube structure wettability before and after oxidation treatment. J Phys Condens Matter 2008;20:474206–474213.
68. Satyanarayana N, Lau KH, Sinha SK. Nanolubrication of poly (methyl methacrylate) films on Si for microelectromechanical systems applications. Appl Phys Lett 2008;93:261906.
69. Dyett BP, Wu AH, Lamb RN. Mechanical stability of surface architecture-consequences for superhydrophobicity. ACS Appl Mater Interfaces 2014;6(21):18380–18394.
70. Bhushan B. Mechanical properties of nanostructures. In: Bhushan B, editor. *Nanotribology and Nanomechanics II*. Berlin Heidelberg: Springer; 2011. p 527–584.
71. Bhushan B. *Modern Tribology Handbook*. Boca Raton, FL: CRC Press; 2001.
72. Fischer-Cripps AC. *Nanoindentation*. Vol. 1. New York: Springer; 2011.
73. Bhushan B. Scale effect in mechanical properties and tribology. In: Bhushan B, editor. *Springer Handbook of Nanotechnology*. Berlin Heidelberg: Springer; 2010. p 1023–1054.
74. Nix WD, Gao H. Indentation size effects in crystalline materials: A law for strain gradient plasticity. J Mechan Phys Solids 1998;46:411–425.
75. Fleck NA, Muller GM, Ashby MF, Hutchinson JW. Strain gradient plasticity: Theory and experiment. Acta Metall Mater 1994;42:475–487.
76. Aston DE, Bow JR, Gangadean DN. Mechanical properties of selected nanostructured materials and complex bionano, hybrid and hierarchical systems. Int Mater Rev 2013;58:167–202.
77. Zhanga T-Y, Xu W-H. Surface effects on nanoindentation. J Mater Res 2002;17:1716.
78. Doerner MF, Nix WDA. Method for interpreting the data from depth-sensing indentation instruments. J Mater Res 1986;1.
79. Oliver WC, Pharr GM. Improved technique for determining hardness and elastic modulus using load and displacement sensing indentation experiments. J Mater Res 1992;7:1564–1583.
80. Joslin DL, Oliver WC. A new method for analyzing data from continuous depth-sensing microindentation tests. J Mater Res 1990;5:123–126.
81. Bhushan B, Li X. Nanomechanical characterisation of solid surfaces and thin films. Int Mater Rev 2003;48:125–164.
82. Palacio MLB, Bhushan B. Depth-sensing indentation of nanomaterials and nanostructures. Mater Charact 2013;78:1–20.
83. Charitidis CA. Nanomechanical and nanotribological properties of carbon-based thin films: A review. Int J Refract Met Hard Mater 2010;28:51–70.
84. Sneddon IN. Boussinesq's problem for a flat-ended cylinder. Proc Cam Philos Soc 1946;42:29–39.
85. Bhushan B. Nanotribology, nanomechanics and nanomaterials characterization. Philos Trans R Soc A 2008;366:1351–1381.

86. Bhushan B. Surface forces and nanorheology of molecularly thin films. In: Bhushan B, editor. *Springer Handbook of Nanotechnology*. Berlin Heidelberg: Springer; 2010. p 857–911.
87. Kim H-J, Kim D-E. Nano-scale friction: A review. Int J Prec Eng Manuf 2009;10:141–151.
88. Skarmoutsou A, Charitidis CA, Gnanappa AK, Tserepi A, Gogolides E. Nanomechanical and nanotribological properties of plasma nanotextured superhydrophilic and superhydrophobic polymeric surfaces. Nanotechnology 2012;23(50):505711.
89. Zeniou A, Ellinas K, Olziersky A, Gogolides E. Ultra-high aspect-ratio Si nanowires fabricated with plasma etching: Plasma processing, mechanical stability analysis against adhesion and capillary. Nanotechnology 2014;25:035302.
90. Zhang J, Li J, Han Y. Superhydrophobic Ptfe surfaces by extension. Macromol Rapid Commun 2004;25(11):1105–1108.
91. van de Grampel RD, Ming W, Gildenpfennig A, Laven J, Brongersma HH, de With G, van der Linde R. Quantification of fluorine density in the outermost atomic layer. Langmuir 2004;20(1):145–149.
92. Brown L, Koerner T, Horton JH, Oleschuk RD. Fabrication and characterization of poly(methylmethacrylate) micro-fluidic devices bonded using surface modifications and solvents. Lab Chip 2006;6(1):66–73.
93. Zhao H, Park K-C, Law K-Y. Effect of surface texturing on superoleophobicity, contact angle hysteresis, and “robustness”. Langmuir 2012;28(42):14925–14934.
94. Mesarovic S, Dj MCCM, Bahr DF, Radhakrishnan H, Richards RF, Richards CD, McClain D, Jiao J. Mechanical behavior of a carbon nanotube turf. Scr Mater 2007;56:157–160.
95. Pathak S, Goknur Cambaz Z, Kalidindi SR, Gregory Swadener J, Gogotsi Y. Viscoelasticity and high buckling stress of dense carbon nanotube brushes. Carbon 2009;47:1969–1976.
96. Misra A, Greer JR, Daraio C. Strain rate effects in the mechanical response of polymer-anchored carbon nanotube foams. Adv Mater 2008;21:334–338.
97. Waters JF, Guduru PR, Jouzi M, Xu JM, Hanlon T, Suresh S. Shell buckling of individual multiwalled carbon nanotubes using nanoindentation. Appl Phys Lett 2005;87:103109.
98. Chaiwong C, Rachtanapun P, Wongchaiya P, Auras P, Boonyawan P. Surf Coat Technol 2010;204:2933–2939.
99. Sheu M-S, Hoffman AS, Feijen J. A glowdischarge treatment to immobilize poly(ethylene oxide)/poly(propylene oxide) surfactants for wettable and non-fouling biomaterials. J Adhes Sci Technol 1992;6:995.
100. Cheng YT, Cheng CM. What is indentation hardness? Surf Coat Technol 2000;133–134:417–424.
101. Leyland A, Matthews A. Surf Coat Technol 2004;177–178:317–324.
102. Koumoulos EP, Charitidis CA, Papageorgiou DP, Papathanasiou AG, Boudouvis AG. Nanomechanical and nanotribological properties of hydrophobic fluorocarbon dielectric coating on tetraethoxysilane for electrowetting applications. Surf Coat Technol 2012;206(19–20):3823–3831.

10

METHODS FOR TESTING DUSTINESS

K. A. JENSEN

Danish Centre for Nanosafety, National Research Centre for the Working Environment, Copenhagen, Denmark

M. LEVIN

Danish Centre for Nanosafety, National Research Centre for the Working Environment, Copenhagen, Denmark; Department of Micro- and Nanotechnology, Technological University of Denmark, Lyngby, Denmark

O. WITSCHGER

Laboratoire de Métrologie des Aérosols Département Métrologie des Polluants, Institute National de Recherche et de Sécurité, Nancy, France

10.1 INTRODUCTION

Dustiness is a term addressing the ability of a material (e.g., loose, granulated, or pelletized powder) to generate dust during agitation [1]. It is important to note that the level of dustiness and dust characteristics (size-distribution and level of agglomeration) is not an intrinsic physical or chemical defined property of a material. In fact, the property of dustiness is not yet fully understood, and it is, therefore, currently not possible to predict the dustiness of a material from physicochemical characteristics. It is known that the level of dustiness and dust size-distributions can vary with factors such as physical and chemical properties of the tested powder, the energy applied to the material in a given test method, and environmental conditions during storage and testing [2–8].

Traditionally, the level of dustiness is given in milligram dust per kilogram powder of the health-relevant inhalable, thoracic, and respirable dust size fractions (Fig. 10.1).

Nanomaterial Characterization: An Introduction, First Edition. Edited by Ratna Tantra.

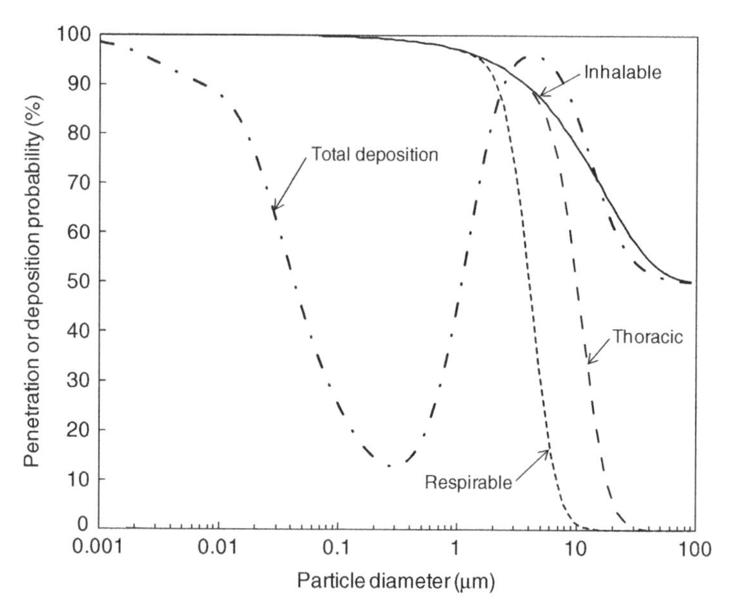

Figure 10.1 Standardized particle size-selective criteria (inhalable, thoracic, and respirable) for health-related aerosol sampling according to the EN 481 (1994) and ISO 7708 (1995) standards, and the total aerosol deposition curve in the human respiratory tract according to the ICRP model [9]. Calculations assume a spherical particle of density $\rho = 1\ g/cm^3$ and a standard worker according to the ICRP [9].

Inhalable dust is particles that can be inhaled through nose or mouth. Thoracic and respirable dust can reach the bronchial and the alveolar region (of the lung), respectively. Dustiness of environmental relevance may be given in, for example, PM_1, $PM_{2.5}$, and PM_{10}, which are size fractions typically monitored for particulate matter (PM) in ambient air pollution (see, e.g., review by Gill et al. [10]). The numbers in the subscript of PM_1, $PM_{2.5}$, and PM_{10} are the aerodynamic diameters of particles that can pass through a size-selective inlet with a 50% efficiency cutoff at 1, 2.5, and 10 µm, respectively. In other words, PM_{10} refers to particulate matter with aerodynamic diameters smaller than 10 µm. For the sake of comparison, the 50% cut-point for respirable dust is ca. 4.0 µm; respirable dust could also be referred to as $PM_{4.0}$. In regard to this and the cut-points of the mentioned dust sampling standards, it is interesting to note that both PM_1 and $PM_{2.5}$ underestimate the particles than can reach and deposit in the alveolar region of the airways (as indicated in Fig. 10.1).

As dustiness levels and characteristics are expected to vary considerably with the test method, dustiness testing must be performed using well-defined methods and procedures to enable reliable comparisons or ranking of materials. Dustiness indexes or material groups (with associated default dustiness values) are already in use for regulatory exposure estimation for conventional materials [11]. The use of dustiness data for exposure assessment has also been recommended for nanomaterials by the Organisation for Economic Co-operation and Development (OECD) [12] and are already in use as an input parameter in some control banding tools for nanomaterials (e.g., [13–16]). Control banding tools are precautionary methods for risk management, which can be applied when information on material hazard and/or exposure is limited (ISO TS/12901-2, 2014).

Despite the obvious needs for standardized methods and the widespread potential use of dustiness data for risk assessment, there is currently only one document standard on dustiness testing in the world, namely the European EN 15051-1 (2013). This standard describes two fundamentally different test methods: The rotating drum (RD) method (EN 15051-2, 2013) and the continuous drop (CD) method (EN 15051-3, 2013). Both methods are intended to mimic dust release during powder work processes. In brief, the RD method performs repeated pouring or agitation of the same material, whereas the CD method simulates continuous feed of "undisturbed" powders. It is up to the user or assessor to decide which of the two methods are most suitable to indicate the release potential for the work process in question.

For ease of communication, the EN 15051 standard specify that the dustiness levels can be classified into four dustiness categories ("very low," "low," "moderate," or "high"). These (or comparable) dustiness categories have been established to group materials with the same characteristics or emission potentials. Such categories can allow some idea of the dustiness levels of new powders with unknown dustiness indexes. This has been used in the ECETOC TRA and EMKG-Expo tools (as recommended by ECHA [11]) for exposure assessment. However, recent compilation of dustiness data shows that categorizing of dustiness based on material type or general appearance is not a reliable approach, as fine powders if categorized as one group do not exhibit similar dustiness levels. Such findings have been highlighted in several previous publications [3, 4, 17–19].

There is no doubt that material-specific dustiness data may be of indispensable help, for example, to accelerate the ability to perform exposure assessments for nanomaterials and keep pace with the technological evolution. However, the sole measurement of mass-based dustiness indexes (as used for conventional powders) may not be applicable when it comes to nanomaterials. The reason behind this stems from the fact that the primary particle size of a nanomaterial is inherently smaller than microparticulate powders. Therefore, the dust generated by handling of a nanomaterial is likely to consist of finer particles with higher surface areas than observed for microparticle powder. This effect was clearly demonstrated by Schneider and Jensen [17] who showed that the dust size-distributions of nanosize TiO_2 (primary particles ca. 20 nm) was much finer (with a peak size-mode around 200 nm) as compared with a pigment-size TiO_2 (primary particle size ca. 220 nm and with no detected dust particles smaller than 500 nm). At the same time, the inhalable dustiness index of the nanosize TiO_2 was found to be about 300 times higher than for the pigment-size TiO_2. Overall, nano-TiO_2 had much higher dustiness level and generated much finer dust (with a much higher surface area than the pigment-size TiO_2). Such an observation has important implications for risk assessment, because the deposition efficiency in the airways varies with particle size (as shown in Fig. 10.1). Several previous studies have highlighted that the inhalation hazard of dust appears to be, at least in part, related to fine particle sizes and high specific surface area doses rather than the total mass-based exposure dose [20–25]. This highlights the importance of considering particle size-distributions and specific surface area for risk assessment as well. This in turn has prompted new requirements for exposure measurements and subsequently for dustiness data to also include number-concentration and number-size distribution data. In relation to airborne particle surface area, its measurement is still associated with high uncertainty [26–30] and, therefore, not included for general purpose at this point in time.

In response to the special requirements associated with testing and measuring dustiness of nanomaterials, a number of recent studies have already aimed at either modification of the two EN 15051 methods or to develop new approaches. In addition to adding real-time aerosol measurements for particle numbers and size-distributions, the motivation of these works have been to:

- enable testing of much smaller amounts (from few tens of milligrams to grams) of nanomaterials that are either potentially very toxic and/or costly [4, 17, 31];
- develop smaller setups that can be used in ventilated enclosures or fume hoods [17, 32];
- establish setups that are easy, simple, and compact to be used by a larger number of laboratories and industries [33];
- establish test methods that are more energetic than the EN 15051 RD and CD methods to potentially describe a worst-case scenario in a workplace [4, 34].

Availability of time-resolved number and size-distribution measurements also enables new procedures for data analysis and new data for reporting [5, 17, 35, 36]. The inclusion of such new data in data reporting, in addition to mass-based dustiness indexes, can facilitate use of dustiness data in more specific emission or exposure assessments. This is demonstrated by, for example, Tsai et al. [37], who studied the comparability between EN 15051 RD dustiness characteristics and workplace measurements when working with nano-SiO_2, carbon black, and nano-$CaCO_3$, and Levin et al. [19], who demonstrated the use of dustiness levels, dustiness kinetics, and filter penetration data to select the best candidate among four pharmaceutical active ingredients by assessing the potential worker exposure and product cross-contamination in a factory.

One issue associated with dustiness testing that needs resolving is associated with the different methods and protocols for measurements. It is likely that the lack of a harmonized approach concerning the measurement strategies, metrics, and size ranges and the procedures of data analysis (and reporting) severely limits the ability to make reliable comparisons between the different dustiness methods. It was already shown from research in the European Union Seventh Framework Programme (FP 7) NANODEVICE (http://www.nano-device.eu/index.php?id=249) and the NANOGENOTOX projects (http://www.nanogenotox.eu/) that testing and establishment of potential interrelationships between some of the above-mentioned dustiness methods (as they exist now) is not trivial [32]. In response to the need for harmonized methods, a CEN project was established in order to deal with this issue

[38]. The project was initiated in response to the European Commission mandate M/461 to CEN (Comité Européen de Normalisation, i.e., European Committee for Standardisation) requesting development of standardization activities regarding nanotechnologies and nanomaterials. The aim of the currently ongoing CEN project is thus to develop a harmonized approach for evaluating dustiness of bulk nanomaterial powders when considering four fundamentally different concepts and test methods.

In this chapter, we present the four different methods, which are considered in the above-mentioned CEN project. We discuss some of the findings from using these methods, as illustrated by examples of previously generated data. We will perform a first small comparison between the test methods using these available data and also demonstrate the potential use of dustiness data for the assessment of exposure and dust transfer in a pharmaceutical company. It should be noted that dustiness data and their application have a strong practical industrial, occupational health and safety, and regulatory orientation. Consequently, many of the dustiness data and their applications may never reach the public domain and scientific literature. Lastly, the examples given in this chapter are likely more indicative for potential applications of dustiness data rather than the current level of use by practitioners.

10.2 CEN TEST METHODS (UNDER CONSIDERATION)

Currently, four different nanomaterial dustiness test methods are under consideration for standardization in CEN. The methods are modifications of the small rotating drum (RD), continuous drop (CD) methods (as detailed in EN 15051 standard), the small rotating drum (SRD), and the vortex shaker (VS) methods.

10.2.1 The EN 15051 Rotating Drum (RD) Method

The rotating drum method was first described by the British Occupational Hygiene Society (BOHS) in 1985 [39]. A principle sketch of the EN 15051-2 RD setup modified for measurements of dustiness of nanomaterials is shown in Figure 10.2. The standard test apparatus (as shown) consists of a stainless steel drum (inner diameter = 300 mm) with conical ends equipped with eight longitudinal lifter vanes (230 × 25 mm). The (earthed) drum rotates at 4 rpm during which the lifter vanes repeatedly *lift and let fall* the test material. To determine the mass fractions, the emitted dust cloud is drawn by a vacuum pump at a flow rate of 38 l/min to a sampling system.

As described by EN 15051, the sampling system consists of two particle size-selective polyurethane foam (PUF) stages in series followed by a backup filter. This allows the determination of emitted inhalable, thoracic, and respirable dust mass (as well as the respective dustiness indexes). However, the need for further characterization of the emitted dust has led to two different sampling outlets, where one of them allows the standard gravimetric test and the other attachment of real-time particle sizing instruments such as Condensation Particle Counter (CPC), Scanning

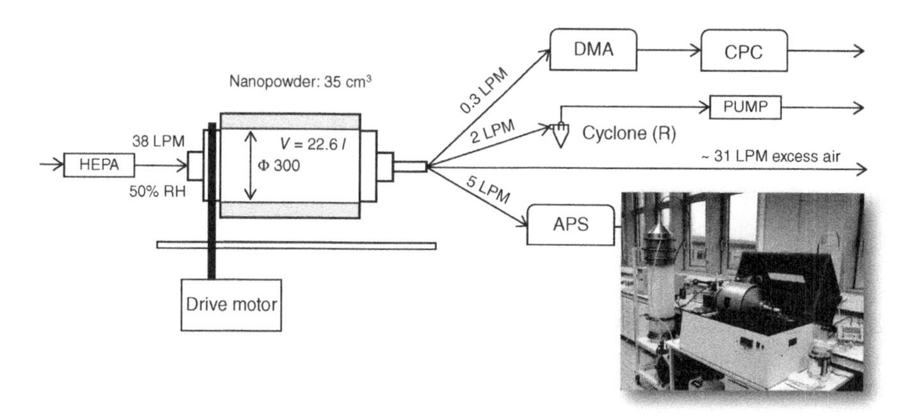

Figure 10.2 Principle sketch and photograph of the EN 15051 standard rotating drum with a modified sampling train. In the illustration given here, real-time size-distribution measurements were acquired using a Differential Mobility Particle Sizer (DMPS) and an Aerodynamic Particle Sizer (APS). Previously, an Electrical Low-Pressure Impactor (ELPI) has also been used to acquire real-time size-distribution measurements in a similar setup. The inset photograph shows a commercial version of the EN 15051 rotating drum.

Mobility Particle Sizer (SMPS), Aerodynamic Particle Sizer (APS), Fast Mobility Particle Sizer (FMPS), and Electrical Low Pressure Impactor (ELPI) [3, 40].

A dustiness test, following the EN 15051 standard method, is conducted by placing 35 cm^3 of test material (with predetermined moisture content and bulk density) into the drum on the "upside" of a lifter vane placed in bottom position. An air flow of preconditioned air (50 ± 10% RH, 21 ± 3 °C) is then turned on and when the conditions are stabilized, the drum is rotated for a duration of 1 min and 5 s at 4 rpm. After 5 s rotation, the 1-min sampling is started during which the lifter vanes *lift and drop* the powder 32 times during rotation. After completion, the foams and filter are carefully removed, acclimatized, and then weighed to determine the emitted dust mass and material dustiness index. For measurements of the size-distributions of nanomaterial dusts, the volume of powder tested often needs to be reduced as compared with the amount used for gravimetric test to avoid instrumental errors.

Table 10.1 presents some examples of published data on dustiness using the rotating drum (the focus here is on testing of nanomaterial powders). It was noted that several studies have included the use of modified sampling trains, to allow for particle counting and sizing, and in some cases the foams for dustiness indexes are forgone to focus solely on other types of characterization [7, 37]. It is apparent from the table that materials of the same chemical composition can vary considerably in dustiness depending on other parameters, such as the chemical surface modification [3]. From the table, several important findings can be highlighted. Tsai et al. [40] reported that the particle number size-distribution measurement of TiO_2 and ZnO during the dustiness testing resulted in a bimodal particle size-distributions (as measured by SMPS and APS). Furthermore, TiO_2 had a dominating peak at 300 nm and a secondary peak

TABLE 10.1 Examples of Dustiness Data and Specific Surface Areas of Nine Nanomaterials with Indication on the Type of Real-Time Measurements that were Conducted for the Dust Characterization

Material	Specific Surface Area (m^2/g)	Inhalable DI mg/kg (Rank)	Thoracic DI mg/kg (Rank)	Respirable DI mg/kg (Rank)	Additional Characterization
$CaCO_3$[a]	12–14	305 (Low)	37 (Low)	2 (Very low)	SMPS, FMPS, CPC, ELPI
$CaCO_3$ coated for rubber[a]	28–33	13,845 (High)	3,323 (High)	331 (High)	SMPS, FMPS, CPC, APS, ELPI
$CaCO_3$ coated for adhesives[a]	21–23	4,419 (Moderate)	1,256 (High)	160 (Moderate)	SMPS, FMPS, CPC, APS, ELPI
TiO_2[b]	NA	610 (Low)	60 (Low)	20 (Low)	
TiO_2[c]	NA	6,713 (High)	576 (Moderate)	15 (Low)	SMPS, APS
ZnO[c]	NA	142 (Very low)	72 (Low)	11 (Low)	SMPS, APS
Al_2O_3[d]	NA	8,178 (High)	–	1209 (High)	
$BaSO_4$[d]	NA	258 (Low)	–	49 (Low)	
$CaCO_3$[d]	NA	2,400 (Moderate)	–	617 (High)	

[a]Burdett et al. [3].
[b]Mark et al. [41].
[c]Tsai et al. [40].
[d]Bach et al. [42].
NA: Not available.

at 1–2 μm. The ZnO sample was dominated by a 2 μm peak with a smaller peak at 200 nm; here, measurement of particle size-distribution by number was acquired. Burdett et al. [3] showed number size-distributions for seven variations of $CaCO_3$, all of which gave bimodal distributions; peaks appeared at 150–200 nm as measured by FMPS and 0.8–1.3 μm by APS. However, a large variation in modal concentrations was observed for the different variations of the material.

10.2.2 The EN 15051 Continuous Drop (CD) Method

The continuous drop (CD) method was originally designed for the EN 15051-3 standard and intends to simulate a dust generation process where powder is falling without any major mechanical stresses. It has been slightly modified in order to be used with

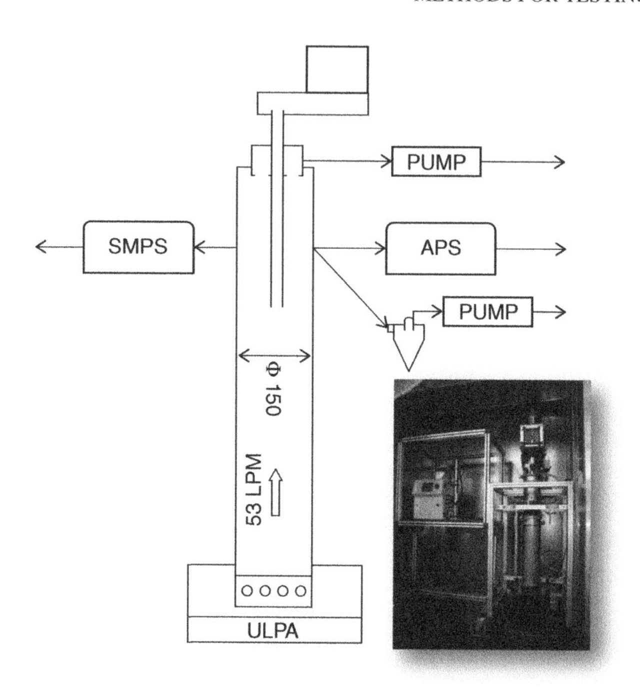

Figure 10.3 Principle sketch and photograph of the EN 15051 continuous drop method. In this illustration, real-time size-distribution measurements using a Scanning Mobility Mobility Particle Sizer and an Aerodynamic Particle Sizer, are shown. The inset photograph shows a version of the EN 15051 continuous drop method.

nanomaterial powders [18]. As shown in Figure 10.3, the CD method consists of a vibration feeder tank from which the test powder is continuously poured through a drop pipe, then into a vertical pipe with an inner diameter of 150 mm. Within the vertical pipe, ULPA-filtered upper airflow is pulled at a rate of 53 l/min. At this volume flow and configuration, the tube acts as a vertical elutriator in which finer particles (below ≈40 μm in aerodynamic diameter) are separated from larger particles. The aerosol is measured via various sampling tubes located at the same height (≈80 cm) from the bottom.

When compared with the description in the EN 15051 standard the CD method for testing nanomaterials has been subject to several modifications. First, the entire system in the CD apparatus is made of stainless steel, and real-time aerosol monitors, such as SPMS and APS, are added for the measurement of particle number size-distribution concentrations (particle size to range between ca. 10 nm and 20 μm).

According to the standard (EN 15051), a preliminary activity (before the test procedure itself) involves filling the sample tank with the test material with predetermined moisture content and bulk density. The amount of sample material (typically 500 g) has to be adjusted together with the vibration of the metering device to obtain a drop mass rate within the range 6–10 g/min. The test procedure starts by turning on the main air flow of preconditioned air (50 ± 10% RH, 21 ± 3 °C) followed by the sampling devices and a metering device (that has been previously set at the proper rate). Duration of the test procedure is typically 10 min. At the end of the test, filters are carefully removed and weighed using an analytical balance (in accordance with best laboratory practices). The test procedure is usually repeated four times.

Within a German project, NanoCare [43], a total of 19 different nanomaterial powders have been investigated using the continuous drop method [18]. Findings showed that the measured number concentrations cover the range between about 10^3 and 10^7 particles/cm^3. In relation to particle number size-distribution data acquired, the findings show clear differences in the particle number size-distribution data associated with released aerosols and in some cases, the size-distributions appeared bimodal.

10.2.3 The Small Rotating Drum (SRD) Method

Figure 10.4 shows a principle sketch of the SRD and a photograph of the system (inset) under current evaluation for standardization. The first version of the SRD method was developed and demonstrated by the Schneider and Jensen [17], in which the test method included both a single-drop and rotating drum test, using real-time measurement with a FMPS and APS and filter sampling for inhalable dust. Subsequent modifications to the system have enabled sampling of respirable dust along with real-time number-concentration and size-distribution measurements using other equipment, such as a CPC and the ELPI [44].

The SRD is, in principle, a miniaturized version of the EN 15051-2 RD. As with the EN 15051-2 RD, the SRD consists of a cylindrical part and 45° truncated conical ends. The length of the cylindrical part is 23 cm and the inner diameter is 16.3 cm

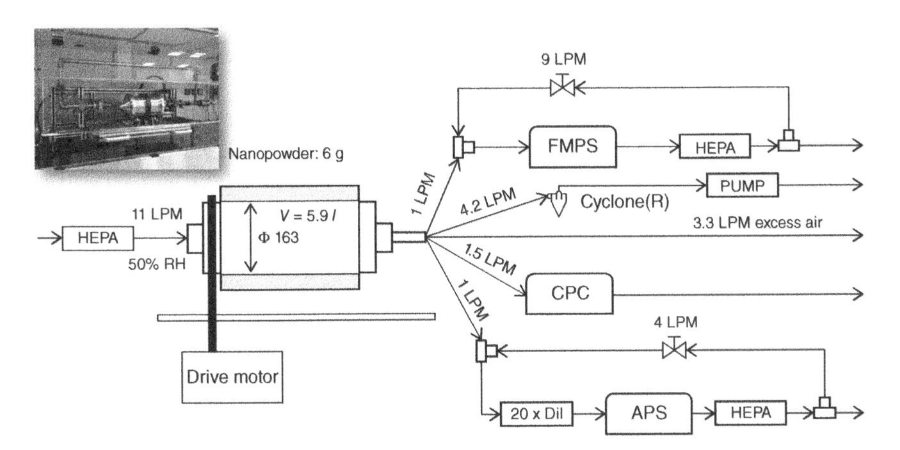

Figure 10.4 Principle sketch and photograph of the small rotating drum (SRD). In this example, real-time size-distribution measurements involved the use of a FMPS, an APS, and a CPC (to measure particle number concentration). The (inset) photograph shows the latest version of SRD, which currently is under evaluation for CEN standardization.

(4.80 l). The conical ends of the drum have a center depth of 6.3 cm (2 × 1.13 l). These dimensions give a total volume of 5.93 l, which is ca. 1/3 of the total volume of the EN 15051 RD. The cylindrical part of the drum contains three lifter vanes (2 × 22.5 cm) 120° apart, used to *lift and drop* the powder during the dustiness test.

In the prototype design developed by Schneider and Jensen [17], a small surplus of 11 l/min humidity-controlled air (50%RH) was fed to the loose connected drum inlet from which 11 l/min was drawn into the drum and the sampling train (from which dust could be collected on filters and also characterized by different real-time aerosol monitors, such as the CPC, FMPS, and APS). In later studies, the dustiness tests were conducted using the ELPI instead of the APS and FMPS [44]. Size-selective dust sampling can be made using either cyclones, inline filter samplers, or impactors. In the new CEN standard design (inset photograph in Fig. 10.4), the air volume flow to the drum is passed through a closed connection, to reduce the risk of accidental release into the fume-hood during testing. In any configuration, the air volume flowing through the drum is maintained by sampling pumps and the real-time aerosol monitors.

The test procedure for the SRD is also based on the EN 15051-2 RD procedure (with regard to conditions and test requirements). However, so far, slightly different SRD procedures have been applied in different studies. In one setup, the SRD was used for a single-drop test (measured for 60 s) followed by 60 s of rotating drum testing followed by 120 s measurement [5, 17, 19]. Hence, this procedure tested the effect of 1 single drop followed by 33 powder agitations. In other studies, the protocol involved conducting only the rotating drum test with 60 s rotation (33 powder agitations) and a total of 180 s sampling [44–47]. In all cases, up to 6 g sample was used in each run and at least three replicate measurements were acquired, to determine the mean value with standard deviations. In some cases, where very high dustiness levels were indicated from the saturation test, the amount of test material was reduced down to ca. 1 g powder. This was necessary to avoid oversaturation problems associated with (highly sensitive) real-time measurement devices such as the APS. Thus, standardization of the test volume is currently being considered.

Table 10.2 lists some examples of dustiness data produced using the SRD method. Results showed that mass-based dustiness levels can vary considerably from sample to sample. Respirable dust was shown to vary from 2 to 1058 mg/kg and the particle by number data ranged from ca. 3×10^5 to 3×10^7 particles per mg powder. Results indicate a high dynamic range in the test system. Moreover, it has been observed that materials of the same chemical composition do not have the same dustiness levels. This clearly shows that the exposure potential of a nanomaterial cannot reliably be grouped according to their chemical makeup.

In addition to the dustiness indexes, the SRD system can also give information about the particle size-distributions. Figure 10.5 gives an example of the FMPS and APS number size-distributions of different TiO_2 samples (associated with tests materials NM-101 to NM-105) in Table 10.2. These NM TiO_2 nanomaterial samples were all included on the OECD Working Party on Manufactured Nanomaterials Test Programme. It is worth noting that the particle size-distributions of the different TiO_2 nanomaterials vary considerably when measured using these techniques. NM-103

TABLE 10.2 Examples of Dustiness Data Using the SRD Obtained on 13 Nanomaterials Associated Material

Material	XRD Phases	Specific surface area (m^2/g)	Respirable DI (mg/kg)	Inhalable DI (mg/kg)	Particle Number (n/mg)
NM-101[a]	Anatase	229–316	24	728	1.10 E6[d]
NM-102[a]	Anatase	78–82	15	268	0.30 E6[d]
NM-103[a]	Rutile	51	323	9185	18.0 E6[d]
NM-104[a]	Rutile	56–57	38	3911	0.41 E6[d]
NM-105[a]	Anatase–rutile	46–55	28	1020	0.32 E6[d]
NM-200[b]	SAS	189	293	6459	6.16 E6[d]
NM-201[b]	SAS	140	218	6034	5.82 E6[d]
NM-202[b]	SAS	187–204	91	4988	4.13 E6[d]
NM-203[b]	SAS	190–204	354	5800	6.30 E6[d]
NM-204[b]	SAS	131–136	1058	24969	8.25 E6[d]
$CaCO_3$[c]	Calcite	12–14	2	NA	0.55 E6[e]
$CaCO_3$ coated for rubber[c]	Calcite	28–33	288	NA	32.7 E6[e]
$CaCO_3$ coated for adhesives[c]	Calcite	21–23	135	NA	10.8 E6[e]

[a]Rasmussen et al. [47].
[b]Rasmussen et al. [46].
[c]KA Jensen and M Levin (previously unpublished data from the National Research for the Working Environment).
[d]CPC data.
[e]FMPS+APS data.

has a clear presence of a size-mode smaller than 100 nm as well as a size-mode at ca. 4 μm. The dust of NM-104 and NM-105 is dominated by particles in the FMPS size range and has only a limited contribution from particles in the coarser APS size range. For NM-102, the micrometer size fraction dominates over the FMPS size fraction. Such differences in size-distributions have significant implications for the estimated doses deposited in the different regions in the lungs (refer to Fig. 10.1, which shows the size-dependent particle deposition efficiencies in the airways) as well as specifications required for filter performance.

10.2.4 The Vortex Shaker (VS) Method

The VS method is a completely new method with regard to standard dustiness testing, and several different configurations of this method have been proposed over time. In this section, the background and historical development of the VS test system will be the main focus.

The VS method comes from an original concept developed first by Baron et al. [34]. As part of a field study devoted to evaluate the aerosol release during

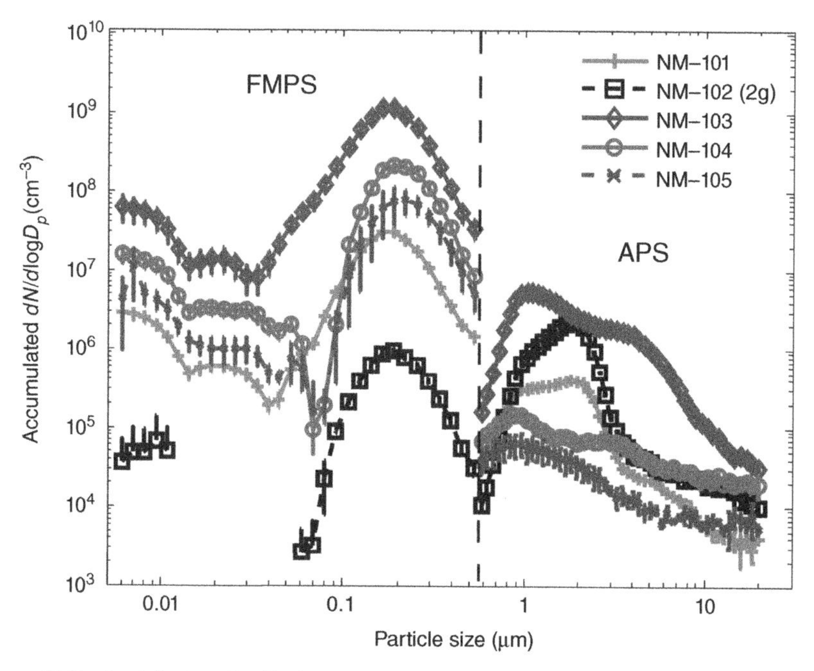

Figure 10.5 Particle size-distributions of five TiO_2 samples acquired from the OECD Working Party on Manufactured Nanomaterials test programme. The FMPS and APS particle size ranges are given in terms of electrical mobility and aerodynamic particle sizes, respectively.

the handling of single-walled carbon nanotube (SWCNT) material, the authors investigated the release of fine airborne particles (by agitating samples of SWCNT powders inside a Pyrex centrifuge tube). Depending on the degree of agitation, particles in the aerodynamic size range between 1 and 10 μm have been measured. In addition, there was some evidence that one of the two SWCNT materials tested led to measurement of particles around 200 nm. Few additional tests indicated that the rate of generating dust particles from the SWCNT materials was approximately two orders of magnitude below that for fumed alumina (for similar volumes of material).

Later, this VS concept was further developed by Ogura et al. [48], to make it suitable for the characterization of aerosol release from different nanomaterial powders, including carbon materials (nanotubes and fullerenes) and metal oxides. Similar to the work by Baron et al. [34], the VS was used to characterize the release of aerosols. The entire particle size range, from about 10 nm to about 20 μm, was measured. Particle-size selectors (with a cutoff diameter of, respectively, 15 and 2.5 μm) were added to the sampling train and positioned upstream of the aerosol instruments. For most of the powders investigated, 0.25, 0.5, and 1 cm^3 of powder samples was employed, with the VS being operated for a few hours at a time. No obvious changes in the particle size-distributions of the released aerosols were observed, regardless of the level of rotation speed (not specified) and the amount (6–360 μg) of the tested nanomaterials. Findings show modal diameters that range between 300 nm and 3 μm. In addition, a small fraction below 100 nm was detected for most of the dusts generated. The dustiness number indexes estimated amongst the samples were shown to differ by more than two orders of magnitude. Furthermore, it was reported that the volume of the powder sample had no effect on the final value.

Since 2008, other experimental works have been performed using the VS method but with different objectives (i.e., other than evaluating the dustiness). The objectives vary, for example, to generate aerosol to test capabilities of portable instruments (e.g., for measuring airborne CNTs [33], or to produce test aerosols for inhalation toxicological studies [49, 50]).

Figure 10.6 shows a schematic illustration of the VS method, as further developed by Witschger et al. [32]. This design is based on the experimental work by Baron et al. [34] and Ogura et al. [48]. The design consists of a stainless steel tube (volume $\approx$100 cm^3) that is continuously shaken in a circular orbital motion (orbit 4 mm, rotation speed 2000 rpm), and in which a small volume (0.5 cm^3) of the nanomaterial is placed for testing. High-efficiency particulate arrestance (HEPA) filtered air, controlled at 50% RH, passes through the tube at 4.2 l/min. This flow rate is necessary not only for the emission of airborne particles from the powder sample but also to transfer of the aerosol formed to the sampling and measurement section. The setup developed by Witschger et al. [32] has not only been used for measuring respirable number concentration and its corresponding particle size-distribution but also for defining the dustiness indexes for the respirable dustiness number index.

In order to rank the dustiness of nanomaterial powders, concentrations in terms of respirable mass (i.e., DI_{RM}) and respirable number dust indexes (i.e., DI_{RN}) can be used. Here, DI_{RM} is acquired from measurements taken from the setup illustrated in Figure 10.6b, whereas the corresponding DI_{RN} is determined using the setup illustrated in Figure 10.6a. DI_{RN} is expressed in 1/mg (Fig. 10.6a), whilst DI_{RM} is expressed in mg/kg. Figure 10.6a also shows the ability of the setup to collect dust particles for electron microscopy analysis (subsequently to characterize or measure particle size-distribution of the released aerosol). Overall, the dustiness protocol associated with VS method can consist of (i) the use of a respirable selection (before taking any aerosol measurement/sampling, (ii) CPC (as reference instrument for number concentration measurement), (iii) MiniParticle Sampler (MPS, Ecomesure, France) as TEM grid holder, and (iv) ELPI (normal or +, Dekati, Finland) as a size-resolved aerosol measurement technique (to measure the entire dust particle size range).

Figure 10.7 presents the two dustiness indexes DI_{RN} and DI_{RM} for 15 nanomaterial powders obtained from the list of nanomaterials in OECD Working Party on Manufactured Nanomaterials test program. The powder selection included several carbon nanotubes, synthetic amorphous silica (SiO_x), and TiO_2 nanomaterials. The findings clearly show a large range of values over several orders of magnitude [32, 46, 47, 51]. A similar observation testing different powders was made by Tsai et al. [7]. They reported significant differences in particle number size-distributions of the released aerosols. However, size-distribution data for all nanomaterials tested appear bimodal

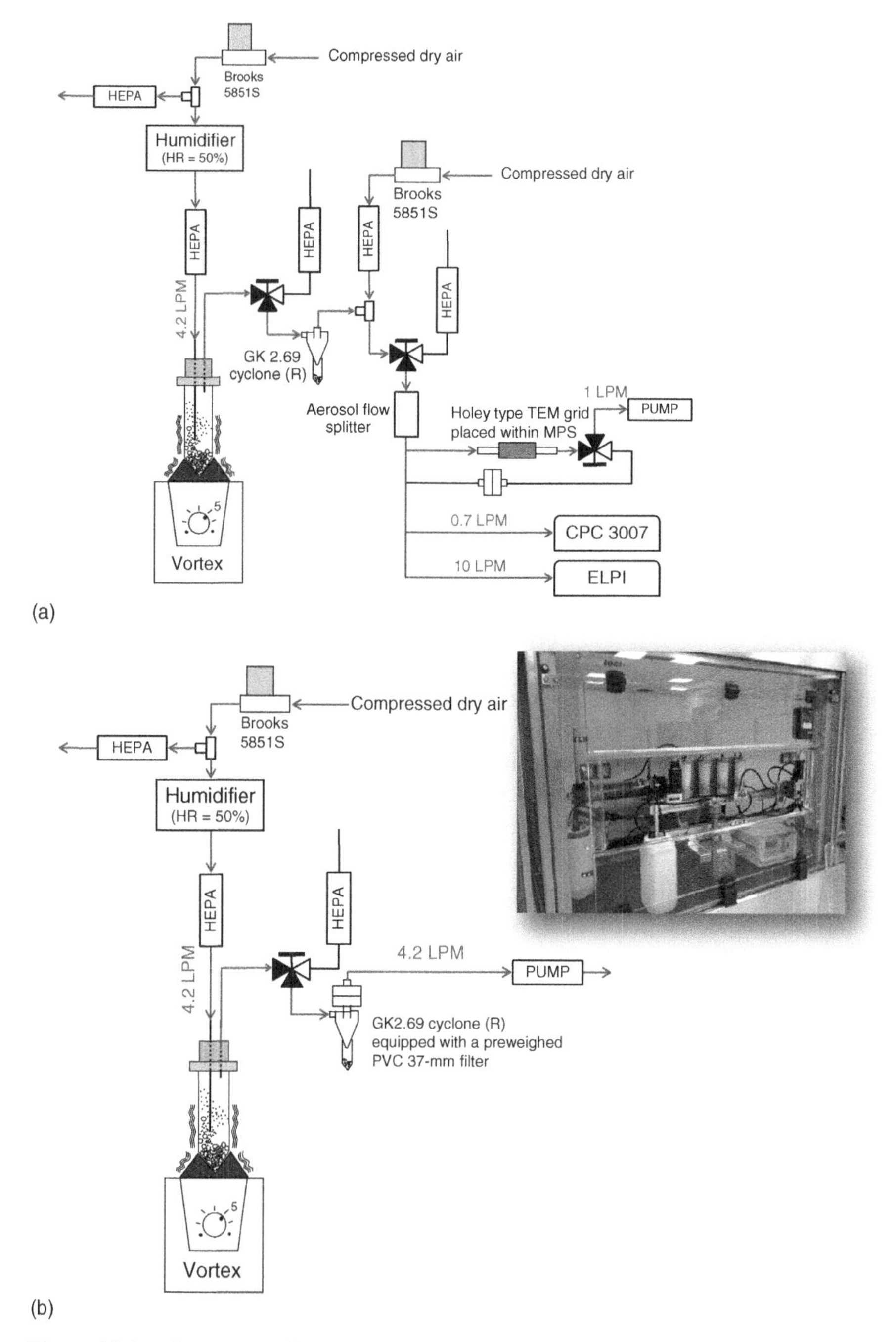

Figure 10.6 The two experimental configurations used in the vortex shaker method according to Witschger et al. [32]: (a) for measuring respirable number concentration and its corresponding particle size-distribution, and for collecting airborne particles for subsequent electron microscopy observations; (b) for collecting respirable mass fraction of the emitted aerosol.

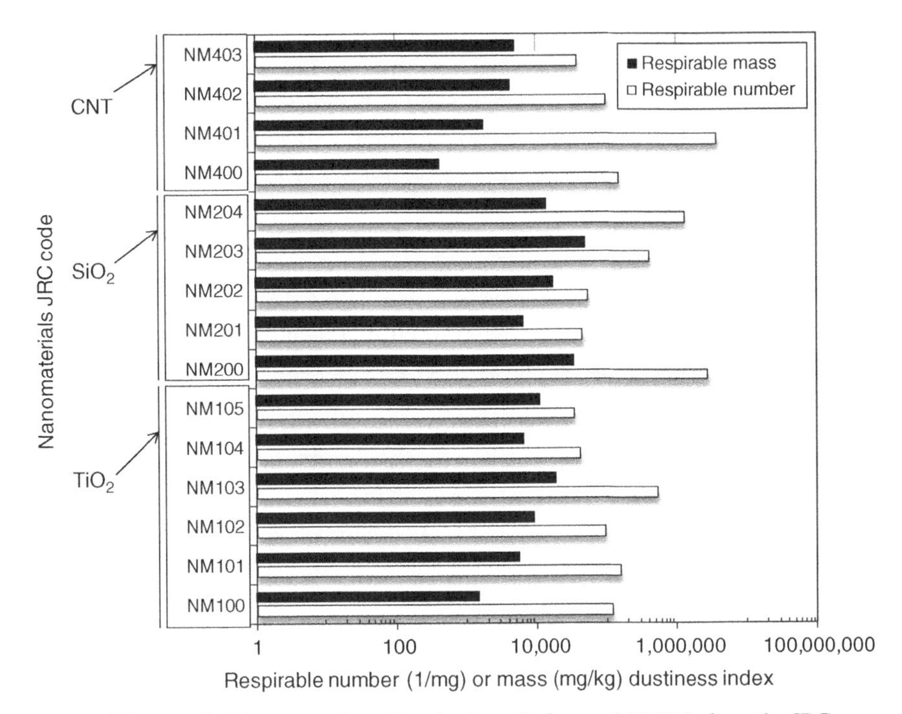

Figure 10.7 Respirable mass and number dustiness **indexes** of 15 NMs from the JRC nanomaterial repository tested with the VS method (data from Witschger et al. [32]).

(with the ratio between the micron and submicron fractions being dependent on the type of the nanomaterial.

10.2.5 Dustiness Test: Comparison of Methods

It is evident that there are important differences between the various dustiness test methods described herein. Yet, to date, comparability between the methods has not been investigated, that is, no data exist across all four methods. In future, however, it is envisaged that this type of comparison will be established in the ongoing CEN standardization work. Having said this, there are a few cases in which dustiness data exist where pairs of methods have been used for the same nanomaterial. These data enable a comparison for the RD versus SRD as well as between SRD and VS, as is discussed in this section.

The most elaborate previous comparative study that has been reported is the work conducted by Tsai et al. [7]. They have investigated the emission characteristics of nanopowders including TiO_2, ZnO, and SiO_2 using the RD and the VS. Tsai et al. expressed the dustiness indexes in terms of particle number concentrations per mass of powder sample ($1/cm^3/g$), the particle number, and mass size-distributions; the particle number data were obtained by SMPS and APS. In both the RD and VS methods used, the mass fractions of dust with sizes smaller than 100 nm were relatively very low, which is also in general agreement with other findings when CD, RD, SRD, and VS are used. In relation to the real-time measurement data, a significant decrease in particle concentration was observed as function of time for all three powders in both test systems, but the number concentrations and size-distributions were observed to differ between the methods. The decrease in particle concentration with time is, however, not a general feature of powders. Different types of temporal evolution of particle concentration patterns (ranging from instant to constant dust release rates) have been reported in experiments conducted using the SRD [5, 17, 19]. Tsai et al. [7] proposed that the VS method causes a higher drag force on the powder particles than the RD method. This could result in a higher dispersion energy and thereby potentially generation of smaller dust particles and higher particle number concentrations.

The major difference between EN 15051 RD and the SRD test procedures is associated with the volume of powder required for the test (35 cm^3 in the RD vs. $\leq$6 g in the SRD). In addition, due to differences in drum diameters, the powder drop-height should be lower in the SRD than in the RD. Furthermore, the sampling lines in the two methods are very different. Dust for gravimetric measurements is collected by sandwiched PUF filters in the RD method, whereas a respirable cyclone with Teflon filter and/or an inline inhalable membrane filter is used for SRD. Lastly, unlike the RD method, the SRD system has a fixed sampling line that not only allows simultaneous monitoring using several different real-time measurement devices but also allows the simultaneous sampling to obtain gravimetrical data. The differences in the designs and sampling principles between two methods could cause some differences in the observed data. However, initial comparison based on three respirable dustiness indexes on carbonates (shown in Tables 10.1 and 10.2) suggests relatively high data comparability between the SRD and RD methods (Fig. 10.8). Having said this, a much larger data set is needed if a reliable comparison is to be made.

A much larger data set for comparison between the SRD and the VS has been acquired on nanomaterials from the OECD Working Party on Manufactured Nanomaterials testing program as part of the NANOGENOTOX project [32]. In particular, TiO_2 (assigned as NM-10#) and synthetic amorphous silica (NM-20#) were generated using the SRD and VS methods. In agreement with previous data from Tsai et al. [7], the NANOGENOTOX data show that the VS generates a much higher mass of respirable dust per kg powder than the SRD. However, a scatter-plot between the respirable dustiness indexes from the two methods shows a general linear relationship (with a general factor of 100), but with a relatively high scatter and a few clear "outliers" (Fig. 10.8). Similar observation has been recorded for the number-based dustiness indexes, but with even more scatter in the data. Some of these differences are ascribed to different sampling times, but it is likely that the different dustiness indexes produced by these two methods are primarily due to the fundamentally different design of the agitation mechanisms. However, further research is needed to clarify these relationships, but this requires harmonization of sampling lines and investigation of the effects of different sampling times in the VS.

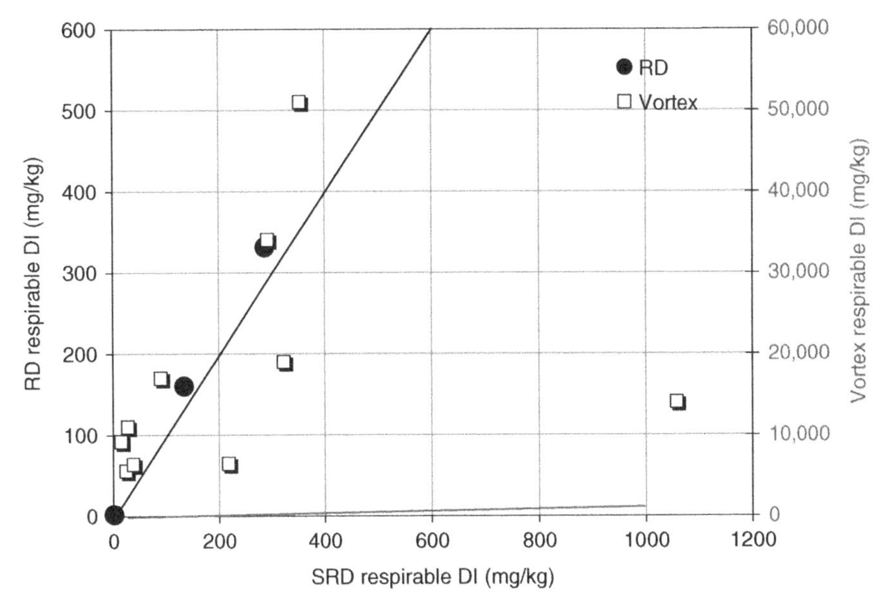

Figure 10.8 Scatter-plot of respirable dustiness data generated by the RD (left *y*-axis) and VS (right *Y*-axis) plotted against the respirable dustiness data generated by the SRD. The black line shows the 1:1 relationship for the RD and SRD data, while the gray line shows the 1:1 relationship between the VS and SRD data. Data from Burdett et al. [3]; Witschger et al. [32]; K.A. Jensen and M. Levin (previously unpublished data from the National Research for the Working Environment).

10.3 CASE STUDIES: APPLICATION OF DUSTINESS DATA

In Section 10.2, we briefly mentioned some of the previous applications of dustiness data. Till now, most work has been devoted to investigate different methods to perform and measure dustiness data on nanomaterials. Schneider and Jensen [52] and Schneider et al. [53] discussed the use of dustiness and dust size-distribution data for exposure modeling. However, dustiness is also used as a key input parameter for exposure assessment or scaling in some control banding tools such as the NanoSafer (http://nanosafer.i-bar.dk/) [14] and Stoffenmanager Nano (https://nano.stoffenmanager.nl/) [15].

To date, we found only one study that currently has demonstrated the possibilities for a more elaborate use of the different dustiness parameters [19]. In the work reported by Levin et al., they assessed and ultimately selected the most suitable molecular active ingredient powder candidate for the production of a pharmaceutical product. In their assessment, dustiness data were acquired for four different nanomaterial powders (referred as Pharma #1–#4) using the SRD. Pharma #1 (already in use) was employed as the test reference material. The dustiness data included dustiness indexes for inhalable dust as well as temporal concentrations and size-distribution data using real-time measurements with an FMPS and an APS.

The inhalable dustiness indexes were used to model the potential exposure levels using the near-field/far-field exposure model from the Advanced Reach Tool [54]. The number size-distributions and number concentrations data were used to assess filter penetration efficiency through damaged filters, which could lead to product cross-contamination. The authors found that the inhalable dustiness indexes of the four different powders were different, ranging from 1036 mg/kg (Pharma #3; granular powder) to 14,501 mg/kg (Pharma #2; fine powder). All powders generated dusts with three size-modes located between 66 and 86 nm, 230 and 270 nm, and between 2.2 and 2.9 μm. The granular powder (Pharma #3) had the lowest dustiness level, but was shown to release most of the dust in a short burst, as compared with the other Pharma powders. They concluded that Pharma #3 or alternatively the already-in-use Pharma #1 were the most favorable powder ingredients; this was based on the data associated with dustiness indexes, as well as the assessed exposure levels and particle generation rates.

10.4 SUMMARY

Powder dustiness is defined as the ability of powders to release dust during handling. In the context of occupational health risk assessment, the dustiness level is traditionally given in milligram per kilogram for specific health-related size fractions, such as inhalable, thoracic, and respirable dust. The dustiness levels can be used to rank the dustiness of powders or more specific exposure assessments (when using control banding or exposure assessment tools, as recommended by among others ECHA and the OECD). Currently, only one standard (EN 15051) exists, which describes determination of gravimetric dustiness indexes using the so-called rotating drum (RD) and continuous drop (CD) methods. The RD and CD methods are also being considered for CEN standardization in a modified design for dustiness testing of nanomaterials along with a miniaturized SRD and the VS methods. In addition to gravimetric dustiness levels, all nano-specific dustiness test methods also include measurement of particle number concentrations and size-distributions. This enables reporting of the number of dust particles generated per mass unit, the temporal dust particle generation rate, and the size-distribution.

The four dustiness test systems proposed for the new CEN standard on nanomaterial dustiness testing are fundamentally very different. The RD and SRD systems generate dust by repeated *lift and drop* agitation of a powder sample. The CD produces dust from continuous feed and drop of "undisturbed" powder, while the VS generates dust by continuous vibration of the powder. To date, there is still insufficient data available to enable a thorough comparison between the different methods. However, analysis using a small data set suggests a linear relationship between the RD and SRD respirable dustiness indexes and some overall linear correlation between the dustiness indexes (as determined by the SRD and the VS methods). However, the dustiness indexes produced by the VS are normally much higher using the measurement protocols that have been hitherto applied.

The main issue still is the absence of harmonization of methods or fixed protocols for nanomaterial dustiness testing. Most studies have focused on developing methods and assessing dustiness characteristics (when using different methods). So far, only a single study has used dustiness data to establish a more detailed assessment of the comparability between particle size-distributions (arising from dustiness testing and at the workplace). Another study used gravimetric dustiness indexes as well as particle generation rates and size-distributions for exposure assessment modeling, in order to identify the best suitable powder candidate for a molecular active ingredient relevant for a pharmaceutical company.

It is envisaged that future work in this area will focus on reaching a final decision on specific test designs and protocols, toward the development of a CEN dustiness standard for nanomaterials. Furthermore, a reliable comparison between the four different methods (associated with the upcoming CEN standard) is expected to shed further light in regards to performance comparability and the potential applications of the different test results.

ACKNOWLEDGMENTS

KAJ and ML gratefully acknowledge the financial support from the Danish Centre for Nanosafety (grant agreement No. 20110092173/3) funded by the Danish Working Environment Research Fund that made writing this chapter possible.

REFERENCES

1. Liden G. Dustiness testing of materials handled at workplaces. Ann Occup Hyg 2006;50(5):437–439.
2. Breum NO. The rotating drum dustiness tester: Variability in dustiness in relation to sample mass, testing time, and surface adhesion. Ann Occup Hyg 1999 Nov;43(8):557–566.
3. Burdett G, Bard D, Kelly A, Thorpe A. The effect of surface coatings on the dustiness of a calcium carbonate nanopowder. J Nanopart Res 2012;15(1):1–17.
4. Evans DE, Turkevich LA, Roettgers CT, Deye GJ, Baron PA. Dustiness of fine and nanoscale powders. Ann Occup Hyg 2013;57(2):261–277.
5. Jensen KA, Koponen IK, Clausen PA, Schneider T. Dustiness behaviour of loose and compacted Bentonite and organoclay powders: What is the difference in exposure risk? J Nanopart Res 2009;11(1):133–146.
6. Levin M, Rojas E, Vanhala E, Vippola M, Liguori B, Kling KI, et al. Influence of relative humidity and physical loading during storage conditions on dustiness of inorganic nanomaterials – Implications for testing and risk assessment. J Nanopart Res 2015;17:337.
7. Tsai CJ, Lin GY, Liu CN, He CE, Chen CW. Characteristic of nanoparticles generated from different nano-powders by using different dispersion methods. J Nanopart Res 2012;14(4).
8. Plinke MAE, Leith D, Boundy MG, Loffler F. Dust generation from handling powders in industry. Am Ind Hyg Assoc 1995;56(3):251–257.

9. ICRP. *Human Respiratory Tract Model for Radiological Protection.* ICRP Publication 66; 1994. Report No.: 24 (1–3).
10. Gill TE, Zobeck TM, Stout JE. Technologies for laboratory generation of dust from geological materials. J Hazard Mater 2006;132(1):1–13.
11. *Chapter R.14: Occupational Exposure Estimation v. 2.1.* Helsinki, Finland: European Chemicals Agency; 2012. Report No.: ECHA-2010-G-09-EN.
12. OECD. *Important Issues on Risk Assessments of Manufactured Nanomaterials.* Paris, France: OECD; 2012. Report No.: 33, ENV/JM/MONO(2012)8.
13. Paik SY, Zalk DM, Swuste P. Application of a pilot control banding tool for risk level assessment and control of nanoparticle exposures. Ann Occup Hyg 2008;52(6):419–428.
14. Kristensen HV, Hansen SB, Holm GR, Jensen KA, Koponen IK, Saber AT, et al. *Nanopartikler i arbejdsmiljøet - Viden og inspiration om håndtering af nanomaterialer.* Copenhagen, Denmark: Industriensens Branchearbejdsmiljøråd; 2010. Report No.: ISBN: 978-87-92141.
15. Van Duuren-Stuurman B, Vink SR, Verbist KJM, Heussen HGA, Brouwer DH, Kroese DED, et al. Stoffenmanager nano version 1.0: A web-based tool for risk prioritization of airborne manufactured nano objects. Ann Occup Hyg 2012;56:525–541.
16. Riediker M, Ostiguy C, Triolet J, Troisfontaine P, Vernez D, Bourdel G, et al. Development of a control banding tool for nanomaterials. J Nanomater 2012.
17. Schneider T, Jensen KA. Combined single-drop and rotating drum dustiness test of fine to nanosize powders using a small drum. Ann Occup Hyg 2008;52(1):23–34.
18. Dahmann D, Monz C. Determination of dustiness of nanostructured materials. Gefahrstoffe – Reinhaltung der Luft 2011;71(11/12):481–487.
19. Levin M, Koponen IK, Jensen KA. Exposure assessment of four pharmaceutical powders based on dustiness and evaluation of damaged HEPA filters. J Occup Environ Hyg 2014;11(3):165–177.
20. Jacobsen NR, Moller P, Jensen KA, Vogel U, Ladefoged O, Loft S, et al. Lung inflammation and genotoxicity following pulmonary exposure to nanoparticles in ApoE(−/−) mice. Part Fibre Toxicol 2009;6.
21. Kobayashi N, Naya M, Endoh S, Maru J, Yamamoto K, Nakanishi J. Comparative pulmonary toxicity study of nano-TiO_2 particles of different sizes and agglomerations in rats: Different short- and long-term post-instillation results. Toxicology 2009;264(1–2):110–118.
22. Oberdörster G, Oberdörster E, Oberdörster J. *Nanotoxicology: An Emerging Discipline Evolving from Studies of Ultrafine Particles.* 113 ed. National Institute of Environmental Health Sciences; 2005. p 823–839.
23. Saber AT, Jensen KA, Jacobsen NR, Birkedal R, Mikkelsen L, Møller P, et al. Inflammatory and genotoxic effects of nanoparticles designed for inclusion in paints and lacquers. Nanotoxicology 2012;6(5):453–471.
24. Suttiponparnit K, Jiang J, Sahu M, Suvachittanont S, Charinpanitkul T, Biswas P. Role of surface area, primary particle size, and crystal phase on titanium dioxide nanoparticle dispersion properties. Nanoscale Res Lett 2011;6.
25. Wittmaack K. Search of the most relevant parameter for quantifying lung inflammatory response to nanoparticle exposure: Particle number, surface area, or what? Environ Health Perspect 2007;115(2):187–194.

26. Ku BK, Kulkarni P. Comparison of diffusion charging and mobility-based methods for measurement of aerosol agglomerate surface area. J Aerosol Sci 2012;47:100–110.
27. Bau S, Witschger O, Gensdarmes F, Rastoix O, Thomas D. A TEM-based method as an alternative to the BET method for measuring off-line the specific surface area of nanoaerosols. Powder Technol 2010;200(3):190–201.
28. Asbach C, Fissan H, Stahlmecke B, Kuhlbusch TAJ, Pui DYH. Conceptual limitations and extensions of lung-deposited nanoparticle surface area monitor (NSAM). J Nanopart Res 2009;11(1):101–109.
29. Leavey A, Fang J, Sahu M, Biswas P. Comparison of measured particle lung-deposited surface area concentrations by an Aerotrak 9000 using size distribution measurements for a range of combustion aerosols. Aerosol Sci Tech 2013;47(9):966–978.
30. Bau S, Witschger O, Gensdarmes F, Thomas D. Evaluating three direct-reading instruments based on diffusion charging to measure surface area concentrations in polydisperse nanoaerosols in molecular and transition regimes. J Nanopart Res 2013;14(1217):1–17.
31. O'Shaughnessy PT, Kang M, Ellickson D. A novel device for measuring respirable dustiness using low-mass powder samples. J Occup Environ Hyg 2012;9(3):129–139.
32. Witschger O, Bianchi B, Bau S, Levin M, Koponen IK, Jensen KA. *Key intrinsic Physicochemical Characteristics of NANOGENOTOX Nanomaterials – Deliverable 4.6: Dustiness of NANOGENOTOX Nanomaterials Using the NRCWE Small Rotating Drum and the INRS Vortex Shaker*. French Agency for Food, Environmental and Occupational Health & Safety: Paris, France; 2012.
33. Hashimoto N, Ogura I, Kotake M, Kishimoto A, Honda K. Evaluating the capabilities of portable black carbon monitors and photometers for measuring airborne carbon nanotubes. J Nanopart Res 2013;15(11):1–17.
34. Baron PA, Maynard AD, Foley M. *Evaluation of Aerosol Release During the Handling of Unrefined Single Walled Carbon Nanotube Material*. Cincinatti, OH, USA: National Institute of Occupational Safety and Health; 2002. Report No.: DART-02-191.
35. Ogura I, Sakurai H, Mizuno K, Gamo M. Release potential of single-wall carbon nanotubes produced by super-growth method during manufacturing and handling. J Nanopart Res 2011;13(3):1265–1280.
36. Witschger O. Monitoring Nanoaerosols and Occupational Exposure. In: Houdy P, Lahmani M, Marano F, editors. *Nanoethics and Nanotoxicology*. Springer: Berlin Heidelberg; 2011. p 163–199.
37. Tsai CJ, Huang CY, Chen SC, Ho CE, Huang CH, Chen CW, et al. Exposure assessment of nano-sized and respirable particles at different workplaces. J Nanopart Res 2011;13(9):4161–4172.
38. Witschger O, Jensen KA, Brouwer DH, Tuinman I, Jankowska E, Dahmann D, et al. DUSTINANO: A CEN pre-normative research project to harmonize dustiness methods for manufactured nanomaterial powders. Abstracts ed. 2014. pp. T230A09.
39. BOHS. *Dustiness Estimation Methods for Dry Materials: Part 1, Their Uses and Standardization; Part 2, Towards a Standard Method*. Leeds, UK: British Occupational Hygiene Society Technology Committee, Working Group on Dustiness Estimation; 1985. Report No.: 4.
40. Tsai CJ, Wu CH, Leu ML, Chen SC, Huang CY, Tsai PJ, et al. Dustiness test of nanopowders using a standard rotating drum with a modified sampling train. J Nanopart Res 2009;11(1):121–131.

41. Mark D, Bard D, Thorpe A, Wake D. *Some Considerations for the Measurement of the Dustiness of Nanopowders*. Taipei, Taiwan: ; 2007.
42. Bach S, Eickmann U, Schmidt E. Comparison of established systems for measuring the dustiness of powders with the UNC dustiness tester developed especially for pharmaceutical substances. Ann Occup Hyg 2013;57(8):1078–1086.
43. Nau K, Krug HF. The NanoCare project: A German initiative on health aspects of synthetic nanoparticles. J Phys 2009;170(1):012038.
44. Koivisto AJ, Jensen ACØ, Levin M, Kling KI, Dal Maso M, Nielsen SH, et al. Testing the near-field/far-field model performance for prediction of particulate matter emissions in a paint factory. Environ Sci 2015;17:62–73.
45. Singh C, Friedrichs S, Carlander D, Levin M, Jensen KA, Infante HG, et al. *Cerium Dioxide, NM-211, NM-212, NM-213. Characterisation and Test Item Preparation*. Luxembourg: Publications Office of the European Union; 2014.
46. Rasmussen K, Mech A, Mast J, De Temmermann P-J, Waeggeners NVSF, Pizzolon JC, et al. *Synthetic Amorphous Silicon Dioxide (NM-200, NM-201, NM202, NM-203, NM-204): Characterisation and Physico-Chemical Properties*. Luxembourg: European Union; 2013. Report No.: EUR 26046 EN.
47. Rasmussen K, Mast J, De Temmermann P-J, Verleysen E, Waegeneers N, Van Steen F, et al. *Titanium Dioxide, NM-100, NM-101, NM-102, NM-103, NM-104, NM-105: Characterisation and Physico-Chemical Properties*. Luxembourg: European Union; 2013. Report No.: EUR 26046 EN.
48. Ogura I, Sakurai H, Gamo M. Dustiness testing of engineered nanomaterials. J Phys 2009;170(012003):1–4.
49. Ku BK, Deye G, Turkevich LA. Characterization of a vortex shaking method for aerosolizing fibers. Aerosol Sci Tech 2013;47(12):1293–1301.
50. Le Bihan OLC, Ustache A, Bernard D, Aguerre-Chariol O, Morgeneyer M. Experimental study of the aerosolization from a carbon nanotube bulk by a vortex shaker. J Nanomater 2014;2014(193154):1–11.
51. Rasmussen K, Mast J, De Temmermann P-J, Verleysen E, Waegeneers N, Van Steen F, et al. *Multi-Walled Carbon Nanotubes, NM-400, NM-401, NM-402, NM-403: Characterisation and Physico-Chemical Properties*. Luxembourg: European Union; 2014. Report No.: EUR 26796 EN.
52. Schneider T, Jensen K. Relevance of aerosol dynamics and dustiness for personal exposure to manufactured nanoparticles. J Nanopart Res 2009.
53. Schneider T, Brouwer DH, Koponen IK, Jensen KA, Fransman W, Van Duuren-Stuurman B, et al. Conceptual model for assessment of inhalation exposure to manufactured nanoparticles. J Exposure Sci Environ Epidemiol 2011;21(5):450–463.
54. Fransman W, van Tongeren M, Cherrie JW, Tischer M, Schneider T, Schinkel J, et al. Advanced reach tool (ART): Development of the mechanistic model. Ann Occup Hyg 2011;55(9):957–979.

11

SCANNING TUNNELING MICROSCOPY AND SPECTROSCOPY FOR NANOFUNCTIONALITY CHARACTERIZATION

D. Fujita

Advanced Key Technologies Division, National Institute for Materials Science, Tsukuba 305-0047, Japan

11.1 INTRODUCTION

Analytical techniques under the scanning probe microscopy (SPM) umbrella rely on the use of a physical probe (sometimes referred to as a tip) in order to raster across a specimen. It is the interaction between the probe and the surface of the specimen (as a function of position) that is monitored to produce an image on the basis of theoretical/experimental assumptions of such interactions. The way the image is obtained can be seen as "indirect" in nature, as it employs a probe to "see," much in the same way as a blind person reading Braille. Currently, there are more than 20 well-established SPM techniques [1], including atomic force microscopy (AFM), scanning ion conductance microscopy (SICM), tip-enhanced Raman spectroscopy (TERS), and so on.

As previously reported [2, 3], SPM has proved useful for characterizing nanomaterials. SPM is not only suited to three-dimensional topography measurements but also to the characterization of novel properties and functionalities of surface nanostructures on the sub-nanometer scale. An extremely powerful SPM technique is the scanning tunneling microscopy (STM). This technique was introduced by Binning and Rohrer at IBM Zurich. In 1981, they provided the first observation of vacuum tunneling between a sharp tip and a platinum surface.

Nanomaterial Characterization: An Introduction, First Edition. Edited by Ratna Tantra.

STM measurements are based on quantum mechanical phenomenon called quantum tunneling. Consider a potential barrier and an electron with energy smaller than the potential barrier. According to classical mechanics, the electron cannot travel across the barrier. However, if the thickness of the barrier is in the nanoscale, then quantum mechanics allows a small number of electrons to traverse through the barrier (referred to as quantum tunneling). Thus, STM is based on the concept of the quantum mechanical tunneling of electrons. When a conductive sharp tip is brought to the proximity of a conductive sample's surface, a bias voltage applied between two electrodes to allow electrons to tunnel through the insulating gap between them. The tunneling current is a function of tip-surface separation, applied voltage, and the local density of states (LDOS) of the sample surface. The high vertical resolution and sensitivity of STM is thus governed by the exponential dependence of tunneling current on the tip–sample separation. The use of an atomically sharpened tip allows the production of molecular or atomic resolution of STM images.

STM image can be acquired in two modes: constant-current and constant-height modes. In the constant-current operation, the height image represents a constant charge density contour of the sample surface. In the constant-height operation, the current image scan can be related to charge density. The scanning tunneling spectroscopy (STS) is one of the spectroscopic measurement modes of STM, which provides information on the electronic structure of the sample surface by probing the current as a function of energy or applied bias voltage.

STM is a useful method for nanomaterial characterization and an especially powerful tool for analyzing the surfaces of conductive nanometer-scale materials [4]. Furthermore, various nanometer-scale materials with low-dimensional structures have recently been fabricated by using STM nanoprobe technology and where STM/STS methods have been applied to exploring their unique properties [5–7]. Also, it has been shown that with STS, it is possible to probe the site-specific LDOS of the top-most surfaces on the atomic scale.

Nanomaterials are often exploited commercially because of their interesting nano-features and novel nanofunctionality. They can exhibit unique properties, which can be attributed to their small size and low dimensionality compared with their bulk counterparts. For example, quantum mechanical effects can manifest at the nanoscale. Although such effects can manifest even at room temperature, they can be made more obvious under extreme environments. Hence, to make STMs powerful enough to characterize novel surface nanomaterials, it is important to achieve so-called true atomic-scale analyses under such extreme environments (such as low temperatures (LTs), high-magnetic fields, and ultrahigh vacuums (UHVs)).

There are several important advantages when STM is conducted under extreme environments. First, at low-temperature (LT) environments, thermal disturbances such as phonon-scattering events decrease. The consequent is an increase in the mean free path and the coherence length of electrons, which subsequently enhances the manifestation of various quantum effects (where the wave–particle duality of electrons is involved). Mesoscopic-scale quantum effects such as standing-wave formation in a two-dimensional electron gas (2DEG) [6], the Kondo resonance around single magnetic atoms [7], the single-electron tunneling (SET) effect in metal nanoparticles [8], and the inelastic tunneling effect in molecules [9] are typical quantum effects at the nanoscale level. Another important nanoscale quantum-mechanical phenomenon appearing at LTs is superconductivity in classical and high-temperature superconductors [10]. In relation to STM/STS imaging, the LT conditions will mean that thermal drift of the STM instrument and the thermal smearing of the LDOS are significantly reduced with decreasing temperature. Subsequently, this will result in high-energy-resolution STS imaging on the atomic scale and the feasibility to acquire the image under relatively long duration (>hours) time. Second, when the STM is operated under a high-magnetic-field (HMF) environment, such an extreme environment has been shown to play important role in controlling the spin states of magnetic-material surfaces [11] and atomic-scale structures [12], the manipulation and measurement of superconducting states [13], and the observation of Landau quantization in 2DEG [14, 15]. Third, the use of UHV environments, where the pressure is kept at less than 1×10^{-7} Pa, is advantageous as such condition is a pre-requisite for standard surface-chemical analysis. It is generally more favorable to have UHV environments with a pressure of the order of 10^{-9} Pa or even below for detailed STM/STS imaging analyses in which a sufficiently prolonged duration of analysis is required. Furthermore, the UHV environment (with minimized residual gas) is required to ensure high purity and allows extremely clean surfaces [16] to preserve clean atomistic structures prepared on sample surfaces. Figure 11.1 depicts an illustration of an STM system working under combined extreme conditions of UHV, very low temperature, and very high magnetic field.

This chapter presents state-of-the-art STM/STS when the instrument is operated under extreme conditions (of LT, HMF, and UHV). The chapter focuses on various topics associated with this state-of-the-art technology, including a brief history, how STM is able to measure the LDOS of a material, and several examples showing the usefulness of such a technology in different applications.

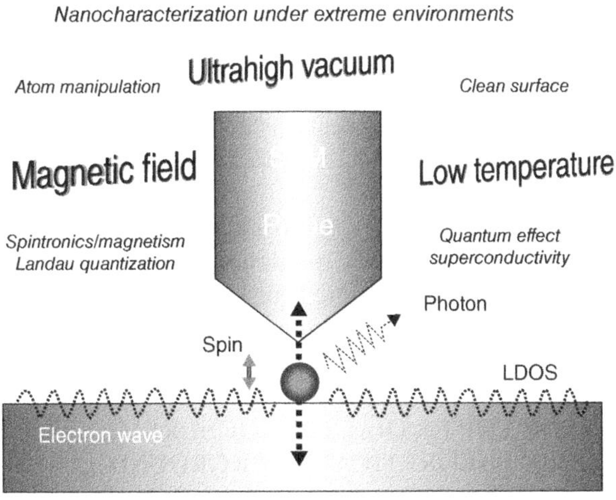

Figure 11.1 Schematic representation of STM nanoscale characterization under extreme environments for novel nanofunctionality research.

11.2 EXTREME FIELD STM: A BRIEF HISTORY

The instrumentation for LT-STM with an externally applied HMF was initially promoted in the late 1980s to investigate superconducting states [13]. It was reported that the application of external magnetic fields between two critical fields of H_{C1} and H_{C2} to a type-II superconductor induces an Abrikosov flux lattice to emerge with marked spatial variations in the surface electronic structures. With an increased magnetic field, the vortex spacing decreases. Due to the pronounced vortex–vortex interactions, the array of vortices is formed into a hexagonal lattice. The first direct STM observation of the Abrikosov flux in $NbSe_2$ was achieved by Hess et al. at 1.8 K [13].

In the early 1990s, further efforts toward constructing atomic-resolution STM/STS in more extreme environments were made, where lower temperatures ($T <\sim 1$ K) and higher magnetic fields in a UHV or cryogenic vacuum were aimed at. The first high-energy-resolution STS measurements below 1 K were achieved again by Hess et al. using a helium three (^{3}He) refrigerator [17]. The STM was operated at 300 mK with an energy resolution improved to 100 mV, which clarified the detailed structures of Abrikosov vortex cores in $NbSe_2$. One year later, they developed an ultralow temperature STM operated at 50 mK using a ^{3}He–^{4}He dilution refrigerator [18]. Stimulated by the pioneering achievements of Hess et al., other groups followed and constructed novel STMs with lower temperatures, higher magnetic fields, higher spatial resolutions, and lower base pressures. For example, Schulz and Rossel developed a UHV compatible LT (7 K) and an HMF (8 T) STM based on a ^{4}He cryostat with an *in situ* tip/sample exchange facility in a UHV [19].

In the late 1990s, Davis et al. developed a high spatial and energy resolution STM/STS operated in very-low-temperature ($T = \sim 240$ mK), variable-magnetic-field (up to 7 T), and cryogenic-UHV environments using ^{3}He refrigeration [20]. Highly improved imaging and spectroscopic quality were proved by atomic-resolution imaging on superconducting materials. However, a cryogenic vacuum is only suitable for samples whose clean surfaces can be prepared by *in situ* cleaving. Just after a truly UHV-compatible LT-HMF-STM using a ^{3}He refrigerator had been constructed by Kugler et al. [21], Wiebe et al. completed the construction of an LT-HMF-UHV STM system using a ^{3}He refrigerator. The system contained a standard surface-analysis tool for preparing and characterizing tips and samples, which could achieve spin-resolved STS measurements [22].

Although ultralow temperatures ($T < \sim 100$ mK) can be attained by using dilution refrigeration, the addition of a true UHV to a dilution-refrigerated STM is difficult to achieve (since its operation is much more complicated). However, Fukuyama et al. had succeeded in constructing a UHV-compatible LT (30 mK)-HMF (6 T)-STM [23, 24].

11.3 STM/STS FOR THE EXTRACTION OF SURFACE LOCAL DENSITY OF STATES (LDOS): THEORETICAL BACKGROUND

Energy-resolved STM or STS has been a powerful tool in mapping the LDOS in the topmost layer of conductive materials. The theoretical approach to describing

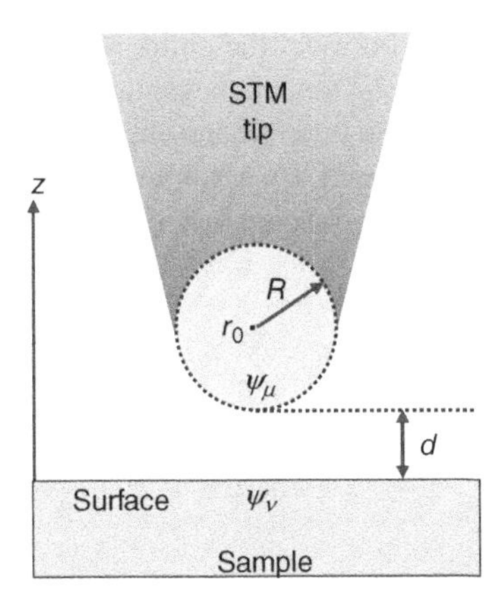

Figure 11.2 Schematic representation of geometry of STM tip in Tersoff–Hamann model. Tip apex with distance d from sample surface is assumed to have a hemispherical shape with a curvature radius R.

the electron-tunneling process within first-order time-dependent perturbation theory was first described by Bardeen in 1961 for a planar metal–insulator–metal structure [25]. Bardeen's transfer Hamiltonian approach was applied to an STM configuration by Tersoff and Hamman using a simple model for the tip [26]. The geometry of the STM tip and sample surface is schematically shown in Figure 11.2. The tunneling current, I, with an applied bias voltage, V, of the sample electrode relative to the tip electrode can be evaluated by summing over all the relevant states, expressed as the following equation:

$$I = \frac{2\pi e}{\hbar}\sum_{\mu,\nu} f(E_\mu)[1 - f(E_\nu + eV)]|M_{\mu\nu}|^2\delta(E_\mu - E_\nu) \tag{11.1}$$

Here, $f(E) = \{1+\exp[(E-E_F)/k_BT]\}^{-1}$ is the Fermi–Dirac distribution function (FDDF), where E_F represents the Fermi energy and k_B represents Boltzmann's constant. $M_{\mu\nu}$ is the tunneling matrix element between the electronic states, ψ_ν, of the sample surface (ν) and the ψ_μ of the tip apex (μ). E_ν and E_μ correspond to the energies of states ψ_ν and ψ_μ.

At LTs, the FDDFs can be approximated by unit step functions. Within the limits of LTs and small applied voltages (<~10 meV for metal–metal tunneling), the tunneling current becomes

$$I = \frac{2\pi e^2}{\hbar}V\sum_{\mu,\nu} |M_{\mu\nu}|^2\delta(E_\nu - E_F)\delta(E_\mu - E_F) \tag{11.2}$$

According to Bardeen's consideration, the tunneling matrix element, $M_{\mu\nu}$, can be expressed as

$$M_{\mu\nu} = -\frac{\hbar^2}{2m}\int dS(\psi_\mu^*\nabla\psi_\nu - \psi_\nu\nabla\psi_\mu^*) \tag{11.3}$$

Integration has to be carried out over the entire surface lying under the vacuum barrier. Within the Tersoff–Hamman model, the tunneling matrix element is evaluated within s-wave approximation for the tip where only s-like electronic states (orbital quantum number $l = 0$) are considered. By assuming that the tip apex and the sample surface have an identical local potential barrier height, ϕ, the tunneling current is then given by

$$I \propto V \cdot \rho_t(E_F) \cdot \exp(2\kappa R) \cdot \sum_{\nu} |\psi_\nu(r_0)|^2 \delta(E_\nu - E_F)$$

$$\text{where} \quad \kappa = \frac{\sqrt{2m\varphi}}{\hbar} \tag{11.4}$$

Here, $\rho_t(E_F)$ is the LDOS of the tip apex at the Fermi level, κ is the decay constant, R is the curvature radius of the tip apex, and r_0 represents the center of curvature of the tip apex. It should be noted that the sum in Equation 11.4 represents the surface LDOS, $\rho_t(E_F, r_0)$, at the Fermi level, located at the center of curvature r_0 of the tip apex.

$$\rho_s(E_F, r_0) = \sum_{\nu} |\psi_\nu(r_0)|^2 \delta(E_\nu - E_F) \tag{11.5}$$

Equations 11.4 and 11.5 suggest that the surface LDOS is constant if the tunneling current is kept constant. Therefore, within the approximations of LTs ($T \cong 0$ K), small bias voltages ($V \ll \phi$), s-like tip states, and identical local barrier heights (LBHs), STM images obtained in the constant current mode (CCM) can be regarded as contour maps of surface LDOS at around the Fermi level, evaluated at the location of the tip apex.

A more general relation between tunneling current I and surface LDOS can be evaluated by converting the sum of Equation 11.4 into the integral over all the relevant states. By assuming that the magnitude of the tunneling matrix element (TME) does not change appreciably in the energy regions of interest and the temperature is sufficiently low to consider the FDF as a step function, tunneling current I can be expressed as

$$I \propto \int_0^{eV} \rho_s(r_0, E_F + E) \cdot \rho_t(E_F - eV + E) dE \tag{11.6}$$

Equation 11.6 indicates that tunneling current I is proportional to the convolution product of the surface LDOS, $\rho_s(r_0)$, and the tip LDOS, ρ_t, over an energy range, eV. It should be noted that the LDOS of the tip and sample contribute to the tunneling current in a symmetric way. To extract the LDOS of the sample with tunneling spectroscopy, a further assumption is required. Provided that the LDOS of the tip apex is almost constant or weakly varying ($d\rho_t/dV \cong 0$), the derivative of Equation 11.6 can be expressed as

$$\frac{dI}{dV}(V) \propto \rho_s(r_0, E_F + eV) \tag{11.7}$$

Thus, the differential tunnel conductance, dI/dV, as a function of applied bias voltage V can be considered as a representation of the energy-resolved LDOS of the sample surface at the location of the tip apex, r_0, and at energy $E_F + eV$. It should be noted that this equation is only valid for small bias voltages.

More general cases including higher bias voltages can be treated by introducing a transmission coefficient, $T(E, V, s)$, where tunneling current I is given by

$$I \propto \int_0^{eV} \rho_s(E_F + E)\rho_t(E_F - eV + E)T(E, V, s)dE \tag{11.8}$$

Here, s denotes the effective distance ($s = d + R$) between the tip and the sample. For a trapezoidal tunnel barrier, the transmission coefficient, T, may be calculated using the Wentzel–Kramers–Brillouin (WKB) method of approximation [27] as

$$T(E, V, s) \cong \exp\left(-2s\left\{\frac{2m}{\hbar^2}\left(\frac{\varphi_s + \varphi_t}{2}\right) + \frac{eV}{2} - E\right\}^{1/2}\right) \tag{11.9}$$

Here, ϕ_t is the work function of the tip and ϕ_s is that of the sample. Based on Equation 11.8 for the tunneling current, the differential tunnel conductance, $dI/dV(V)$, can be expressed as follows (with the assumption that the LDOS of the tip is a constant ($d\rho_t/dV \cong 0$)):

$$\begin{aligned}\frac{dI}{dV}(V) &\propto \rho_s(r_0, E_F + eV)\,\rho_t(E_F)\,T(eV, V, s)\\ &+ \int_0^{eV} \rho_s(E_F + E)\,\rho_t(E_F + E - eV)\,\frac{d}{dV}T(E, V, s)dE\end{aligned} \tag{11.10}$$

Since transmission coefficient T with applied voltage V increases monotonically, the second term of Equation 11.10 generally contributes to a monotonic increase in the background at high bias voltages. Therefore, the structures appearing in $dI/dV(V)$ can be attributed to the structures of the LDOS of the sample expressed by the first term.

To extract the LDOS of the sample from the $dI/dV(V)$ curves containing the effect of the voltage-dependent increase in the transmission coefficient, a good method of approximation was proposed by Feenstra et al. [28]. They used the normalization of differential conductance $dI/dV(V)$ by total conductance I/V. The LDOS of the sample can be approximated by the normalization, which is equivalent to the logarithmic derivative of the tunneling current as

$$\rho_s(E_F + eV) \propto \frac{dI}{dV}\bigg/\frac{I(V)}{V} = \frac{d\ln I}{d\ln V} \tag{11.11}$$

This method can mostly extract the dependence of transmission coefficient T on distance s between the tip and the sample and on applied bias voltage V.

11.4 SCANNING TUNNELING SPECTROSCOPY (STS) AT LOW TEMPERATURES: BACKGROUND

The differential tunnel conductance, $dI/dV(V)$, can directly be measured with a lock-in detection technique, or it can be computed from the I–V curves by numerical differentiation. In the lock-in detection technique, a high-frequency sinusoidal modulation voltage ($V_m \sin \omega t$) is added to the constant bias voltage, V_0, applied between the tip and the sample. Thus, tunnel current I under the total applied bias voltage, V, can be expressed as

$$I(V) = I(V_0 + V_m \sin \omega t) \tag{11.12}$$

Provided that the superimposed modulation voltage, V_m, is much smaller than constant-DC bias-voltage V_0, Equation 11.12 can be Taylor-expanded as

$$\begin{aligned} I(V) &= I(V_0) + \left(\frac{dI}{dV}\right) V_m \sin \omega t + \frac{1}{2}\left(\frac{d^2I}{dV^2}\right)(V_m \sin \omega t)^2 + \Delta \\ &= I(V_0) + \left(\frac{dI}{dV}\right) V_m \sin \omega t + \frac{V_m^2}{4}\left(\frac{d^2I}{dV^2}\right)(1 - \cos \omega t) + \Delta \end{aligned} \tag{11.13}$$

By detecting the amplitude of the in-phase (ω) signal as a function of the applied DC bias voltage using a lock-in amplifier, a spectrum of differential tunnel conductance $dI/dV(V)$ can be extracted, which can be attributed to the surface LDOS at the specified location [29, 30].

A spatially resolved image of the surface LDOS can be obtained simultaneously with constant current imaging by recording the dI/dV signals when the probe is scanned over the surface. It is not difficult to obtain differential tunnel conductance imaging with atomic-scale lateral resolution, only if measured with an optimized tip at LTs in a UHV [31]. Furthermore, by applying a two-dimensional (2D) fast Fourier transform (FFT) to energy-resolved surface LDOS images obtained in real space, it is possible to extract valuable information related to the 2D k-space, such as 2D reciprocal lattices, 2D Fermi surfaces, and 2D band mapping.

The energy resolution of tunneling spectroscopy observed at finite temperatures is mainly limited by the thermal broadening of the FDDF, $f(E)$, and by the amplitude of the modulation voltage, V_m, if the method of lock-in detection is used. Thermal broadening width $\Delta E_{\text{thermal}}$ can be evaluated as the full width at half maximum (FWHM) of the convolution function of the FDDFs for the tip and the sample [32]. The dependence of the convolution function on temperature is plotted in Figure 11.3. Thermal broadening width $\Delta E_{\text{thermal}}$ deduced from the FWHM of the convolution functions is approximately $3.5\,k_B T$, which is consistent with the thermal width evaluated as the FWHM of the differential FDDF [33]. Note that thermal energy $k_B T$ at room temperature (RT) is ~26 meV.

Figure 11.3 suggests that the amount of thermal broadening can be significantly diminished by lowering the temperature. Since $\Delta E_{\text{thermal}}$ can be estimated as low as ~300 μeV at $T = 1$ K, a very LT environment is necessary for high-energy-resolution

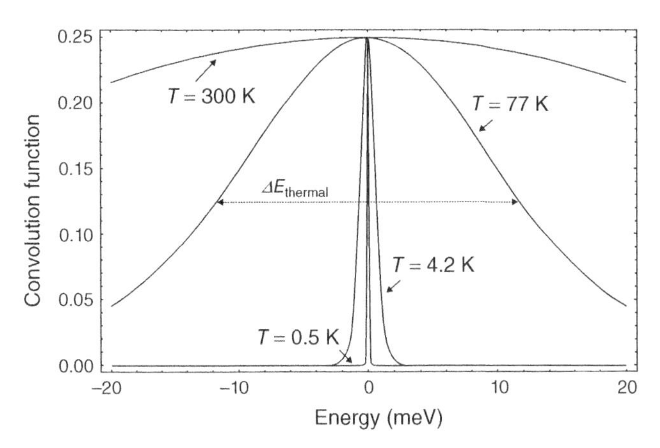

Figure 11.3 Dependence of convolution function on temperature obtained by convoluting FDDF with itself.

tunneling spectroscopy. Such high-energy-resolution STS is a powerful tool especially for detailed studies on the surface electronic states of superconducting materials [10].

At LTs, intrinsic energy resolutions caused by the measurement mode and electronic noise become more dominant. In the measurement of differential tunnel conductance using lock-in detection, the amplitude of the modulation voltage, V_{mod}, contributes to the energy resolution of spectroscopy as $\Delta E_{\text{mod}} \cong eV_{\text{mod}}$. Due to electrical noise in an STS measurement system, uncertainties or fluctuations in V_0 also cause intrinsic energy broadening ΔE_{noise}. Consequently, the total energy resolution, ΔE, can be expressed as

$$\Delta E = \sqrt{\Delta E_{\text{thermal}}^2 + \Delta E_{\text{mod}}^2 + \Delta E_{\text{noise}}^2} \tag{11.14}$$

Therefore, the amplitude of modulation voltage should be as low as ~100 μeV or even lower to enable superconducting energy gaps of about the order of millielectron volts [13, 20] to be precisely measured.

11.5 STM INSTRUMENTATION AT EXTREME CONDITIONS: SPECIFICATION REQUIREMENTS AND DESIGN

In order to keep the surface clean for a sufficiently long duration (for sample transfer, cooling, and measurement), true UHV environments with base pressures of $\sim 10^{-8}$ Pa or better are preferred. The concept of monolayer formation time t_{ml} is helpful in understanding the quality of UHVs [34, 35]. According to kinetic theory, the time to saturate a surface with a monolayer of molecules is a function of the molecular arrival rate, Γ, and the molecular size, d_0. Assuming that the sticking coefficient is unity, the time to form a monolayer is

$$t_{\mathrm{ml}} = \frac{1}{\Gamma d_0^2} = \frac{4}{nvd_0^2} \tag{11.15}$$

where n is the particle density and v is the average velocity. At RT, 1% of a monolayer of air will form in about 250 s at a pressure of 10^{-8} Pa. Fortunately, since the residual gas in UHV is mainly hydrogen, the sticking coefficient is less than unity, resulting in the longer formation time.

In low-temperature environments where various quantum-mechanical effects are expected to be observed, the temperature to be achieved should be that of liquid ^{4}He or lower ($T < \sim 4$ K). To observe the integer quantum Hall effect (QHE) in 2D-electron systems at LTs, magnetic fields higher than several teslas are required. Furthermore, to achieve an atomic-resolution of STS measurement extreme environments of low temperature (LT), HMF, and UHV, are required.

A schematic representation of UHV-LT-HMF STM based on the single-shot ^{3}He refrigerator is illustrated in Figure 11.4 [16]. A sub-kelvin temperature as low as ~350 mK can be achieved by using a one-shot type ^{3}He refrigerator. To generate a variable HMF, a superconducting magnet made of NbTi alloy is usually employed for magnetic fields of less than ~11 T. Although the superconducting critical temperature, T_c (~9 K), and upper critical field H_{c2} (~12 T at 4.2 K) are relatively low, NbTi wire is widely used for superconducting magnets due to its excellent mechanical properties. The superconducting magnet (in Fig. 11.4) is made of an NbTi solenoid with a bore diameter of ~70 mm. It is installed in the bottom of a ^{4}He bath, and can provide homogeneous magnetic fields up to 11 T that are perpendicular to the sample surface at 4.2 K. When a magnetic field higher than ~11 T is required, Nb_3Sn, a superconducting intermetallic compound, is employed. Since the superconducting critical temperature T_c (~18 K) and upper critical field H_{c2} (~25 T at 4.2 K) of Nb_3Sn are superior, it is possible to generate an HMF of ~22 T with a reduced bore diameter. However, the maximum field attained in a UHV-LT-HMF STM is still 12 T at 4.2 K and 14 T at 2.2 K by using a lambda fridge [22]. To increase the maximum field, it is necessary to develop a much more compact STM head that can be installed in a narrower bore space. If a UHV environment is not required, further increase in the magnetic field (higher than 30 T) can be accomplished by using a hybrid magnet, which is composed of an outer superconducting magnet and an inner water-cooled Cu magnet.

A UHV-LT-HMF STM system normally consists of several vacuum chambers for STM/STS, sample preparation, and load-locking. These UHV chambers should be installed on an external vibration-isolation system normally composed of air spring legs. Levitating the system effectively dampens low-frequency vibrations originating from the floor. To damp medium-frequency vibrations originating from the resonances of the refrigerator, chambers, and other attached equipments, the STM head is suspended with CuBe tension springs from the bottom of the ^{3}He pot. To efficiently conduct heat to achieve a rapid thermal equilibrium, fine copper-wire bundles are used to connect the STM head to the ^{3}He pot.

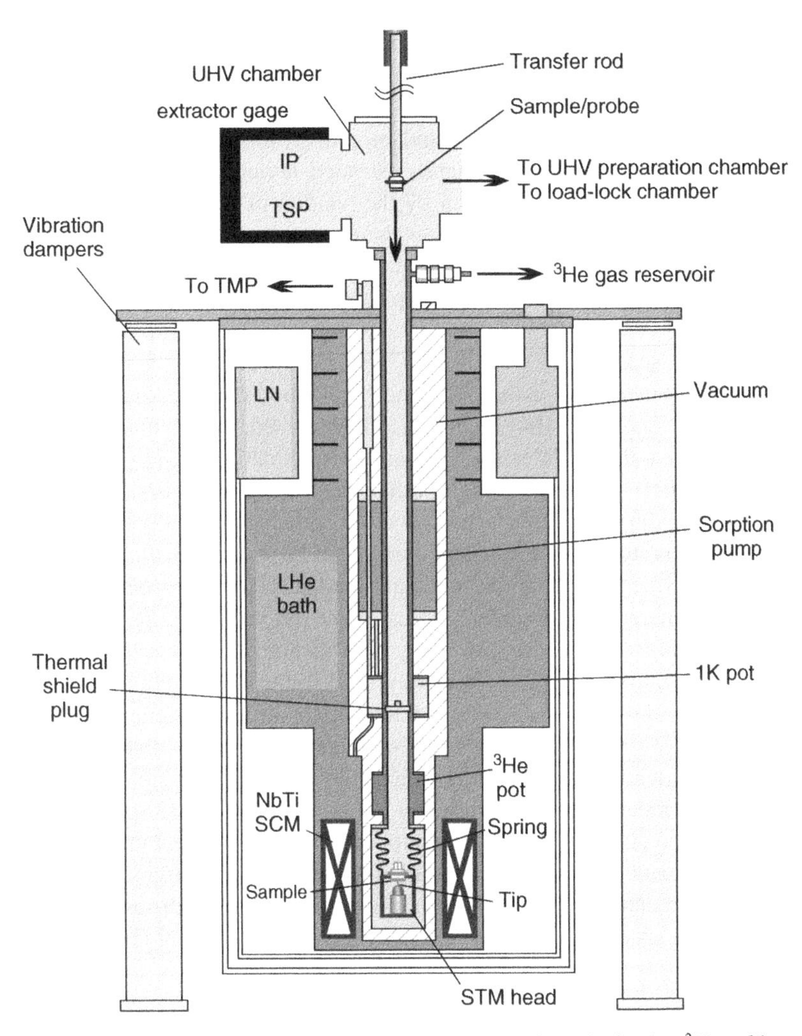

Figure 11.4 Schematic view of UHV-LT-HMF STM based on single-shot ^{3}He refrigeration system. Load-lock and preparation chambers are not shown (adapted from Sagisaka et al. [16]).

The STM head itself should be constructed so that it is rigid and compact to make its resonance frequency as high as possible, which effectively eliminates the high-frequency vibration caused by the resonances of the piezo-elements. By using this double-vibration damping system, routine imaging at atomic level is possible. It is possible to change the sample and tip *in situ* (by using a transfer rod from the preparation chamber without breaking the vacuum). After the sample and tip have been changed, the sample access column is baffled at the level of the 1-K pot using a gold-plated copper plug to block RT radiation to the sample and tip.

The UHV chambers are made of nonmagnetic materials such as 300 series austenitic stainless steels. Types 304 and 316 are the most commonly used stainless steels for vacuum and cryogenic uses. To reduce the outgassing rate from the walls of UHV chambers, the inner wall should be mirror-finished by electro-polishing and degassed in a vacuum at sufficiently elevated temperatures. Pressure P (Pa) in a vacuum system is generally given by the relation of $P = qA/S$, where q is the outgassing rate ($Pa \cdot m^3/m^2 = W/m^2$) of the chamber wall, A is the inner surface area (m^2), and S is the pumping speed (m^3/s) of the vacuum pump. The outgassing rate of stainless steel after normal baking is typically $\sim 10^{-8}$ W/m^2 [34]. The load-lock chamber (typically $A \sim 0.1$ m^2) pumped by a turbo molecular pump ($S \sim 0.25$ m^3/s) can achieve a pressure of $\sim 10^{-8}$ Pa.

The UHV requirements for the STM and preparation chambers are much more rigid. A base pressure of $\sim 10^{-9}$ Pa is typically expected. To maintain a sufficient pumping speed in the UHV region, titanium sublimation pumps (TSPs) with liquid nitrogen (LN) shrouds and high-performance ion pumps (IPs), are normally employed as the main pumps. The use of nonevaporable getter (NEG) pumps with the assistance of IPs is suitable for a system that should be operated without a magnetic field. A thorough degassing of the materials used in UHV is especially effective in reducing the increase in outgassing rates. After the vacuum system is properly baked, it is possible to achieve an extremely high vacuum ($P < \sim 1 \times 10^{-9}$ Pa).

Although ionization gauges are normally used in the UHV region, they have a background limit of $\sim 10^{-8}$ Pa at which the ionization particle current is equal to the background photocurrent caused by soft X-rays generated by bombarding the grid with electrons. Extending the operating region below 10^{-8} Pa, the use of an extractor gauge is recommended, which has an X-ray limit of 2×10^{-10} Pa [35].

The preparation chamber should be equipped with various apparatuses for cleaning tips, cleaning and annealing sample surfaces, depositing thin films, and monitoring deposition amounts. Finally, tip preparation optimized for high-resolution imaging and tunneling spectroscopy is an especially important factor for consideration.

11.6 STM/STS IMAGING UNDER EXTREME ENVIRONMENTS: A REVIEW ON APPLICATIONS

11.6.1 Atomic-Scale STM Imaging

When STM works in extreme environments of low temperature and high magnetic fields, researchers have shown that atomic-resolution has been achieved on clean surfaces of various materials such as metals, semiconductors, semimetals, and superconductors [16]. Two typical examples are shown in Figure 11.5. Figure 11.5a shows the feasibility of achieving atomic on a reconstructed Si(001) surface, when STM imaging was acquired under very LTs, that is, below 1 K. The reconstructed Si(001) surface is shown to be composed of a buckled dimer structure. True atomic resolution in STM imaging is required to clarify the buckled dimer structure in detail. One of the difficulties in the high-resolution STM imaging of a semiconductor surface is the

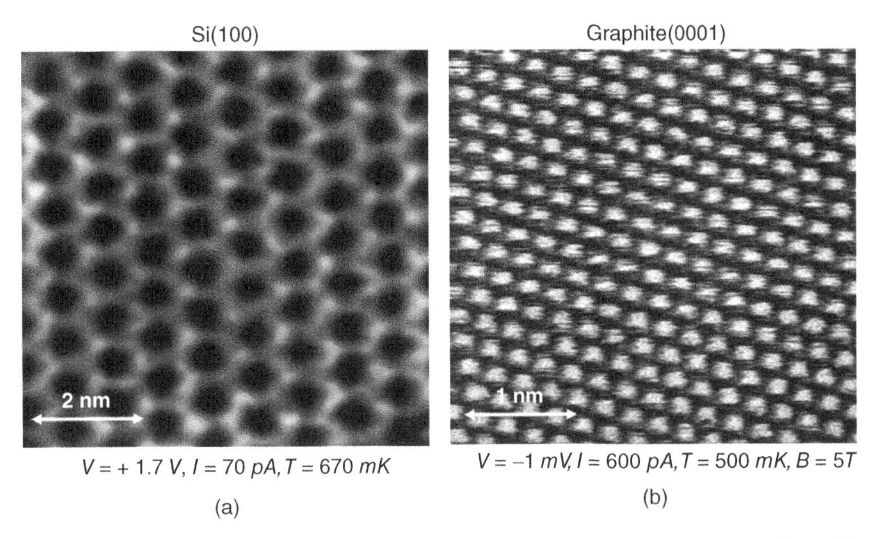

Figure 11.5 Atomic-resolution STM images taken by UHV-LT-HMF STM using ^{3}He refrigeration. (a) Constant current STM image of reconstructed Si(001) surface at 670 mK, exhibiting single phase of c(4×2) reconstruction. (b) Constant current STM image of highly oriented pyrolytic graphite (0001) surface at 500 mK and 5 T.

significant decrease in conductivity due to carrier freeze at LTs [36]. It consequently requires low-noise STM operation with a small tunneling current. Another difficulty is to achieve a true UHV environment. To prepare a low defect density Si(001) surface, a clean environment with a low vacuum pressure maintained below 1×10^{-8} Pa during flashing at ~1420 K is required [37]. Such difficulties have meant that the true ground state of the Si(001) surface at LTs has been a matter of controversy for decades, and has attracted significant interest from experimental and theoretical surface physicists [38–44]. The proposed surface structures for the ground state are static (2×1), dynamic (2×1), c(4×2), and p(2×2). As shown in Figure 11.5a, the surface of an n-type Si(001) wafer at 670 mK exhibits c(4×2) periodicity, which enabled us to conclude that c(4×2) structure is the most stable phase at very LTs.

Figure 11.5b shows a constant-current topography STM image of a (0001) surface of highly oriented pyrolytic graphite (HOPG). Atomically resolved imaging was achieved at a very LT ($T = 500$ mK) and in an HMF of 5 T (applied normal to the surface). It should be noted that atomic-resolution imaging on the HOPG (0001) surface could be attained even by using a very low bias voltage of −1 mV. This indicates that the detailed LDOS analysis of a 2D electronic system around the Fermi level is highly feasible even in the combined extreme environments of VLT, HMF, and UHV.

To date, there have been few reports on the atomically resolved STM imaging of metal surfaces below 1 K in UHV; this is mainly due to their relatively small corrugations [16]. To evaluate the imaging quality of the VLT–STM system, an Au(111) surface was selected because it is the only face-centered-cubic (fcc) metal whose (111) surface reconstructs itself. Figure 11.6 shows an atomically resolved image of a

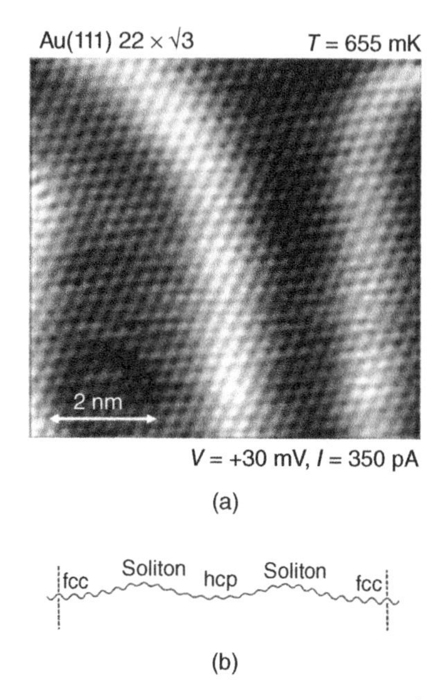

Figure 11.6 (a) Atomic resolution imaging of Au(111) $22 \times \sqrt{3}$ surface at 650 mK using VLT-UHV STM with ^{3}He refrigeration system. (b) Schematic representation of cross-sectional profile of reconstruction.

reconstructed Au(111) surface at 655 mK in UHV. The image clearly resolves the surface gold atoms aligned in a $22 \times \sqrt{3}$ reconstruction and the two bright stripes within the reconstructed unit cell. These features are consistent with the proposed model for the reconstructed Au(111) surface [45]. The observed ridges, called "soliton walls," are interpreted as transition regions consisting of bridge-site atoms, which separate fcc-type stacking (ABC ...) and hexagonal-closed-packed (hcp) stacking (ABA ...) regions. The formation of zigzag patterns is characteristic of herringbone reconstruction, and can be explained by the spontaneous formation of "stress domains" to reduce the uniaxial surface stress [46].

11.6.2 Interference of Low-Dimensional Electron Waves

Low-dimensional surface states may exist on some surfaces of metals, semiconductors, and alloys [47–49]. For example, Shockley-type surface states have been observed on the close-packed surfaces of noble metals such as Au(111), Ag(111), and Cu(111) by band mapping using angle-resolved photoelectron spectroscopy [50]. These surface state bands in the center of the surface Brillouin zone have shown parabolic dispersions in the gaps of the projected bulk states. Therefore, the surface-state electron system can be considered as a 2D nearly free electron gas. These free electrons in the 2D surface states may be scattered by potential barriers such as adatoms, point defects, and surface monatomic steps. Quantum mechanics

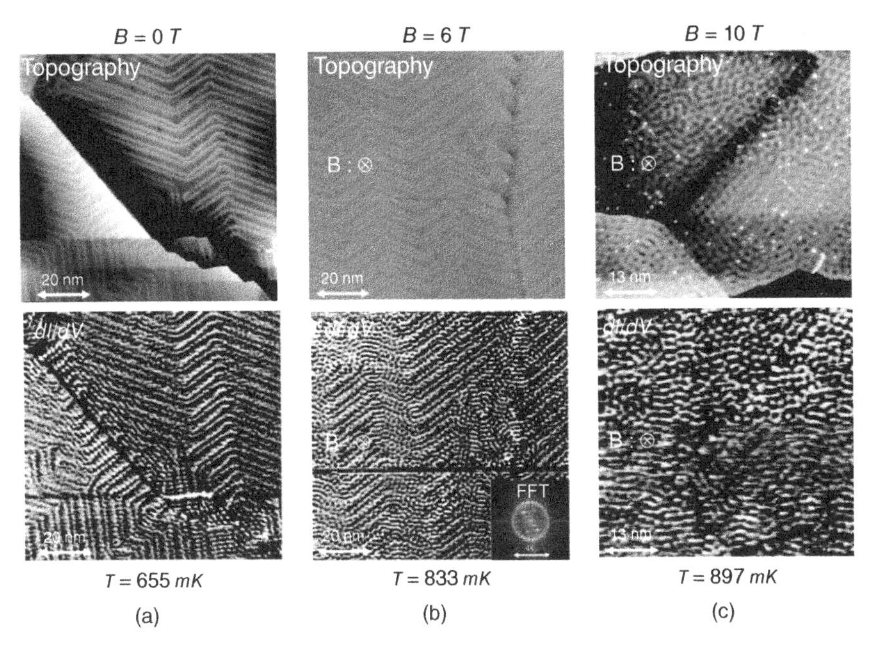

Figure 11.7 Constant-current STM images and simultaneously obtained *dI/dV* images of reconstructed Au(111) surface at very low temperatures and under different perpendicular magnetic fields. (a) $T = 650$ mK and $B = 0$ T ($V = +10$ mV, $I = 100$ pA). (b) $T = 833$ mK and $B = 6$ T. Inset 2D FFT of *dI/dV* image. (c) $T = 897$ mK and $B = 10$ T ($V = +10$ mV, $I = 150$ pA).

suggests that the interference of incident and scattered electron waves at around potential barriers results in standing-wave formation, or spatial modulation of the LDOS of the surface-state electrons. The direct observation of the spatial modulation of the LDOS caused by 2D wave interference in the vicinity of defects was first demonstrated on a Cu(111) surface by Crommie et al. with a UHV-LT STM [49]. Following their findings, various surface states on metal surfaces, such as Au(111), Ag(111), and Be(0001), were studied under UHV and LT conditions [33, 51–53].

A constant-current topography and a corresponding *dI/dV* image of Au(111) surfaces in a very LT environment at 650 mK under a zero magnetic field is shown in Figure 11.7a. The STM topography image shows a dominant long-range superstructure and an overlapping wavy structure caused by standing wave formation of the surface state electrons. Since the STM imaging was operated at around the Fermi level ($V = 10$ mV), the *dI/dV* imaging enabled real-space visualization of the surface LDOS. This is why the *dI/dV* imaging can visualize the spatial modulation of LDOS much more clearly. The surface 2D wave vector, $k_{//} = (k_x{}^2 + k_y{}^2)^{1/2}$, is related to the 2D wavelength, $\lambda = 2\pi/k_{//}$.

The spatially modulated LDOS, $n(k_{2D}, x)$, at a distance from a scattering center such as a monatomic step is expressed [54] as

$$n(k_{//}, x) = n_0[1 - J_0(2k_{//}x)] \tag{11.16}$$

Here, J_0 is the zeroth-order Bessel function, and n_0 is the LDOS of a 2DEG in the absence of scattering. By applying a 2D FFT to the STM image containing the standing wave patterns near the Fermi level, k-space mapping of the surface 2D wave vectors can be extracted [55]. The FFT operation of a constant-current or dI/dV image visualizes a ring with a radius of 2 $k_{//}$ in the case of a 2D nearly free electron gas, due to its isotropic wave propagation. If the bias voltage is very close to the Fermi level ($V \sim 1$ mV), then k-space imaging can extract 2D Fermi contour mapping. In the case of Au(111), $k_{//}$ is equal to the 2D Fermi wave vector k_F (=1.73 nm^{-1}). Applying a variable high-magnetic field to the 2D free electron gas on the reconstructed Au(111) surface may affect the trajectory of charged particles due to Lorentz force and change the standing wave patterns, as shown in Figure 11.7b,c. The 2D FFT of the LDOS (dI/dV) image under an HMF (6 T) has a circular 2D Fermi contour, suggesting that there is no significant effect on isotropic 2D wave propagation or the magnitude of the Fermi vector. The origin of the apparent change in contrast in the LDOS images under HMFs is not yet well understood since it also depends on the electronic condition of the tip [56].

11.6.3 INTERESTING PHENOMENA RELATED TO HIGH-MAGNETIC FIELDS

To understand the effects of high-magnetic fields on STM/STS imaging, we need to understand the motion of low-dimensional electrons in a magnetic field. In the presence of a magnetic field, B, that is applied perpendicularly to the x–y plane, the resulting Lorentz force causes 2D electrons with an effective mass, m_{eff}, and a kinetic energy, E, to move in circular cyclotron orbits perpendicular to the field, which is called cyclotron motion. The angular frequency, ω_c (cyclotron frequency), and the radius, R_c (cyclotron radius), of the circular orbit are expressed as

$$\omega_c = \left| \frac{eB}{m_{\text{eff}}} \right|, \qquad R_c = \frac{v}{\omega_c} = \frac{\sqrt{2m_{\text{eff}}E}}{|eB|} \tag{11.17}$$

where v is the constant magnitude of the velocity.

It is possible to reduce cyclotron radius R_c in the HMF down to the nanometer scale, which is comparable to the wavelength of 2D surface state electrons. The plane-wave-like motion of the electron waves in the x–y plane may be affected at the nanometer scale. Consequently, standing wave patterns will be formed by the superposition of waves with different propagation directions.

In the presence of a high-magnetic field parallel to the z axis, the formation of cyclotron orbits in the x–y plane changes the nature of wave functions from a plane wave extending over the whole x–y plane to a localized one on a length scale of the order of the cyclotron radius. As a result, the continuous DOS for the 2DEG in the absence of the magnetic field changes into a series of δ-functions called Landau levels, as shown in Figure 11.8. The energy of the nth Landau levels can be given as

$$\varepsilon_n = \left(n + \frac{1}{2}\right)\hbar\omega_c \qquad n = 0, 1, 2, 3, \ldots \tag{11.18}$$

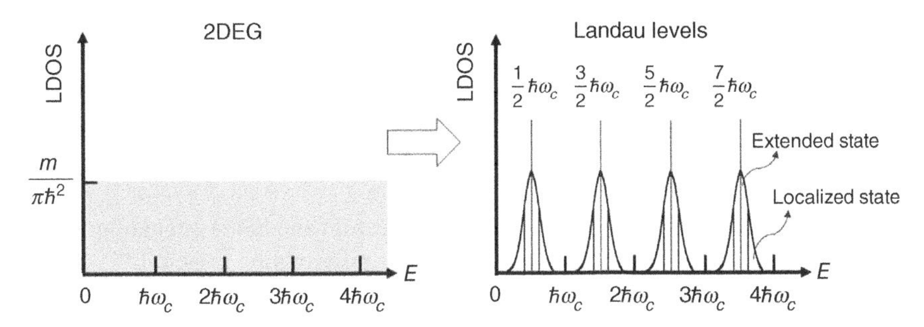

Figure 11.8 Schematic representation of Landau quantization of 2D electron system. By applying high-magnetic field perpendicular to x–y plane, continuous DOS collapses from constant for 2D system to series of discrete levels called Landau levels.

The energy intervals, $\hbar\omega_c$, of the adjacent Landau levels are dependent on the effective mass, m_{eff}, of electrons and the applied magnetic field, B. The length scale, l_B, called the magnetic length, sets the scale of the spread of the wave function of the lowest Landau level, which is given by

$$l_B = \sqrt{\frac{\hbar}{m_{\text{eff}}\omega_c}} = \sqrt{\frac{\hbar}{|eB|}} \tag{11.19}$$

Magnetic length l_B depends only on the magnetic field and not on effective mass m_{eff} of the 2D electrons. In the presence of an HMF, for example, $B = 10$ T, magnetic length l_B is approximately 8 nm.

If the temperature is sufficiently low and the magnetic field is sufficiently high, it is possible to observe Landau quantization by using LT-HMF-UHV STM for the 2DEG with a relatively small effective mass. Wildör et al. chose a clean InAs(110) surface to illustrate this point. This surface was chosen because of its small effective mass at the conduction band edge, $m_{\text{eff}} = 0.023m_e$, where m_e is the static mass of the electrons [14]. They demonstrated the quantization of conduction electrons into Landau levels in an external magnetic field. Morgenstern et al. observed Landau quantization on an adsorbate-induced 2D electron system at a cleaved InAs(110) surface with submonolayer iron deposition [57]. Theoretically predicted drift states [58], with a width of about the magnetic length, were observed in real space.

The sharp δ-function-like DOS of the Landau levels can be virtually achieved only in an ideal system where 2D electrons are never scattered. In practice, electrons are scattered by other electrons or by imperfections in materials such as impurities, phonons, and defects. Therefore, the levels are broadened as shown in Figure 11.8. The states close to the peaks represent mobile electrons, which are called extended states because they are believed to extend throughout the plane. The states close to the troughs are called localized states and represent localized nonconducting electrons. The width of the Landau levels depends on a finite lifetime, τ, between the scattering events ($\Gamma = \hbar/\tau$). If the energy separation of the Landau levels ($\hbar\omega_c$) does not sufficiently exceed the width, only a small change in the DOS may be observable.

Thus, to observe Landau quantization, the condition of $\hbar\omega_c > \Gamma$ should be satisfied. Since effective mass $m_{\text{eff}} = 0.27m_e$ is one-order larger than that of InAs in the case of a 2D electron system of metal surfaces such as Au(111), it might be more difficult to observe the Landau quantization of a metal 2D electron system. To measure general 2D electron systems with normal effective masses, high-resolution STS in a higher magnetic field is required.

Recently, Landau quantization of a quasi-2D electron and hole system of graphite was observed by ULT-HMF STM using a dilution refrigeration system [59]. Moreover, coexistence of both massless and massive Dirac Fermions in graphite(0001) surface was recently observed by Li and Andrei, which are a unique electronic behavior of monolayer and bilayer graphene [60].

Other than Landau quantization, novel quantum-mechanical phenomena can be elucidated by using an LT-HMF-UHV STM. For example, Heinrich et al. demonstrated the ability to measure the energy required to flip the spin of single adsorbed atoms [61]. Furthermore, they recently succeeded in probing the interactions between spins in individual atomic-scale magnetic structures [12].

11.7 SUMMARY AND FUTURE OUTLOOK

This chapter has been devoted to state-of-the-art STM for nanomaterial characterization, in which high-resolution STM/STS measurements have been shown to be useful when analyses were carried out under extreme environments of LT, HMF, and UHV conditions. The ability to conduct STM/STS analysis under extreme conditions has meant the ability to probe into novel nanofunctionalities (that cannot be observed otherwise). For example, high-resolution STS was shown to be useful for the detailed analyses of surface LDOS (that can only be accomplished at very LTs). Such high-resolution STS has been shown to achieve clear visualization of the electronic states of novel low-dimensional materials in real space at the nanoscale level. Furthermore, the combination of atom manipulation, atomic-resolution imaging, and spin excitation spectroscopy on the atomic scale can be used to assemble and probe novel magnetic nanostructures.

Undoubtedly, STM/STS is an indispensable tool for exploring novel quantum effects in low-dimensional nanostructures. The development of key technologies for atomic-resolution STM/STS at very LTs ($T < \sim 0.5$K), in an HMF ($B > \sim 10$ T), and in an extremely high vacuum ($P < \sim 10^{-9}$ Pa) are required because the exploration of novel functionalities for nanomaterials will become increasingly important with advances in nanotechnology. In the future, it is envisaged that research will aim to integrate state-of-the-art STM/STS with nanofabrication, which will result in an extremely powerful platform to characterize nanofunctionality of nanomaterial.

ACKNOWLEDGMENTS

The author appreciates the support from Japan Science and Technology Agency (JST) and the collaboration with the MARINA Project.

REFERENCES

1. Wiesendanger R. *Scanning Probe Microscopy and Spectroscopy, Methods and Applications*. Cambridge: Cambridge University Press; 1994.
2. Bhushan B. *Scanning Probe Microscopy in Nanoscience and Nanotechnology 3 (NanoScience and Technology)*. Springer; 2012.
3. Kalinin SV, Gruverman A. *Scanning Probe Microscopy of Functional Materials: Nanoscale Imaging and Spectroscopy*. Springer; 2010.
4. Fujita D, Sagisaka K. Active nanocharacterization of nanofunctional materials by scanning tunneling microscopy. Science and Technology of Advanced Materials 2008;9:013003 (9 pp).
5. Sagisaka K, Fujita D. Quasi-one-dimensional quantum well on Si(100) surface crafted by using scanning tunneling microscopy tip. Applied Physics Letters 2006;88:203118 (3 pp).
6. Heller EJ, Crommie MF, Lutz CP, Eigler DM. Scattering and absorption of surface electron waves in quantum corrals. Nature 1994;369:464–466.
7. Manoharan HC, Lutz CP, Eigler DM. Quantum mirages formed by coherent projection of electronic structure. Nature 2000;403:512–515.
8. Ohgi T, Fujita D. Consistent size dependency of core-level binding energy shifts and single-electron tunneling effects in supported gold nanoclusters. Physical Review B 2002;66:115410–115414.
9. Ho W. Single-molecule chemistry. Journal of Chemical Physics 2002;117:11033–11061.
10. Fischer Ø, Kugler M, Maggio-Aprile I, Berthod C, Renner C. Scanning tunneling spectroscopy of high-temperature superconductors. Reviews of Modern Physics 2007;79:353–419.
11. Pietzsch O, Kubetzka A, Haude D, Bode M, Wiesendanger R. A low-temperature UHV scanning tunneling microscope with a split-coil magnet system and a rotary motion stepper motor for high spatial resolution studies of surface magnetism. Review of Scientific Instruments 2000;71:424–430.
12. Hirjibehedin CF, Lutz CP, Heinrich AJ. Spin coupling in engineered atomic structures. Science 2006;312:1021–1024.
13. Hess HF, Robinson RB, Dynes RC, Valles JM, Waszczak JV. Scanning-tunneling- microscope observation of the abrikosov flux lattice and the density of states near and inside a fluxoid. Physical Review Letters 1989;62:214–216.
14. Wildöer JWG, Harmans CJPM, Van Kempen H. Observation of Landau levels at the InAs(110) surface by scanning tunneling spectroscopy. Physical Review B 1997;55:R16013–R16016.
15. Morgenstern M, Haude D, Gudmundsson V, Wittneven C, Dombrowski R, Wiesendanger R. Origin of Landau oscillations observed in scanning tunneling spectroscopy on n-InAs(110). Physical Review B 2000;62:7257–7263.
16. Sagisaka K, Fujita D, Kido G, Koguchi N. Scanning tunnelling microscopy in extreme fields: very low temperature, high magnetic field, and extreme high vacuum. Nanotechnology 2004;15:S371–S375.
17. Hess HF, Robinson RB, Waszczak JV. Vortex-core structure observed with a scanning tunneling microscope. Physical Review Letters 1990;64:2711–2714.
18. Hess HF, Robinson RB, Waszczak JV. STM spectroscopy of vortex cores and the flux lattice. Physica B 1991;169:422431.

19. Schulz RR, Rossel C. Beetle-like scanning tunneling microscope for ultrahigh vacuum and low-temperature applications. Review of Scientific Instruments 1994;65:1918–1922.
20. Pan SH, Hudson EW, Davis JC. ^{3}He refrigerator based very low temperature scanning tunneling microscope. Review of Scientific Instruments 1999;70:1459–1463.
21. Kugler M, Renner C, Fischer Ø, Mikheev V, Batey G. A 3He refrigerated scanning tunneling microscope in high magnetic fields and ultrahigh vacuum. Review of Scientific Instruments 2000;71:1475–1478.
22. Wiebe J, Wachowiak A, Meier F, Haude D, Foster T, Morgenstern M, Wiesendanger R. A 300 mK ultra-high vacuum scanning tunneling microscope for spin-resolved spectroscopy at high energy resolution. Review of Scientific Instruments 2004;75:4871–4879.
23. Matsui T, Kambara H, Fukuyama H. Development of a new ULT scanning tunneling microscope at University of Tokyo. Journal of Low Temperature Physics 2000;121:803–808.
24. Kambara H, Matsui T, Niimi Y, Fukuyama H. Construction of a dilution refrigerator based ultra-low temperature scanning tunneling microscope. Japanese Journal of Applied Physics 2006;45:1909–1912.
25. Bardeen J. Tunnelling from a many-particle point of view. Physical Review Letters 1961;6:57–59.
26. Tersoff J, Hamann DR. Theory of the scanning tunneling microscope. Physical Review B 1985;31:805–813.
27. Ukraintsev VA. Data evaluation technique for electron-tunneling spectroscopy. Physical Review B 1996;53:11176–11185.
28. Feenstra RM, Stroscio JA, Fein AP. Tunneling spectroscopy of the Si(111)2×1 surface. Surface Science 1987;181:295–306.
29. Binnig G, Frank KH, Fuchs H, Garcia N, Reihl B, Rohrer H, Salvan F, Williams AR. Tunneling spectroscopy and inverse photoemission: Image and field states. Physical Review Letters 1985;55:991–994.
30. Becker RS, Golovchenko JA, Hamman DR, Swartzentruber BS. Real-space observation of surface states on Si(111) 7×7 with the tunneling microscope. Physical Review Letters 1985;55:2032–2034.
31. Sagisaka K, Fujita D. Standing waves on Si(100) and Ge(100) surfaces observed by scanning tunneling microscopy. Physical Review B 2005;72:235327, (8 pp).
32. Sarid D. *Exploring Scanning Probe Microscopy with Mathematica*. Wiley-VCH Verlag GmbH; 2007.
33. Fujita D, Amemiya K, Yakabe T, Nejoh H, Sato T, Iwatsuki M. Anisotropic standing-wave formation on an Au(111)-23×$\sqrt{3}$ reconstructed surface. Physical Review Letters 1997;78:3904–3907.
34. O'Hanlon JF. *A User's Guide to Vacuum Technology*. Hoboken, NJ: Wiley-Interscience; 2003.
35. Watanabe F. New X-ray limit measurements for extractor gauges. Journal of Vacuum Science & Technology A 1991;9:2744–2746.
36. Dickstein RM, Titcomb SL, Anderson RL. Carrier concentration model for n-type silicon at low temperatures. Journal of Applied Physics 1989;66:2437–2441.
37. Hata K, Kimura T, Ozawa S, Shigekawa H. How to fabricate a defect free Si(001) surface. Journal of Vacuum Science & Technology 2000;18:1933–1936.

38. Wolkow RA. Direct observation of an increase in buckled dimers on Si(001) at low temperature. Physical Review Letters 1992;68:2636–2639.
39. Kondo Y, Amakusa T, Iwatsuki M, Tokumoto H. Phase transition of the Si(001) surface below 100 K. Surface Science Letters 2000;453:L318–L322.
40. Yokoyama T, Takayanagi K. Anomalous flipping motions of buckled dimers on the Si(001) surface at 5 K. Physical Review B 2000;61:R5078–R5081.
41. Hata K, Yoshida S, Shigekawa H. p(2×2) phase of buckled dimers of Si(100) observed on n-type substrates below 40 K by scanning tunneling microscopy. Physical Review Letter 2002;89:286104, (4 pp).
42. Sagisaka K, Fujita D, Kido G. Phase manipulation between c(4×2) and p(2×2) on the Si(100) surface at 4.2 K. Physical Review Letters 2003;91:146103, (4 pp).
43. Shirasawa T, Mizuno S, Tochihara H. Electron-beam-induced disordering of the Si(001)-c(4×2) surface structure. Physical Review Letters 2005;94:195502, (4 pp).
44. Kawai H, Narikiyo O, Matsufuji K. Structural phase transition between c(4×2) and p(2×2) structures on Si(001) surface under observation by scanning tunneling microscopy. Journal of the Physical Society of Japan 2007;76:034602, (7 pp).
45. Harten U, Lahee AM, Toennies JP, Wöll C. Observation of a soliton reconstruction of Au(111) by high-resolution helium-atom diffraction. Physical Review Letters 1985;54:2619–2622.
46. Narasimhan S, Vanderbilt D. Elastic stress domains and the herringbone reconstruction on Au(111). Physical Review Letters 1992;69:1564–1567.
47. Zangwill A. *Physics at Surfaces*. Cambridge: Cambridge University Press; 1988.
48. Desjonquêres MC, Spanjaard D. *Concepts in Surfaces Physics*. Berlin Heidelberg: Springer-Verlag; 1993.
49. Mönch W. *Semiconductor Surfaces and Interfaces*. 3rd ed. Berlin Heidelberg: Springer-Verlag; 2001.
50. Kevan SD, Gaylord RH. High-resolution photoemission study of the electronic structure of the noble-metal (111) surfaces. Physical Review B 1987;36:5809–5818.
51. Crommie MF, Lutz CP, Eigler DM. Imaging standing waves in a two-dimensional electron gas. Nature 1993;363:524–527.
52. Li J, Schneider WD, Berndt R, Crampin S. Electron confinement to nanoscale Ag islands on Ag(111): A quantitative study. Physical Review Letters 1998;80:3332–3335.
53. Sprunger PT, Petersen L, Plummer EW, Lægsgaard E, Besenbacher F. Giant friedel oscillations on the beryllium(0001) surface. Science 1997;275:1764–1767.
54. Avouris P, Lyo IW, Walkup RE, Hasegawa Y. Real space imaging of electron scattering phenomena at metal surfaces. Journal of Vacuum Science & Technology B 1994;12:1447–1455.
55. Petersen L, Sprunger PT, Hofmann P, Lægsgaard E, Briner BG, Doering M, Rust HP, Bradshaw AM, Besenbacher F, Plummer EW. Direct imaging of the two-dimensional Fermi contour: Fourier-transform STM. Physical Review B 1998;57:R6858–R6861.
56. Petersen L, Hofmann P, Plummerc EW, Besenbacher F. Fourier transform–STM: Determining the surface Fermi contour. Journal of Electron Spectroscopy and Related Phenomena 2000;109:97–115.
57. Morgenstern M, Klijn J, Meyer C, Wiesendanger R. Real-space observation of drift states in a two-dimensional electron system at high magnetic fields. Physical Review Letters 2003;90:056804, (4 pages).

58. Joynt R, Prange RE. Conditions for the quantum Hall effect. Physical Review B 1984;29:3303–3317.
59. Matsui T, Kambara H, Niimi Y, Tagami K, Tsukada M, Fukuyama H. STS observations of Landau levels at graphite surfaces. Physical Review Letters 2005;94:226403, (4 pp).
60. Li GH, Andrei EY. Observation of Landau levels of Dirac fermions in graphite. Nature Physics 2007;3:623–627.
61. Heinrich AJ, Gupta JA, Lutz CP, Eigler DM. Single-atom spin-flip spectroscopy. Science 2004;306:466–469.

12

BIOLOGICAL CHARACTERIZATION OF NANOMATERIALS

A. Jemec, T. Mesarič, M. Sopotnik, K. Sepčić, and D. Drobne

Biotechnical Faculty, Department of Biology, University of Ljubljana, Ljubljana 1000, Slovenia

12.1 INTRODUCTION

12.1.1 Importance of Nanomaterial Characterization

The safety of nanotechnology products is seen as a crucial element in ensuring that the benefits of nanotechnology outweigh the potential risks of nanomaterials (NMs). A large number of nanosafety-related projects have been launched in the past. For example, only in the EU about 50 projects are either completed or running and represent a total RTD investment of €137 M under FP6 (13 projects, €31 M) and FP7 (34 projects, €106 M) (http://www.nanosafetycluster.eu). Recently, a list of NM physicochemical properties most relevant for NM preparation and safety testing has been suggested by experts from the EU Nanosafety Cluster Working Group 10 on integrated approaches to testing and assessment [1]. These properties are (i) composition, (ii) impurities, (iii) size/shape/and size distribution, (iv) surface area, (v) surface chemistry/crystallinity/reactivity/coating, (vi) pH, and (vii) solubility/dissolution.

In the past decade, it has become clear in nanotoxicology research that the reporting of physicochemical characteristics of NMs is necessary to aid the hazard identification of NMs [2–4]. The ability to establish the relationship between the properties of NMs and the observed biological responses will enable the grouping of hazard NMs according to their specific properties. However, establishing this correlation is not trivial, and the associated challenges have been previously discussed by several

Nanomaterial Characterization: An Introduction, First Edition. Edited by Ratna Tantra.

researchers [5–7]. Particle size and surface area are considered important factors for determining the toxicity of particles; such observation was reported for algae exposed to SiO_2 [8]. Another property of relevance to consider is the surface area, which is not surprising, as there is a correlation between the two properties, that is, when the particle size decreases, the surface area increases. Therefore, NMs with a small particle size (and hence a larger surface area) are often expected to provoke a higher toxicity. In addition to these two properties, the surface texture and the crystallinity of NMs can also play a role. For example, the toxicity of large textured ZnO nanoparticles (NPs) to inflammatory cells (macrophage-derived cells RAW 264.7 cell line) is higher than that of much smaller sized (5 nm), low crystallinity nanoparticles [9]. Furthermore, the shape of the NM has shown to affect the toxicology response such as in Ag NPs with the Gram-negative bacteria *Escherichia coli* [10]. Although several researchers in the past have identified various parameters that can be correlated to a biological response, the larger picture may be that the toxicity can only be explained by the integrated effect of multiple properties, that is, a set of secondary properties that have been referred to as "extrinsic NMs' characteristics in the biological system" [1, 11, 12].

12.1.2 Extrinsic NMs Characterization

The most prominent trait of NMs is that they are not static when entering a biological environment and that their subsequent modifications will result in them acquiring new "extrinsic" properties. NMs acquire new extrinsic properties, which result in different forms of material. These changes can occur either instantly or during their lifetime, upon them entering the biological environments; the latter phenomenon is referred to as the aging of NMs [13]. Several interactions between the NMs and the environment can occur as a result of a combination of several different factors rather than just a single factor. Several factors governing the interactions include ionic strength, pH, and media composition (e.g., presence of natural organic matter, polysaccharides, proteins, specific counter ions, such as Ca^{2+} and Mg^{2+}) [13, 14], colloidal stability, which in turn will be governed by the coatings on the NMs and stabilizing agents [3]. Furthermore, aggregation, flocculation, redox reactions, dissolution, reaction with reduced sulfur species or chloride, photooxidation, photoreduction, adsorption of polymers or natural organic matter, and interaction with essential metals, such as Ag NPs with selenium [15] can also occur [16]. The different factors listed here can result in the NM acquiring extrinsic properties.

A well-known phenomenon of NMs is their readiness to interact with biomolecules [17], resulting in the formation of noncovalent bonds between them [7]. In biological fluids, NMs are known to interact with phospholipids, proteins, DNA, and small metabolites [18] and there is increasing evidence of rapid formation of the so-called "protein corona," that is, the coating of protein molecules around the NMs [7, 19]. The formation of NMs' biocomplexes originating from the adsorption of various components in a biological environment, independent of proteins, has also been recorded; due to salts, these nano–bio complexes originated from the test medium [20].

Due to the inherent complexity of NMs interactions in the biological medium and the subsequent modifications of the NMs, the physicochemical characterization of NMs *in situ* is a difficult analytical challenge, which requires a multimethod approach [21, 22]. The complicated interactions and dynamic changes of NMs mean that it is difficult to make a meaningful characterization of NMs' physicochemical properties [19]. In addition, Tantra et al. [23] discussed difficulties in making a reliable measurement with the current instrumentation, specifically when the NMs are in complex media. The main sources of potential experimental errors identified by Tantra et al. [23] included the (high) polydispersity of NMs and the complex environment the NM is in. Subsequently, this often resulted in employing methods that did not fit the purpose. According to Baalousha and Lead [24], most of the materials tested in nanotoxicology research are considered far too polydisperse to be appropriate for current analytical techniques [24]. Another issue that arises when the NMs are in complex media is the creation of an unstable environment, resulting in NMs to agglomerate and/or sediment. Under such conditions, the measurements may be very unreliable. Furthermore, in relation to bioassay measurements, a number of other interferences that may result in making an unreliable measurement may appear, for example, the contamination of NMs, interference with the assay readout, and variations in dispersion protocols [23].

12.1.3 The Proposal for Measuring "extrinsic" Properties

It is clear that a high level of overarching property to describe NM interactions with biological environment is much needed [25]. In this chapter, such new approaches, based on biological characterization and the measurement of extrinsic properties, are presented. The idea is based on the knowledge that NM properties, such as the size, surface chemistry, crystallinity, and hydrophobicity, govern the adsorption potential of NMs, which is reflected in the interaction of NMs with the biological system, specifically, a protein biosensor system that is able to quantify these interactions is presented.

The chapter is divided into several sections: (i) a general introduction on the existent approaches to describe the integrated surface properties; (ii) the quantification of interactions between NMs and proteins, as novel proposed tools; (iii) an experimental case study that employs the use of acetylcholinesterase (AChE)-based biosensor (used to rank different carbon-based NMs based on their adsorption and interactions with this enzyme).

12.2 MEASUREMENT METHODS

12.2.1 Review of Existing Approaches

An attempt to use extrinsic properties to describe NM interaction with biological media has been discussed by Xia et al. [19]. In particular, a biological surface adsorption index (BSAI) based on computer simulations has been proposed to describe the surface adsorption energy of NMs under biologically relevant aqueous conditions. The BSAI is an integrative measure of the surface adsorption energy of NMs and is derived from five diverse nanodescriptors representing molecular forces of NM interaction with biological system: lipophilicity, hydrogen-bond basicity and acidity, dipolarity/polarizability, and lone-pair electrons. The limitation of this model is the assumption of idealized biological conditions, where the interaction is based on the estimation of how biomolecules may interact with the surface of NMs, rather than a direct experimental measurement of their interactions.

Another approach employed to characterize the extrinsic surface properties of NMs is the quantification of the interactions between NMs and proteins [26]. As discussed earlier, such interactions are complex and dynamic. Upon the interactions of NMs with proteins, a number of phenomena can be monitored to measure the extent of such interactions: (i) the binding affinity of NMs [18], (ii) the NM–protein binding kinetics, followed by surface plasmon resonance [18, 27], (iii) the stoichiometry of interaction [28], (iv) the identification of the binding sites of NMs [29], (v) protein conformational changes [30–34], (vi) protein stability, and (v) proteins function changes [31–34]. The latter refers to changes in enzyme activity and the adsorption of NMs onto the protein. This approach has been previously successfully used to investigate the effects of carbon-based NMs and it is presented in the following experimental case study [31]. In this work, the recombinant enzyme AChE purified from the fruit fly (*Drosophila melanogaster*) was used as the model protein.

12.2.2 Introducing Acetylcholinesterase as a Model Biosensor Protein

AChE (E.C.3.1.1.7) belongs to the family of cholinesterases, which are carboxylic ester hydrolases that break down esters of choline. In vertebrates, AChE is mainly found at neuromuscular junctions and cholinergic synapses in the central nervous system, where it is responsible for the hydrolysis of acetylcholine into choline and acetate after the activation of acetylcholine receptors at the postsynaptic membrane and thus essential for the proper functioning of the central and the peripheral nervous systems [35, 36]. It is also found in red blood cell membranes and in a number of other non-neuronal cells/tissues [37].

The hydrolysis of acetylcholine happens at the catalytic site of the enzyme, which is deeply buried, located at the bottom of a 20 Å long and narrow cavity, called the active-site gorge. The gorge is covered by as many as 14 conserved aromatic residues covering by over 70% of its surface [38]. In addition, all known AChEs include a secondary substrate-binding site, referred to as the peripheral anionic site. This site is involved in several functions, such as the modification of catalytic activity, the mediation of interaction of AChE with inhibitors and the noncatalytic role, namely β-amyloid fibril formation [39].

In general, AChEs are specifically inhibited by organophosphates (OPs) and carbamates (CAs) [36, 40]. The proposed mechanism of inhibition is binding to the serine hydroxyl group at the active site, and this binding is much stronger than the binding of acetylcholine. In addition, AChEs are also inhibited by other types of pollutants, such as metals [41–43], other pesticides [44], polycyclic aromatic hydrocarbons, detergents [36,45], and more recently NMs [31,32]. In relation to metal inhibition, the proposed mechanism of action for metal inhibition lies in the alteration of AChE's binding properties for acetylcholine [46]. It has also been suggested that the conformation of AChE is highly responsive to even subtle changes in ionic composition. The influence of ions on AChE conformation might arise from ion association with the peripheral anionic site, which seems to interact with the anionic site that resides inside the AChE active site gorge [47].

As pointed out by Lionetto et al. [36], the successful application of AChEs as biomarkers for use in the occupational and environmental risk assessment can be attributed to the ease of measurement, the dose-dependent response to pollutants, and the high level of sensitivity and links at an organism level. These properties are also the reason for the wide application of AChE in biosensors. As a result, numerous cholinesterase-based biosensors with various enzyme sources, detection, and immobilization methods have already been developed [48,49].

12.3 EXPERIMENTAL CASE STUDY

12.3.1 Introduction

Carbon-based NMs have emerged in recent years as promising candidates for drug delivery systems, cellular imaging, biosensor matrices, and other biomedical applications. Carbon-based NMs comprise a variety of different NMs with very different properties. Among them are fullerenes (C_{60}), graphene-family NMs, and carbon black. Graphene is a single-atom thick, two-dimensional sheet of hexagonally arranged carbon atoms isolated from its three-dimensional parent material, graphite. Related materials include few-layer-graphene, ultrathin graphite, graphene oxide (GO), reduced graphene oxide, and graphene nanosheets. GO has unique structural, mechanical, and electronic properties and is used in biodevices [50,51]. Fullerenes are carbon allotropes similar in structure to graphene but rolled up to form closed-cage, hollow spheres. The C_{60} fullerene is a remarkably stable compound consisting of 60 carbon atoms [52]. They have been mass-produced for many applications in recent decades, including targeted drug delivery, polymer modifications and cosmetic products, energy storage, sensors, and semiconductors [53]. Carbon black is produced from incomplete combustion and is an amorphous carbon material with a high surface-area-to-volume ratio. The three NMs differ in their hydrophobicity; while GO is amphiphilic due to the presence of epoxide and hydroxyl groups on the surface, CB and C_{60} are hydrophobic.

Adsorption of proteins on carbon-based NMs has been extensively studied, and it was shown that these interactions can affect both the protein structure and function [30,54]. Hydrophobic interactions, $\pi-\pi$ stacking interactions, and electrostatic interactions are reported to play key roles in the binding of proteins to NMs [54]. Recent molecular dynamics (MD) studies have, for example, shown that both carbon nanotubes and graphene have the capability to disrupt α-helical structures of short peptides and that graphene may possess even a higher capability to break α-helices due to its more favorable surface curvature [30].

The aim of this study is to present an example of the interaction between carbon-based NMs (carbon black (CB), graphene oxide (GO) and fullerenes (C_{60})) and recombinant AChE (from fruit fly (*D. melanogaster*)). This particular AChE was chosen as the model system, since it has a well-known structure. The adsorption and inhibition of AChE activity reported will present measures of its interaction with carbon-NMs. The data presented here are based on the data previously reported by Mesarič et al. [31].

12.3.2 Method: Assay of AChE Activity

The measurement of AChE activity was done according to the most widely applied method by Ellman et al. [55], adapted for microplates. AChE hydrolyzes the substrate acetylthiocholine chloride to produce thiocholine and acetate. The thiocholine in turn reduces the color indicator (5,5′-dithiobis-(2-nitrobenzoic acid)) liberating 3-thio-6-nitrobenzoate. The formation of this chromogenic product is followed at 405 nm and the rate of formation is considered to be related to the activity of the AChE. Figure 12.1 shows a schematic of the overall reaction.

Figures 12.2 and 12.3 show an illustration of different steps associated with adsorption and inhibition experiments, respectively. Please note that there are differences between the two types of experiments. In the case of inhibition experiments, the activity of total AChE (the NMs-bound and "free" AChE) was measured, because the reagents were added to the AChE–NMs incubation mixture, before the separation of AChE–NMs complexes. On the contrary, in the case of adsorption experiments, the activity of "free" AChE is measured to evaluate the adsorbed share of AChE.

The inhibition experiment was set up in the following way: the *D. melanogaster* AChE (50 µl of AChE dissolved in 100 mM phosphate buffer; pH 8.0; 0.06 U/ml) was first incubated with 10 µl of NM suspension in the same buffer (final concentrations in

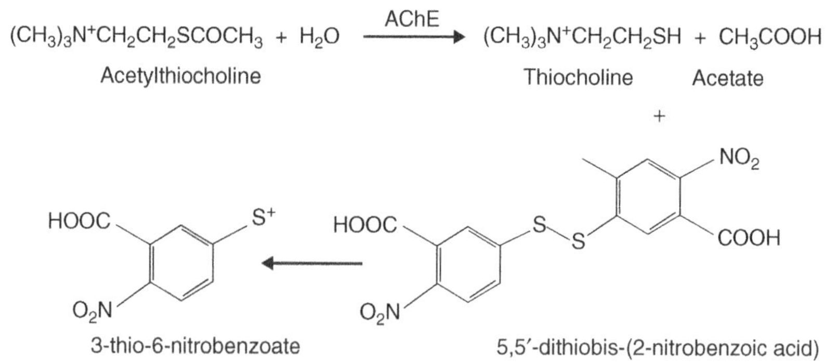

$$(CH_3)_3N^+CH_2CH_2SCOCH_3 + H_2O \xrightarrow{\text{AChE}} (CH_3)_3N^+CH_2CH_2SH + CH_3COOH$$

Acetylthiocholine — Thiocholine — Acetate

+

HOOC, S^+, O_2N ⟵ HOOC, S–S, O_2N, NO_2, COOH

3-thio-6-nitrobenzoate — 5,5′-dithiobis-(2-nitrobenzoic acid)

Figure 12.1 The principle of the acetylcholinesterase (AChE) reaction according to Ellman et al. [55].

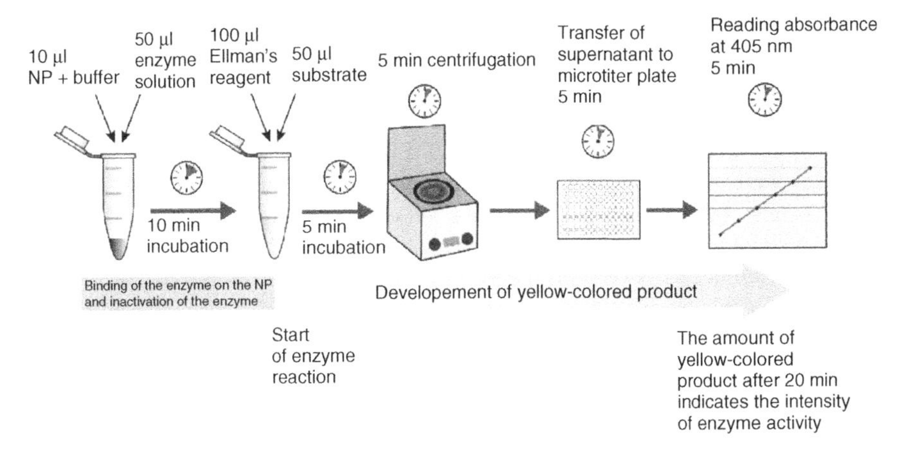

Figure 12.2 Experimental setup for measurement of AChE inhibition.

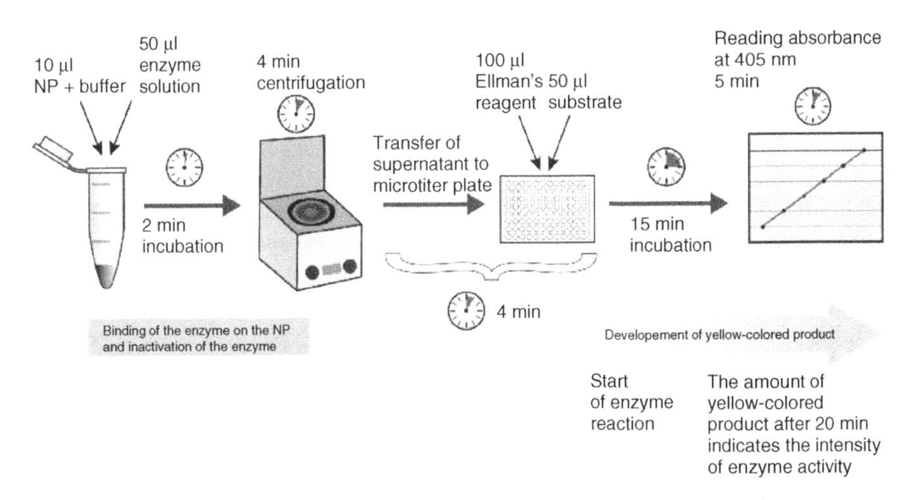

Figure 12.3 Experimental setup for measurement of AChE adsorption.

the range of 0–1 mg/ml). After 10 min of incubation between AChE and NMs, 100 μl of Ellman's reagent and 50 μl of the substrate acetylthiocholine chloride (2 mM) were added. This reagent mixture was incubated for another 5 min. The NM–AChE complexes were then separated by centrifugation (5 min at 12,000×*g*). The supernatants (210 μl) were pipetted onto a microtiter plate and the absorbance was measured at 405 nm exactly 20 min after the addition of the substrate and the Ellman's reagent to the reaction mixture, using the automatic VIS microplate reader (Dynex technologies, USA). For every sample that contained NMs, an appropriate blank was prepared, where the enzyme was replaced by 50 μl of 100 mM phosphate buffer pH 8.0 (Fig. 12.2).

The adsorption procedure was as follows: AChE was incubated with the NMs for 2 min using the same volumes of AChE and NMs as in the case of inhibition experiments. Afterward, the sample was centrifuged (4 min at 12,000×*g*) and the supernatants (60 μl), containing the nonadsorbed "free" enzyme, were pipetted onto the microtiter plate. The reagents (100 μl of Ellman's reagent and 50 μl of 2 mM substrate) were added only to the supernatant. In this case, the activity of free, non-NMs-adsorbed AChE was read at 405 nm exactly 20 min after the addition of the substrate and Ellman's reagent to the reaction mixture (Fig.12.3).

The interference of NMs with the reaction product (3-thio-6-nitrobenzoate) was evaluated. For this purpose, 5,5′-dithiobis-(2-nitrobenzoic acid) was reduced with a minimal volume of 2-mercaptoethanol to obtain the 3-thio-6-nitrobenzoate. It was diluted to give the final value of absorbance at 405 nm, identical to the one obtained in the enzyme reaction without the NM. A 100 μL of such solution was combined with 50 μl of 2 mM substrate, 50 μl of 100 mM phosphate buffer (pH 8.0), and 10 μl of the appropriate NM suspension. The absorbance was read at 405 nm to check for changes in the color; no interference of NMs with the test assay was found.

12.3.3 Results and Discussion

Table 12.1 shows a summary of the results. As mentioned before, the work has been previously published and details can be found elsewhere [31]. Based on the IC_{20} values (this is the concentration that causes 20% of the changes in comparison to control), the NMs can be ranked in the order from the least to the most AChE-adsorptive and inhibitory. The NM with a lower IC_{20} is considered more adsorptive/inhibitory. Results show that GO and CB exhibited similar adsorption and inhibition properties. Although GO shows a slightly higher adsorptive/inhibitory potential than CB, it is clear that they both have a significantly higher adsorption and inhibition than C_{60}.

Table 12.1 shows the inherent differences between the physicochemical properties of the three NMs, which may explain their adsorptive/inhibitory potentials. It is clear that there is no evidence that links particle size with adsorptive/inhibitory potentials. Although GO has similar adsorptive/inhibitory potentials to CB, the particle sizes (both primary and secondary particles) are very different. GO is an 80% single sheet with a size of 0.5–5 μm, whereas carbon black is amorphous and globular, with a primary size of 20 nm. In test media, aggregates of carbon black in the range of 100 nm–1 μm are present.

Out of the different properties noted in Table 12.1, there is some indication that surface curvature may play an important role. Here, GO and CB both exhibit a low surface curvature, while the opposite is true for C_{60}. The high surface curvature of C_{60} could explain its low adsorption/inhibition to the enzyme. Another difference between the materials concerns hydrophobicity. While GO is amphiphilic, that is, possesses both hydrophilic and hydrophobic residues due to the presence of epoxide and hydroxyl groups on the surface, CB and C_{60} are hydrophobic in nature. These differences, however, do not affect their adsorption/inhibition properties, since GO is similarly affected by AChE as CB despite the different hydrophobicity properties. Clearly, the study presented here is preliminary in nature and a clearer link between the observed effects of AChE and the aforementioned properties can be established if

TABLE 12.1 Summary of Data: The Physicochemical Properties of NMs, Their AChE Inhibition and Adsorption Potentials to Recombinant AChE (from *Drosophila melanogaster*). In the case of enzyme activity inhibition the IC20 denotes the concentration of NMs where 20% inhibition of activity in comparison to control was found. In the case of adsorption efficiency the IC20 means the concentration where 20% of the enzyme has adsorbed NMs and is hence inactive

	Physico-Chemical Properties	Inhibition of Activity (mg/l, Time of Incubation is Noted)	Adsorption Efficiency (mg/l; time of Incubation is Noted)
Graphene oxide (GO)	80% single sheet, sheet size varies from 0.5 to 5 μm secondary size in test media: DLS analysis not possible ***low*** surface curvature ***amphiphilic*** nature: presence of epoxide and hydroxyl groups on the surface	10 min $IC_{20} = 0.057 \pm 0.008$	2 min $IC_{20} = 0.005 \pm 0.001$
Carbon black (CB)	amorphous, globular primary size: 13 nm secondary size in test media (100 nm–1 μm range) ***low*** surface curvature ***hydrophobic*** nature	10 min $IC_{20} = 0.15 \pm 0.04$	2 min $IC_{20} = 0.06 \pm 0.01$
Fullerene (C_{60})	primary size distribution (20 nm to several 100 nm) secondary size in test media (250 nm to several μm) ***high*** surface curvature ***hydrophobic*** nature	10 min $IC_{20} = 40 \pm 5$	2 min $IC_{20} = 30 \pm 5$

Further details can be found elsewhere [31].

a larger set of data is acquired, for example, investigating the interaction of different NMs with AChE (of different isoforms and from different sources).

In an attempt to elucidate the reasons behind the observed differences in AChE inhibitions, Mesarič et al. [31] performed computational simulation studies to investigate the probable adsorption site of AChE on the surface of carbon NMs. Results from the simulation study showed that (i) in most of the cases, the interaction site of AChE with carbon-based NMs is far from the active site of the enzyme, (ii) CB seems to form more atomic contacts than GO and C_{60}, (iii) the hydrophobic binding of CB affects the secondary structure of the enzyme. However, the results of the simulation do not seem to give any explanation as to the results reported here. For example,

CB and GO have a quite similar adsorption/inhibition potential (Table 12.1), so the number of atomic contacts does not influence the inhibition.

To date, a similar study has been published by Wang and coworkers [32]. Their study differs in that they have used AChE from another source (i.e., electric eel) and applied a longer incubation time of NMs with AChE (15 min). Their findings indicated a higher adsorption potential of carbon NMs in comparison to metal oxide NMs (SiO_2, TiO_2) (15 min IC_{50} >800 mg/l) [32]. They also suggested that special attention should be paid to those metal oxide NMs where metals dissolve and metal ions are the source of inhibition (e.g., Cu^{2+} in the case of Cu nanoparticles). Their results concur with the modeled adsorption potentials introduced by Xia et al. [19], where carbon NMs were ranked as significantly more adsorptive than metal oxide NMs (SiO_2, Ag-SiO_2, and TiO_2). The model introduced by Xia has been employed for the three NMs in this study, but findings suggest that it does not predict the differences between them.

Finally, it is important to point out that the data presented in this work refer specifically to the recombinant AChE from *D. melanogaster.* Although the differences in the three-dimensional structure of AChE from different species are not significant, some subtle differences are present, which may result in different inhibitor susceptibility [56]. For example, although the comparison of AChE purified from the Pacific electric ray (*Torpedo californica*), the human, and the mouse revealed no conformational differences within the active-site gorge or in the composition of its surface residues, there were differences related to the layers behind those lining the active site [57]. Also, differences between the *D. melanogaster* AChE and AChE from the human, the mouse, and the fish were not found in the overall fold, charge distribution, and deep active-site gorge, but in the external loops and in the tilt of the C-terminal helix [58]. Furthermore, the active-site gorge of the insect enzyme is significantly narrower than that of the *T. californica* AChE and its trajectory is shifted by several angstroms [58]. Marked structural differences are also found between different AChE isoforms, for example, between the erythrocytic and synaptic variant [39]. Overall, this indicates that the interaction (adsorption and inhibition) of NMs with different variants of AChE from different species may vary.

12.3.4 Conclusions

In conclusion, a novel type of NM characterization has been presented, that is, a biological characterization using an enzyme biosensor. Using the current model based specifically on recombinant AChE from the fruit fly (*D. melanogaster*), three carbon-based NMs were ranked according to their adsorptive and inhibitory properties. These results suggest a promising use of the proposed biosensor for ranking NMs with regard to their hazard. The results presented here are preliminary in nature and further work is undoubtedly needed in order to establish a clear link between their properties and the biological response. Furthermore, for a successful uptake of this new tool for characterizing NMs, it is important to validate the method, for example, investigate the effect of the different sources (and isoforms) of AChE, the different incubation periods of NMs with the enzymes, and the interaction with a variety of NMs (with variable properties). Only after an acquisition of a larger set of data on the interaction of different NMs with different forms of AChE has been accomplished, will it be possible to connect the observed effects of AChE on the extrinsic properties of NMs.

12.4 SUMMARY

The characterization of NMs is of crucial importance in ranking NMs according to their hazard. In real biological systems, NMs undergo complex modifications, thus potentially gaining what has been referred to as secondary, "extrinsic" properties. Due to the difficulties in making reliable measurements of NMs when in complex media, the characterization of NMs under such conditions can be a challenging task. In this study, an overarching concept of NM characterization has been put forward, based on a biological characterization approach using an enzyme biosensor. The idea is based on the knowledge that the properties of NMs, such as size, surface chemistry, crystallinity, and hydrophobicity, govern their adsorption potential, which is reflected in their interaction with biological systems, specifically with proteins. It has been shown that the biosensor system is able to quantify these interactions. Results show that AChE is a promising candidate for ranking different NMs according to their adsorptive and inhibitory properties.

ACKNOWLEDGMENTS

This research was funded by the EU FP7 project NanoValid (Development of reference methods for hazard identification, risk assessment, and LCA of engineered nanomaterials; Grant No 263147) and NanoMile (NMP4-LA-2013-310451). We thank the editor for valuable suggestions during the preparation of this manuscript.

REFERENCES

1. Oomen AG, Bos PMJ, Fernandes TF, Hund-Rinke K, Boraschi D, Byrne HJ, Aschberger K, Gottardo S, von der Kammer F, Kühnel D, Hristozov D, Marcomini A, Migliore L, Scott-Fordsmand J, Wick P, Landsiedel R. Concern-driven integrated approaches to nanomaterial testing and assessment – Report of the NanoSafety Cluster Working Group 10. Nanotoxicology 2014;8(3):334–348.
2. Menard A, Drobne D, Jemec A. Ecotoxicity of nanosized TiO2. Review of in vivo data. Environ Pollut 2011;159(3):677–684.
3. Hristozov DR, Zabeo A, Foran C, Isigonis P, Critto A, Marcomini A, Linkov I. A weight of evidence approach for hazard screening of engineered nanomaterials. Nanotoxicology 2014;8(1):72–87.
4. Kühnel D, Nickel C. The OECD expert meeting on ecotoxicology and environmental fate — Towards the development of improved OECD guidelines for the testing of nanomaterials. Sci Total Environ 2014;472:347–353.

5. Kahru A, Dubourguier H-C. From ecotoxicology to nanoecotoxicology. Toxicology 2010;269(2–3):105–119.
6. Oberdörster G, Oberdörster E, Oberdörster J. Nanotoxicology: An emerging discipline evolving from studies of ultrafine particles. Environ Health Perspect 2005;113(7):823–839.
7. Nel A, Xia T, Mädler L, Li N. Toxic potential of materials at the nanolevel. Science 2006;311(5761):622–627.
8. Van Hoecke K, De Schamphelaere KAC, Van der Meeren P, Lcucas S, Janssen CR. Ecotoxicity of silica nanoparticles to the green alga *Pseudokirchneriella subcapitata*: Importance of surface area. Environ Toxicol Chem 2008;27(9):1948–1957.
9. Selim AA, Al-Sunaidi A, Tabet N. Effect of the surface texture and crystallinity of ZnO nanoparticles on their toxicity. Mater Sci Eng C 2012;32(8):2356–2360.
10. Pal S, Tak YK, Song JM. Does the antibacterial activity of silver nanoparticles depend on the shape of the nanoparticle? A study of the gram-negative bacterium Escherichia coli. Appl Environ Microbiol 2007;73(6):1712–1720.
11. Jemec A, Djinović P, Tišler T, Pintar A. Effects of four CeO_2 nanocrystalline catalysts on early-life stages of zebrafish *Danio rerio* and crustacean *Daphnia magna*. J Hazard Mater 2012;219–220:213–220.
12. Lynch I, Weiss C, Valsami-Jones E. A strategy for grouping of nanomaterials based on key physico-chemical descriptors as a basis for safer-by-design NMs. Nano Today 2014;9(3):266–270.
13. Mudunkotuwa IA, Pettibone JM, Grassian VH. Environmental implications of nanoparticle aging in the processing and fate of copper-based nanomaterials. Environ Sci Technol 2012;46(13):7001–7010.
14. Hassellöv M, Readman JW, Ranville JF, Tiede K. Nanoparticle analysis and characterization methodologies in environmental risk assessment of engineered nanoparticles. Ecotoxicology 2008;17(5):344–361.
15. Liu J, Wang Z, Liu FD, Kane AB, Hurt RH. Chemical transformations of nanosilver in biological environments. ACS Nano 2012;6(11):9887–9899.
16. Sharma VK, Siskova KM, Zboril R, Gardea-Torresdey JL. Organic-coated silver nanoparticles in biological and environmental conditions: Fate, stability and toxicity. Adv Colloid Interface Sci 2014;204:15–34.
17. Park S, Hamad-Schifferli K. Nanoscale interfaces to biology. Curr Opin Chem Biol 2010;14(5):616–622.
18. Mu Q, Jiang G, Chen L, Zhou H, Fourches D, Tropsha A, Yan B. Chemical basis of interactions between engineered nanoparticles and biological systems. Chem Rev 2014;114(15):7740–7781.
19. Xia XR, Monteiro-Riviere NA, Mathur S, Song X, Xiao L, Oldenberg SJ, Fadeel B, Riviere JE. Mapping the surface adsorption forces of nanomaterials in biological systems. ACS Nano 2011;5(11):9074–9081.
20. Xu M, Li J, Iwai H, Mei Q, Fujita D, Su H, Chen H, Hanagata N. Formation of nano-bio-complex as nanomaterials dispersed in a biological solution for understanding nanobiological interactions. Sci Rep 2012;2.
21. Fabrega J, Luoma SN, Tyler CR, Galloway TS, Lead JR. Silver nanoparticles: Behaviour and effects in the aquatic environment. Environ Int 2011;37(2):517–531.

22. Domingos RF, Baalousha MA, Ju-Nam Y, Reid MM, Tufenkji N, Lead JR, Leppard GG, Wilkinson KJ. Characterizing manufactured nanoparticles in the environment: Multimethod determination of particle sizes. Environ Sci Technol 2009;43(19): 7277–7284.

23. Tantra R, Oksel C, Puzyn T, Wang J, Robinson KN, Wang XZ, Ma CY, Wilkins T. Nano(Q)SAR: Challenges, pitfalls and perspectives. Nanotoxicology 2014;1–7.

24. Baalousha M, Lead JR. Nanoparticle dispersity in toxicology. Nat Nanotechnol 2013;8(5):308–309.

25. Tantra R, Shard A. We need answers. Nat Nanotechnol 2013;8(2):71–71.

26. Monopoli MP, Åberg C, Salvati A, Dawson KA. Biomolecular coronas provide the biological identity of nanosized materials. Nat Nanotechnol 2012;7(12):779–786.

27. Cedervall T, Lynch I, Lindman S, Berggård T, Thulin E, Nilsson H, Dawson KA, Linse S. Understanding the nanoparticle–protein corona using methods to quantify exchange rates and affinities of proteins for nanoparticles. Proc Natl Acad Sci U S A 2007;104(7):2050–2055.

28. Ipe BI, Shukla A, Lu H, Zou B, Rehage H, Niemeyer CM. Dynamic light-scattering analysis of the electrostatic interaction of hexahistidine-tagged cytochrome P450 enzyme with semiconductor quantum dots. ChemPhysChem 2006;7(5):1112–1118.

29. Calzolai L, Franchini F, Gilliland D, Rossi F. Protein–nanoparticle interaction: Identification of the ubiquitin–gold nanoparticle interaction site. Nano Lett 2010;10(8):3101–3105.

30. Zuo G, Zhou X, Huang Q, Fang H, Zhou R. Adsorption of villin headpiece onto graphene, carbon nanotube, and C60: Effect of contacting surface curvatures on binding affinity. J Phys Chem C 2011;115(47):23323–23328.

31. Mesarič T, Baweja L, Drašler B, Drobne D, Makovec D, Dušak P, Dhawan A, Sepčić K. Effects of surface curvature and surface characteristics of carbon-based nanomaterials on the adsorption and activity of acetylcholinesterase. Carbon 2013;62:222–232.

32. Wang Z, Zhao J, Li F, Gao D, Xing B. Adsorption and inhibition of acetylcholinesterase by different nanoparticles. Chemosphere 2009;77(1):67–73.

33. Käkinen A, Ding F, Chen P, Mortimer M, Kahru A, Ke PC. Interaction of firefly luciferase and silver nanoparticles and its impact on enzyme activity. Nanotechnology 2013;24(34):345101.

34. Šinko G, Vrček IV, Goessler W, Leitinger G, Dijanošić A, Miljanić S. Alteration of cholinesterase activity as possible mechanism of silver nanoparticle toxicity. Environ Sci Pollut Research 2013;21(2):1391–1400.

35. Pohanka M. Cholinesterases, a target of pharmacology and toxicology. Biomed Pap 2011;155(3):219–223.

36. Lionetto MG, Caricato R, Calisi A, Giordano ME, Schettino T. Acetylcholinesterase as a biomarker in environmental and occupational medicine: New insights and future perspectives. BioMed Res Int 2013;2013.

37. Daniels G. Functions of red cell surface proteins. Vox Sang 2007;93(4):331–340.

38. Sussman JL, Harel M, Frolow F, Oefner C, Goldman A, Toker L, Silman I. Atomic structure of acetylcholinesterase from Torpedo californica: a prototypic acetylcholine-binding protein. Science 1991;253(5022):872–879.

39. Grisaru D, Sternfeld M, Eldor A, Glick D, Soreq H. Structural roles of acetylcholinesterase variants in biology and pathology. Eur J Biochem 1999;264(3):672–686.

40. Sultatos LG. Interactions of organophosphorus and carbamate compounds with cholinesterases. In: *Toxicology of Organophosphate and Carbamate Compounds*. Amsterdam: Elsevier Academy Press; 2006. p 209–218.
41. Diamantino TC, Almeida E, Soares AMVM, Guilhermino L. Characterization of cholinesterases from *Daphnia magna straus* and their inhibition by zinc. Bull Environ Contam Toxicol 2003;71(2):0219–0225.
42. Frasco MF, Fournier D, Carvalho F, Guilhermino L. Do metals inhibit acetylcholinesterase (AChE)? Implementation of assay conditions for the use of AChE activity as a biomarker of metal toxicity. Biomarkers 2005;10(5):360–375.
43. Jemec A, Tišler T, Drobne D, Sepčić K, Jamnik P, Roš M. Biochemical biomarkers in chronically metal-stressed daphnids. Comp Biochem Physiol C Toxicol Pharmacol 2008;147(1):61–68.
44. Reddy PM, Philip GH. In vivo inhibition of AChE and ATPase activities in the tissues of freshwater fish, *Cyprinus carpio* exposed to technical grade cypermethrin. Bull Environ Contam Toxicol 1994;52(4):619–626.
45. Guilhermino L, Lacerda MN, Nogueira AJA, Soares AMVM. In vitro and in vivo inhibition of Daphnia magna acetylcholinesterase by surfactant agents: possible implications for contamination biomonitoring. Sci Total Environ 2000;247(2–3):137–141.
46. Guilhermino L, Barros P, Silva MC, Soares AMVM. Should the use of inhibition of cholinesterases as a specific biomarker for organophosphate and carbamate pesticides be questioned? Biomarkers 1998;3(2):157–163.
47. Romani R, Antognelli C, Baldracchini F, De Santis A, Isani G, Giovannini E, Rosi G. Increased acetylcholinesterase activities in specimens of *Sparus auratus* exposed to sublethal copper concentrations. Chem Biol Interact 2003;145(3):321–329.
48. Kucherenko IS, Soldatkin OO, Arkhypova VM, Dzyadevych SV, Soldatkin AP. A novel biosensor method for surfactant determination based on acetylcholinesterase inhibition. Meas Sci Technol 2012;23(6):065801.
49. Wang J, Timchalk C, Lin Y. Carbon nanotube-based electrochemical sensor for assay of salivary cholinesterase enzyme activity: An exposure biomarker of organophosphate pesticides and nerve agents. Environ Sci Technol 2008;42(7):2688–2693.
50. Wang K, Li H-N, Wu J, Ju C, Yan J-J, Liu Q, Qiu B. TiO2-decorated graphene nanohybrids for fabricating an amperometric acetylcholinesterase biosensor. Analyst 2011;136(16):3349.
51. Zhao J, Wang Z, White JC, Xing B. Graphene in the aquatic environment: Adsorption, dispersion, toxicity and transformation. Environ Sci Technol 2014;48(17):9995–10009.
52. Aschberger K, Johnston HJ, Stone V, Aitken RJ, Tran CL, Hankin SM, Peters SAK, Christensen FM. Review of fullerene toxicity and exposure – Appraisal of a human health risk assessment, based on open literature. Regul Toxicol Pharmacol 2010;58(3):455–473.
53. Baughman RH, Zakhidov AA, de Heer WA. Carbon nanotubes--The route toward applications. Science 2002;297(5582):787–792.
54. Mu Q, Liu W, Xing Y, Zhou H, Li Z, Zhang Y, Ji L, Wang F, Si Z, Zhang B, Yan B. Protein binding by functionalized multiwalled carbon nanotubes is governed by the surface chemistry of both parties and the nanotube diameter. J Phys Chem C 2008;112(9): 3300–3307.
55. Ellman GL, Courtney KD, jr. Andres V, Featherstone RM. A new and rapid colorimetric determination of acetylcholinesterase activity. Biochem Pharmacol 1961;7(2):88–95.

56. Massoulié J, Perrier N, Noureddine H, Liang D, Bon S. Old and new questions about cholinesterases. Chem Biol Interact 2008;175(1–3):30–44.
57. Silman I, Sussman JL. Acetylcholinesterase: How is structure related to function? Chem Biol Interact 2008;175(1–3):3–10.
58. Harel M, Kryger G, Rosenberry TL, Mallender WD, Lewis T, Fletcher RJ, Guss JM, Silman I, Sussman JL. Three-dimensional structures of Drosophila melanogaster acetylcholinesterase and of its complexes with two potent inhibitors. Protein Sci 2000;9(6):1063–1072.

13

VISUALIZATION OF MULTIDIMENSIONAL DATA FOR NANOMATERIAL CHARACTERIZATION

J. J. Liu

School of Chemistry and Chemical Engineering, South China University of Technology, Guangzhou 510640, PR China; Institute of Particle Science and Engineering, School of Chemical and Process Engineering, University of Leeds, Leeds LS2 9JT, UK

J. Li

School of Chemistry and Chemical Engineering, South China University of Technology, Guangzhou 510640, PR China

C. Oksel and C. Y. Ma

Institute of Particle Science and Engineering, School of Chemical and Process Engineering, University of Leeds, Leeds LS2 9JT, UK

X. Z. Wang

School of Chemistry and Chemical Engineering, South China University of Technology, Guangzhou 510640, PR China; Institute of Particle Science and Engineering, School of Chemical and Process Engineering, University of Leeds, Leeds LS2 9JT, UK

13.1 INTRODUCTION

Data visualization is often carried out in order to identify patterns and extract useful information that is hidden within a given data set. Different visualization techniques can be used for visual exploration of multidimensional data; multidimensional

Nanomaterial Characterization: An Introduction, First Edition. Edited by Ratna Tantra.

data here can be considered as data that describe an item with more than three attributes. Thus, multidimensional visualization techniques can be used to identify patterns/correlations and detecting clusters/outliers from such data set. They can visually display relationships between multiple variables, handle limited data sets, and allow investigators to interactively make an analysis with the help of visual features such as color. Depending on the dimensionality of the original data, different visualization tools can be used to display large and noisy data in the form of meaningful plots or pictures. For the purpose of visualization, several techniques can be used to handle multidimensional data, such as parallel coordinates, heat-maps, projection, and clustering methods [1–4].

Parallel coordinate transformation is a method that allows the multidimensional data to be analyzed and visualized in a two-dimensional space. It thus induces a nonprojective mapping between *N*-dimensional data on a 2D surface. In a Cartesian coordinate system, orthogonal axes are used to visualize data points effectively up to three dimensions. In parallel coordinates (introduced by Inselberg [3, 5]), the axes are laid parallel to each other so that the number of dimensions that can be visualized is limited only by the horizontal screen resolution. Hence, in parallel coordinates, an *N*-dimensional space is represented as *N* parallel lines, typically vertical and equally spaced. The values of parallel coordinates are that certain geometrical properties in high dimensions can be transformed into a lower dimensional space. Details about parallel coordinate transformation are presented in Section 13.2.2.

Multidimensional data visualization has many important applications and, in particular, can be considered as an important tool in decision-making processes. In the nanotoxicology community, for example, effective data visualization will mean the ability to visualize multidimensional data to discover correlations between nanomaterial physicochemical properties with toxicological effect, that is, to establish what properties nanoscale materials have and how these attributes influence their performance and biological effects. The complexity within nanotoxicology is that no single parameter can describe the properties (e.g., physical, chemical, and toxicological) of engineered nanomaterials (ENMs). In fact, there are various features including physical structure, chemical composition, and surface characteristics that have been suggested to contribute to the biological effects and behaviors of ENMs in different environments. A detailed characterization including the careful assessment of a wide range of characteristics is often required to understand the physical behavior of ENM, to ensure the correct interpretation of the biological activity studies and also to make the intercomparison of studies possible. However, the complete characterization of ENMs can lead to the generation of large amounts of data that need to be analyzed in detail and well understood. Therefore, there exists a need for a simple but yet effective method of converting multidimensional characterization data (corresponding to multivariables or features) into a more efficient format that can be visually explored and examined. Such visualization techniques are necessary in order to get an overall picture of the properties describing individual characteristics of ENMs. This is useful when a large amount of characterization information is involved. The result of effective data visualization in nanotoxicology will mean the ability to help prioritize

ENMs for screening, to identify the key physicochemical parameters that affect toxicity, to provide practical solutions to the risk-assessment-related problems caused by the diversity of ENMs, and to group ENMs (crucial in many aspects, from hazard assessment to knowledge-gap-filling).

In nanotoxicology, it is usual to work with multidimensional data set (taken from experimental or computational information) for subsequent grouping of ENMs (according to their physicochemical or toxicological properties). Validity of findings arising from such data set will not only depend on the reliability of data but the need to have sufficiently large data set. As such, it is imminent that future requirements hinge on the availability of tools that can handle multivariate and large data set. If only a few properties are considered, then the relationship can be explored by plotting them in simple graphs and charts. However, if there is a large amount of characterization data (dimensions >3), then these two- or three-dimensional plots are not efficient to extract meaningful information from. As comprehensive characterization data sets (often) consist of a measurement of more than a few features, more sophisticated visualization tools (to find out how to group ENMs together based on their physicochemical properties) are needed. Dimensionality reduction techniques, such as principal component analysis (PCA) and factor analysis, can be used for representing data in a simpler form. However, these techniques have some drawbacks, for example, some links may be lost during data transformations. On the contrary, direct visualization techniques (e.g., parallel coordinates, radar charts) allow the efficient visualization of multivariate data points without any information loss.

This chapter presents a case study in the field of nanotoxicology and in particular of how large multidimensional data can be visually represented using structure–activity relationship (SAR) approach based on parallel coordinates. The case study introduced here includes data sets of 18 ENMs, each of which was characterized in terms of its physicochemical properties (e.g., particle size, size distribution, surface area, morphology, metal content, reactivity, and free radical generation). In addition, a range of toxicity tests were conducted for the same panel of ENMs to determine their acute *in vitro* toxicity. The main goal of this case study is to show the power of parallel coordinates-based visualization for revealing hidden patterns/features within physicochemical and toxicity data (associated with a panel of ENMs). The focus here is to identify those physicochemical properties that potentially contribute to the toxicity of ENMs.

13.2 CASE STUDY: STRUCTURE–ACTIVITY RELATIONSHIP (SAR) ANALYSIS OF NANOPARTICLE TOXICITY

13.2.1 Introduction

Over the past decade, computational modeling has emerged as a powerful tool to underpin parameters that potentially control properties and effects of chemical substances on the basis of (quantitative) structure–activity relationship (Q)SAR. Such *in silico* models are now being routinely used by researchers, industry, and regulators to estimate physicochemical properties, human health and ecotoxicological effects, and environmental behavior and fate of a wide range of chemical substances. With regard to nanomaterials, researchers have only recently begun to use *in silico* models for similar purpose [6]. The drive in nanotoxicology stems from an urgent need to assess possible risks of manufactured nanomaterials to human health and the environment, due to rapid developments in the manufacturing of such materials. It is clear that there exist a large number of nanoparticles (NPs) of different sizes and coatings, and so on that require testing to assess hazard and risk. The only rational way to achieve a situation where we do not need to test every single ENM (and its variants) in toxicology tests is to employ *in silico* techniques (as with chemical substances) by applying (Q)SAR models in an attempt to relate physicochemical characteristics of ENMs with their toxicity [7, 8]. This is an intelligent way of testing nanomaterials for assessing the hazard and risk related to the end points of toxicological concern. This is not only economically sound but ethical pressures against non-animal testing will mean that expensive animal bioassay is precluded.

Early literature in *in silico* modeling of ENMs toxicity using (Q)SAR includes opinion and perspective articles [8–14]), and several attempts for building real SAR or QSAR models based on nanotoxicity data have been made [15–21]. However, several issues have been highlighted and must be considered when applying a (Q)SAR method for modeling ENM toxicity. One issue is the lack of reliable knowledge about the interactions between ENMs and biological systems. This includes the scarcity of systematic and consistent toxicological data, the lack of verification of *in vitro* findings with *in vivo* observations, as well as the absence of ENM-specific physicochemical descriptors that are able to express novel and size-dependent biological activity of ENMs [22]. It has been also highlighted that besides the obvious, some properties such as aggregation state [23] and surface coatings [24] can affect stability and toxicity of ENMs. Furthermore, ENMs cannot be treated as an equivalent of a single molecule; particle polydispersity (not an issue with chemicals) becomes an issue with ENMs. Overall, the issue is not only related to potential unreliability of the data but that fact that there is insufficient data to derive credible models (with some existing data not containing sufficient information, e.g., missing the measurements of some potentially important descriptors). Therefore, current (Q)SAR studies have relied on "available descriptors" rather than relevant ones. In comparison to chemicals, it is also important to note that current safety assessment of bulk chemicals is supported by a vast amount of data on properties and biological effects, which have been accumulated over several decades. For ENMs, such a knowledge base that can provide a similar level of confidence in safety assessments is currently lacking, and hence, uncertainties with regard to safety of ENMs are quite high.

The issues discussed here make it almost impossible to directly apply some (Q)SAR tools (already widely used for traditional chemical substances) for the purpose of nanotoxicity modeling. For example, in relation to toxicity predictions of chemicals, backpropagation algorithm has been used for neural networks in (Q)SAR modeling [25, 26]. However, this is problematic for ENMs. First, it requires large data sets that are currently not available in nanotoxicology literature. Second, it does not have feature selection capability so, if it is not combined with a descriptor selection method, then all available physicochemical descriptors will be used as inputs. This may result in the inclusion of irrelevant descriptors thus jeopardizing the model's performance (e.g., generalization capability).

In this case study, the parallel coordinates technique [1, 3, 4, 27–31] is employed to study the SAR between toxicity data and physicochemical properties, for a panel of nanoparticles. This technique has been chosen as it can visually display the causal relationships between ENMs' physicochemical descriptors and the toxicity end points, handle limited data sets, and allow investigators to interactively conduct the analysis with the help of the interactive functions and multiple colors (built in a software tool). More information on parallel coordinate method is discussed in Section 13.2.2, so as to give the reader sufficient knowledge to understand the case study. The data set used here to illustrate the use of parallel coordinates in nantoxicology has been taken from past work reported by Wang et al. [21]. As details surrounding the data set used have been previously reported [21], only a summary is presented here.

13.2.2 Parallel Coordinates: Background

The parallel coordinates method was proposed as a device for computational geometry [32–35]. It is a common and very useful method for visualizing multidimensional data. Here, N-dimensional space is represented as N parallel lines, typically vertical, and equally spaced. The value of parallel coordinates is that certain geometrical properties in high dimensions can be easily transformed into a lower 2D space, which breaks the limitation of traditional dimension representation in the Euclidean space. In parallel coordinates, the points used in Euclidean space are represented as series of lines passing through parallel axes, that is, each variable is represented by one parallel axis. Figure 13.1 illustrates the result of transferring a straight line in Cartesian coordinates when converted into parallel coordinates [36].

An interesting feature of parallel coordinates is when overlapping lines between adjacent axes form distinct patterns, representing the relation between variables they

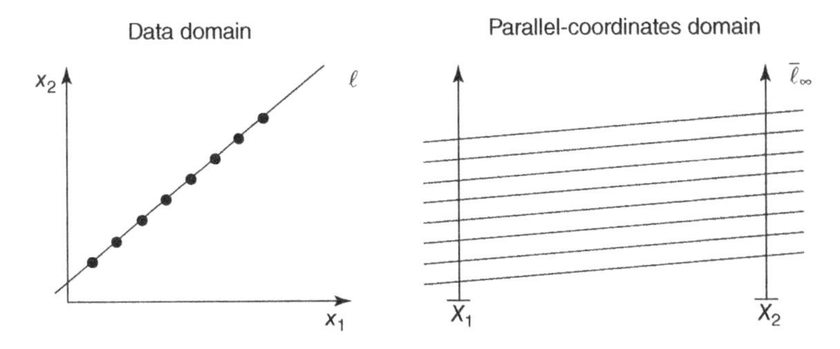

Figure 13.1 An example showing a straight line being translated into parallel coordinates [36].

connect. An advantage of this interactive environment is that it allows the selection of a subset of the plots, thus enabling the operator to highlight the most interesting data, and permuting the axes interactively. The visualization technique using parallel coordinates can transform multidimensional data into 2D patterns and make it possible to visualize clusters and outliers of the data. Therefore, it can be used for data clustering and linking analysis. For nanoparticle toxicity analysis, it can help identify outliers (e.g., particle samples with high toxicity), and aid in finding corresponding responsible physicochemical descriptors (e.g., for the observed high toxicity).

Although there is a large number of papers about parallel coordinates, only a few notable software tools are available to convert databases into parallel coordinates graphs. One of the most sophisticated tool for parallel coordinates transformation is the C Visual Explorer (CVE) software [37], which is used in the case study.

13.2.3 Case Study Data

Table 13.1 lists the 18 nanoparticles used and shows the different ranges of nanomaterials considered here, from carbon-based materials to metals and metal oxides. The data set includes corresponding physicochemical properties and *in vitro* cytotoxicity assays of 18 nanoparticles, which are summarized in Tables 13.1–13.3, respectively.

As previously mentioned, the data sets presented in Tables 13.1–13.3 were taken from a paper published by Wang et al. [21]. According to this paper, physicochemical attributes were characterized using various techniques: (i) thirteen attributes came from TriStar 3000 (Braunner Emmett Teller) BET measurements (to include acquisition of five surface areas based on different definitions, three pore volumes, three pore sizes as well as mean size and particle density); (ii) two attributes came from LEO 1530 scanning electron microscope (SEM)/ Philips CM20 transmission electron microscope (TEM) measurements (to include acquisition of mean size and aspect ratio); (iii) eight attributes came from Mastersizer 2000 measurements of particle size

TABLE 13.1 The 18 Nanoparticles Used in this Study

No	Nanoparticles	No	Nanoparticles	No	Nanoparticles
N1	Carbon black	N7	Polystyrene latex (carboxylated)	N13	Silicon oxide
N2	Diesel exhaust particles	N8	Aluminum oxide (7 nm)	N14	Zinc oxide
N3	Nanotubes	N9	Aluminum oxide (50 nm)	N15	Titanium dioxide (rutile)
N4	Fullerene	N10	Aluminum oxide (300 nm)	N16	Titanium dioxide (anatase)
N5	Polystyrene latex (unmodified)	N11	Cerium oxide	N17	Silver (dry)
N6	Polystyrene latex (amine)	N12	Nickel oxide	N18	Silver (suspension)

TABLE 13.2 Characterization of Physicochemical Properties of the 18 Nanoparticles

Physicochemical Properties
BET surface area and porosity measurements (13 measurements, including 5 surface areas based on different definitions, 3 pore volumes, 3 pore sizes, the mean size, and particle density.)[a]
SEM/TEM mean size and aspect ratio measurements
Particle size and size distribution (size distributions and seven other size properties including mass diameter, uniformity, specific surface area, surface area mean diameter, and three mass diameters)
EPR free radical generation measurements (DMPO and Tempone H measured by electron paramagnetic resonance; EPR)
Particle reactivity (in solution) measurement (DTT consumption)[b]
Metal concentration (Ti, V, Cr, Mn, Fe, Co, Ni, Cu, Zn, and Cd) measurements
Surface charge

[a]Surface area and porosity were measured using TriStar 3000 BET. BET measurements were obtained only for the 14 dry powder samples, not available for three suspensions (N5, N6, and N7) and the silver solution (N18).
[b]Particle reactivity in solution was assessed by the dithiothreitol (DTT) consumption test. Since the DTT consumption test can only be conducted on dry powders, only 14 of the panel of nanoparticles were assessed using this assay.

TABLE 13.3 The Cytotoxicity Assays Performed

In vitro Toxicity Assays	
$DiOC_6$ Mitochondrial Membrane Potential Assay	Hemolysis assay
Lactate dehydrogenase (LDH) release assay	MTT assay
	Cell morphology assay
Proinflammatory effects	Apoptosis/necrosis

distribution, and other measures such as uniformity; (iv) two electron paramagnetic resonance (EPR) measurements of free radicals; (iv) one attribute from reactivity measurement; (v) ten metal content measurements of Ti, V, Cr, Mn, Fe, Co, Ni, Cu, Zn, and Cd; and (vi) others from zeta-potential measurements. BET and reactivity measurements were available for 14 dry powders, while 2 different measures of zeta-potential were available only for 3 polystyrene latex beads (i.e., unmodified, amine, and carboxylated). For convenience, the physicochemical descriptors are subcategorized either "compositional" (referring chemical composition of the nanomaterials and thus is associated with the 10 metal content measurements) or "structural" descriptors (referring to properties such as size, size distribution, and morphology).

Upon initial examination of the data, it was clear that the mean particle size (as measured by Mastersizer 2000 based on laser diffraction technique) is much larger than those reported when SEM/TEM and BET methods were used instead. Such an observation, however, is not surprising and has been attributed to the inherent nature of each measuring technique (rather than the errors associated with the measurement) [38]. The laser diffraction technique, that is, Mastersizer 2000, for example, provides information of aggregated or agglomerated states of the nanomaterials, whereas imaging-based techniques are able to differentiate the sizes of primary particles. As a result, the different size measurements as measured by different techniques are different and all are included in this study. In relation to toxicity assays, the results of hemolysis, MTT (i.e., colorimetric assay for assessing cell viability), cell morphology, and $DiCO_6$ assays that were rationalized to a single value for each assay (e.g., toxicity at a specific dose or area under the curve for a range of doses, as previously reported [21]), are used here.

Wang et al. [21] have used PCA to process the data set and perform a clustering analysis. PCA was applied to the acute cytotoxicity data and the physicochemical descriptors, separately. Using PCA, the authors were able to group (and rank according to their toxicity values) nanoparticles based on the results of LDH release, apoptosis, viability, and necrosis assays. Findings showed that the corresponding 2D and 3D PCA plots that group the particles based on acute cytotoxicity values and physicochemical values are consistent: the particles with low toxicity values are clustered in a single cluster, while the four highly toxic particles are outside of this cluster, and each forms a single class. It was concluded that the most toxic materials screened were zinc oxide (numbered as N14 on Table 13.1), polystyrene latex amine (numbered as N6), followed by Japanese nanotubes (numbered as N3), and nickel oxide (numbered as N12).

Undoubtedly, the work reported by Wang et al. [21] provided a useful reference. The PCA result reported thus gave a strong indication that there exists SARs associated with the nanomaterials. In turn, this prompted the drive behind the work presented here, that is, to identify influencing descriptors potentially responsible for the observed high toxicity values (as observed using PCA and corresponding contribution plots). The visualization of multidimensional data carried out will be achieved using parallel coordinates.

13.2.4 Method

All the raw data were stored in a database. The data were scaled before SAR analysis and this involved by first mean centering the data before range scaling (or normalizing) between 0 and 1 using the minimum and maximum of the data. In relation to attributes associated with structural properties, these steps were carried out for each attribute. For compositional properties (e.g., 10 metal compositions), the scaling steps were performed as a whole.

Since there were no BET and (dithiothreitol) DTT data available for the four wet samples (i.e., the three polystyrene beads (N5, N6, and N7) and the silver solution (N18)), SAR analysis using parallel coordinates was carried out in two ways. First, the 14 dry samples were analyzed, which included the corresponding BET and DTT data. Second, all 18 samples were analyzed that excluded the BET and the DTT test. The explanation as to why data were analyzed this way is further discussed in the following section.

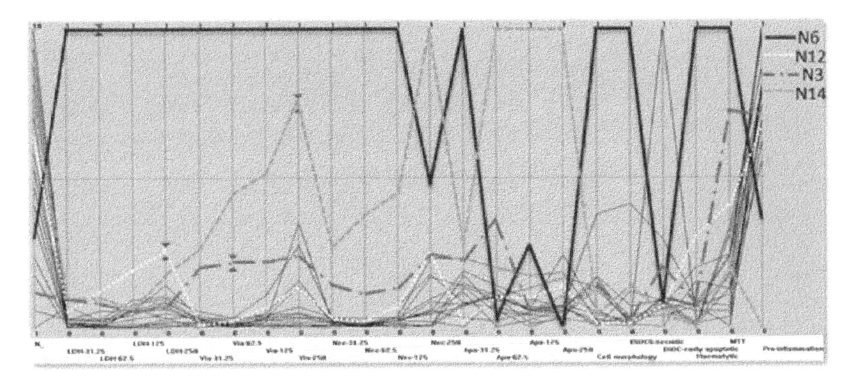

Figure 13.2 Parallel coordinate plot of the cytotoxicity data; four NPs that have high toxicity (N3, N6, N12, and N14) are highlighted.

13.2.5 Results and Interpretation

The cytotoxicity data of the particles listed in Table 13.1 have been measured by different methods and Figure 13.2 shows the corresponding visualization using parallel coordinates on acute cytotoxicity data. If we consider the dense area as the lower toxicity envelope, then any deviation from this area may be considered in the realm of higher toxicity. It is apparent from Figure 13.2 that it is the aminated beads (N6) and zinc oxide (N14) that have the highest toxicity values, followed by nanotubes (N3) and nickel oxide (N12). It should be noted here that, for the remainder of the manuscript, these four relatively high toxic nanoparticles are highlighted with different colors and line styles, for example, thick solid line, dashed line, and so on, while the rest of the nanoparticles that have much lower toxicity values are all shown with thin black (polygonal) lines. Results show that aminated beads (N6) has the highest toxicity values in nearly all assay results (LDH, apoptosis, necrosis, hemolytic, MTT, and cell morphology assays). Zinc oxide (N14) has high toxicity values in LDH, apoptosis, necrosis, and inflammation assays, whereas nanotubes (N3) have high toxicity values in viability and MTT assays. Nickel oxide (N12) has shown high toxicity in LDH and hemolytic assay results. It is worth mentioning that the findings reported here are consistent with the findings reported by Wang et al. [21] when PCA was used [21].

13.2.5.1 Analysis of the 14 Dry Powder Samples Using BET and DTT Data Only BET and DTT measurements require the samples to be dry powders. Therefore, there is no BET and DTT data for the four suspensions, that is, polystyrene latex N5 (unmodified), N6 (amine), and N7 (carboxylated), and N18 (silver suspension). Here, the BET and DTT data were examined. If high toxicity values associated with zinc oxide N14, nanotubes N3, and nickel oxide N12 are not discriminated by the BET and DTT data, then descriptors can be omitted when further analyzing the data set. Figure 13.3 shows the corresponding parallel coordinates plot of BET and DTT data with the 14 dry powder samples. No meaningful clustering was observed from

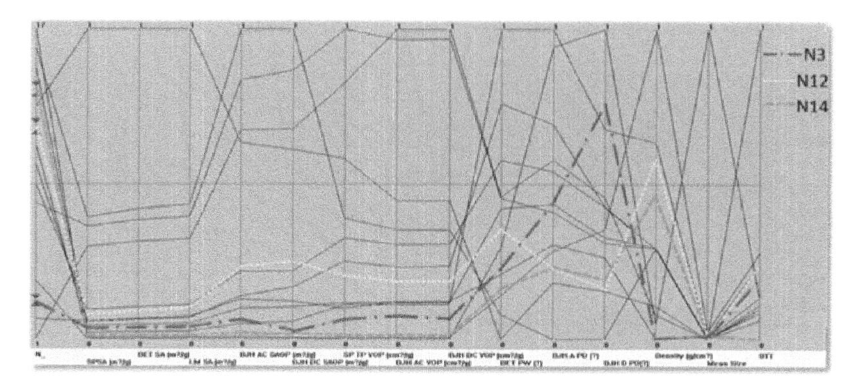

Figure 13.3 Parallel coordinate plot of the BET and DTT data analysis of the 14 dry samples only.

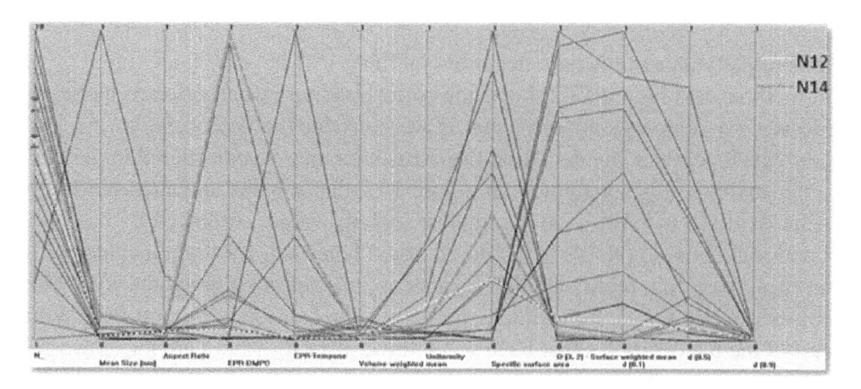

Figure 13.4 Parallel coordinate plot of the structural properties of zinc oxide (N14) and nickel oxide (N12), excluding BET and DTT data, plotted together with the structural properties of low-toxicity particles.

the plots, which suggests that BET and DTT measured descriptors do not play a key role in discriminating the toxicity of the samples. Therefore, for the rest of the chapter, both BET and DTT data have been excluded when analyzing the remaining characterization data.

13.2.5.2 Analysis of the Structural Properties of Zinc Oxide (N14) and Nickel Oxide (N12) (Excluding BET and DTT Data) All the structural descriptors, excluding the compositional (i.e., the metal contents) descriptors and BET and DTT data, were plotted in the parallel coordinates for the 18 nanoparticles. Figure 13.4 shows the parallel coordinate plot of structural properties for zinc oxide (N14) and nickel oxide (N12). The black polygonal lines in Figure 13.4 represent nanoparticles that have low toxicity. As can be seen from this plot, zinc oxide N12 and nickel

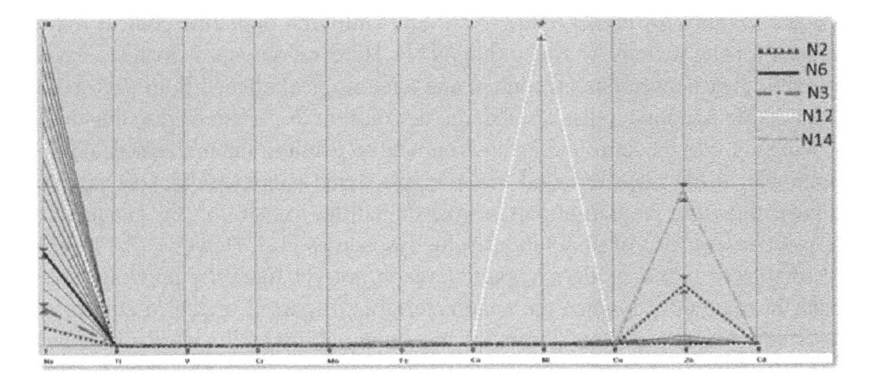

Figure 13.5 Parallel coordinate plot of the metal content analysis of the 18 samples, excluding structural descriptors.

oxide N14 do not show any discriminative patterns from the low-toxicity particles. In other words, the structural descriptors (indicated in Fig. 13.4) do not lead to high toxicity values of zinc oxide N12 and nickel oxide N14 and thus are excluded from this point onward.

13.2.5.3 Metal-Content-Only Analysis of the 18 Samples, Excluding Structural Descriptors The above-mentioned analysis has shown no signs of structural descriptors being responsible for the measured high toxicity of zinc oxide N12 and nickel oxide N14. To further examine the SAR analysis, the metal contents of the 18 samples are plotted (Fig. 13.5). As can be seen from the plot, the majority of the samples stay within a range of low metal content except N12 (nickel oxide), which has a peak value in axis of Ni concentration. It is also noticed that N14 (zinc oxide) shows high zinc concentration.

As there appears to be some relationship between metal content and toxicity response, it is appropriate now to introduce briefly how the metal contents were measured. According to Wang et al. [21], the concentrations of ten metals (Ti, V, Cr, Mn, Fe, Co, Ni, Cu, Zn, and Cd) were measured. Samples were digested for ICP-MS analysis using a CEM MDS-200 microwave system. Particle samples were washed into Teflon-coated composite vessels using 5 ml of 70% nitric acid. The samples were digested using an existing program developed for refractory carbon-based particle matter [39, 40]. The microwave program consisted of a stepped increase in pressure to 80 psi for a period of 20 min, with a corresponding temperature rise to 180 °C. The program lasted for approximately 2.5 h, including warm-up and cool-down periods. Samples were then diluted to a level of 10 μg/ml (depending upon their original weight) using deionized (>18 ΩM) water. Raw data were corrected for blanks and controls accordingly.

Results so far indicate that metallic-based nanomaterials are more toxic than non-metallic nanomaterials. Although it is hard to know the main causative factor in the induction of toxicity, the results suggest that Ni content may be the reason for the high acute toxicity for nickel oxide (N12) and similarly, high zinc content may be the reason for the toxicity of zinc oxide (N14). Here, an attempt is made to explain the possible link between metal content and toxicological effects. Both nickel oxide and zinc oxide nanomaterials can undergo dissolution in certain media, thus potentially leaching into the surrounding environment to result in the toxicological effect. For example, it has been reported that for zinc oxide nanomaterial, free zinc ions have been shown to play an important role in cellular toxicity of this nanomaterial [41], as a consequence of dissolution inside lysosomes [42]. However, the leaching of Ni ions from nickel oxide nanoparticle were not yet found by previous studies to result in toxic effects when the nanomaterial is dispersed in cell in culture, and corresponding aqueous extracts from the nanomaterial dispersion were not shown to induce inflammation when instilled into the lungs of rats [41]. This may be due to that the effective dose of soluble Ni ion (saline (basic)-released) is not sufficient to cause toxicity. Besides, as the nickel oxide nanoparticles are highly acid soluble, it is believed that the acid-releasable Ni ions would be truer reflection of the effective dose of Ni ions than the saline (basic)-released Ni ions.

The potential correlation between metal content and toxicity has prompted us to analyze the results of toxicity of diesel exhaust particles (N2), as shown in Figure 13.6. This is of interest as it is well known that the second largest element detected from diesel fumes is zinc. However, Figure 13.6 clearly indicates that toxicity of diesel exhaust particles (N2) is not so high. The only deviation from the low toxicity region is attributed to its high value of $DiOC_6$ test. As already known, the $DiOC_6$ test is often used as a stamp of early apoptosis, which is an estimate of the collapse of mitochondrial membrane potential. In contrast to the results presented in this study, the assumption that mitochondrial membrane potential collapse precedes apoptosis is not always reliable. As a result, diesel particulate has a unique property in prompting mitochondrial membrane collapse without necessarily inducing apoptosis. However, further research (currently underway) will be needed to verify that this is indeed the case.

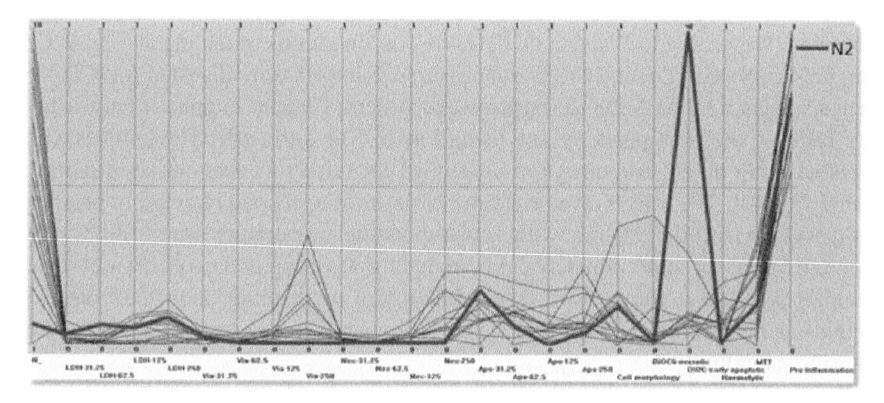

Figure 13.6 Parallel coordinate plot of the toxicity data of diesel exhaust particles (N2), plotted together with data representing lower toxicity.

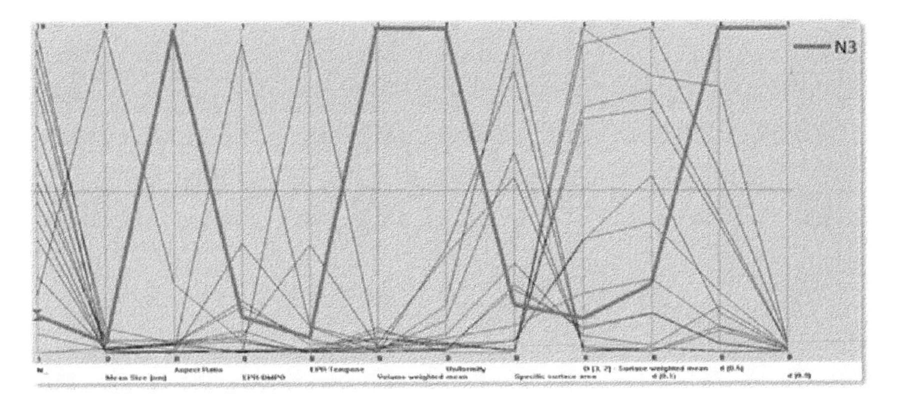

Figure 13.7 Parallel coordinate plot of the structural properties of nanotubes (N3) analysis, without including BET and DTT data, plotted together with structural properties of low-toxicity particles.

13.2.5.4 Analysis of the Structural Properties of Nanotubes (N3) The parallel coordinates plot of the structural descriptors of N3 (nanotubes) and other low-toxicity nanomaterials is shown in Figure 13.7. Note that in order to make the plot neater, not all axes are shown, and the other three high-toxicity materials (i.e., N6, N12, and N14) have not been plotted. Figure 13.7 clearly shows that the variables such as aspect ratio, volume weighted mean, uniformity, D(0.9), and D-PC1 are responsible in differentiating the fold line representing the toxicity of N3 (nanotubes) from low-toxicity samples. In fact, as previously noticed in metal contents analysis of the toxicity data only, the toxicity of N3 cannot be distinguished from the rest of the panel based on metal contents, which implied that metal contents are not responsible for the toxicity of N3.

The ability for the parallel coordinates analysis to pick out aspect ratio as an influencing property for toxicity is an interesting one. In the past, potential toxicity of N3 nanotubes has been associated with their corresponding length and aspect ratio. According to Donaldson et al., length and in particular high aspect ratios of the nanoparticles is important in governing fiber toxicity [43]. This has been previously associated with the fiber pathogenicity paradigm, which according to Donaldson et al. is of relevance to carbon nanotubes (CNTs) [43, 44]. They have indicated that CNTs used are long enough to cause frustrated phagocytosis [44]. They have also shown this to be the case for nickel oxide nanowires [45] and silver nanowires (manuscript in preparation) and concluded that the length of nanofiber has a potential to be the main factor leading to inflammogenicity and toxicity of all high aspect ratio nanomaterials (HARN). As a result, the most likely cause of high toxicity for nanotube (N3) is its high aspect ratio shape.

13.2.5.5 Analysis of the Structural Properties of Aminated Beads (N6) (Excluding BET and DTT Data) In this section, the results for three polystyrene latex (unmodified-N5, aminated-N6, and carboxylated-N7) samples are presented.

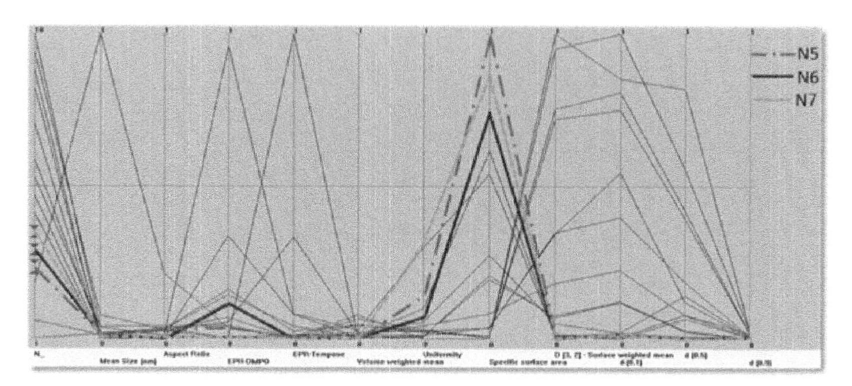

Figure 13.8 Parallel coordinate plot of the structural properties of N6 (aminated beads), N5 (unmodified), and N7 (carboxylated), excluding BET and DTT data, plotted together with structural properties of low-toxicity particles, showing no signs of difference of N6 from N5 and N7.

Obviously, as they are all polystyrene beads, they have very similar structures but according to the toxicity results (Fig. 13.2) it is the aminated-N6 that had the highest toxicity; note that unmodified-N5 and carboxylated-N7 are represented in two of the black lines in the plot that represent low-toxicity particles. As the toxicity of the aminated polystyrene latex is much higher than the other two polystyrene latex samples, the corresponding parallel coordinate plot of structural properties have been plotted (Fig. 13.8). This is to deduce whether any particular physiochemical characteristics have contributed to the toxicity of aminated sample. Result from Figure 13.8 shows that the differentiation of aminated sample based on physico-chemical data are not possible, that is, no clear differentiation between unmodified, aminated, and carboxylated latex beads, based on the measured characteristics. As physchem properties are unable to explain the high toxicity of aminated sample, other possible reasons that explain toxicity specific to the aminated beads were suspected.

According to Wang et al. [21], the toxicity of the three nanoparticles can be explained by their difference in surface properties. They have reported zeta-potential values for N5, N6, and N7 nanoparticles; these have been measured using Malvern Zetasizer. For each sample, three replicate measurements were taken and reported measured zeta-potentials (mV) for N6 (amine) are 37.8, 37.5, and 40.3, for N5 (unmodified) are −36.2, −38.8, and −36.8, and for N7 (carboxylated) are −54.9, −55.3, and −58.6. Results clearly show that N6 has positive zeta-potential values, while N5 and N7 have negative zeta-potentials. According to Verma, et al. [46], any particle with positive charge is likely to interact electrostatically to cell surface and biological membranes, as these surfaces are usually negatively charged under physiological conditions. Potentially, this can cause cell damage and it is most likely that the large positive charge of N6 contributed to its observed high toxicity (despite its structural similarity to N5 and N7).

The significance of surface charge in its ability to cause cellular injury has also been echoed by other workers. Nel et al. [42] described high positive ZP of particles as a factor in destabilizing the phagolysosome after uptake. The positive ZP on nanoparticle (NP) is immediately neutralized on the formation of a corona after deposition in lung lining fluid, but inside the lysosomes the acid and protease/lipase activity may be able to remove this exposed charged surface. This can interact with the lysosomal membrane leading to rupture and release of lysosomal contents with concomitant activation of the NALP3 inflammasome, as has been shown for quartz particles [47]. Another hypothetical mechanism whereby positively charged NP might destabilize membranes implicates the accumulation of chloride ions and water as a consequence of chronic stimulation of the lysosomal membrane proton pump [42].

13.2.6 Conclusion

The case study presented illustrated the benefits of using parallel coordinates methods for SAR analysis of nanomaterials in the field of nanotoxicology. Findings identified physicochemical properties that can potentially govern toxicity for different nanomaterials. For example, the most likely cause of high toxicity in nanotube (N3) is its high aspect ratio. In relation to metal oxide nanomaterials, high toxicity of zinc oxide (N14) and nickel oxides (N12) are most likely due to their high contents of zinc and nickel, respectively. In relation to polystyrene nanoparticles, polystyrene latex N5 (unmodified) and N7 (carboxylated) exhibited low toxicity relative to N6 (aminated) polystyrene. Findings suggest that this may be linked to surface modification, for example, having either a positive or negative surface charge.

Finally, although the case study here shows how parallel coordinates can be useful to visualize patterns, the findings reported here (as well as by Wang and coworkers [21]) are by no means considered to be reliable, as reliable data for robust assessment in nantoxicology is still very much needed.

13.3 SUMMARY

In the field of nanoscience, a wide range of physicochemical properties need to be measured in order to fully characterize and understand the properties of ENMs. Hence, corresponding data collected are often multidimensional in nature (with several variables measured on various ENM samples). Considering the problems of learning in high dimensions, it would be reasonable to use visualization tools that can convert multidimensional characterization data set into a more efficient format for the ease of visual exploration and examination. Such visualization techniques are needed not only for getting an overall picture of the properties describing individual characteristics of ENMs but potentially linking key properties with toxicological effects.

In this chapter, a case study based on experimental characterization and toxicity measurement has been presented to demonstrate the power of a multidimensional visualization technique using parallel coordinates methods, to find the SARs of ENMs. The main advantage of the parallel coordinates methods, compared with other (Q)SAR methods (such as neural networks), is that it can visually display the relationships between physicochemical descriptors and toxicity end points, handle limited data sets, and allow investigators to interactively and visually make analysis. To illustrate this point, observations and outcomes from the analysis of the toxicity data for a panel of 18 nanoparticles have been analyzed with such a method. Our findings show that results are in agreement with previous results obtained by applying PCA to the same data [21]. It should be noted that the main focus of this study was to show how a multidimensional visualization tool can help identify the key physicochemical descriptors that can potentially lead to high toxicity. Having said this, the study is in no way suggesting that the reported findings can be reliably used as predictive model. The main issue complicating the development of computational models in nanotoxicology that hinders the ability to make reliable toxicity predictions is the scarcity of high-quality and useful data on ENM characterization and hazard. In the context of predictive model development, it is not only about the amount of data but also about the variety, quality, consistency, and accessibility of those data that are considered to be vital.

REFERENCES

1. Albazzaz H, Wang XZ. Historical data analysis based on plots of independent and parallel coordinates and statistical control limits. J Process Control 2006;16(2):103–114.
2. Brooks, R.W. and J.G. Wilson, Method and a system for operating a controllable multi-variable process. US Patent No 7916140, 2011.
3. Inselberg A. *Parallel Coordinates: Visual Multidimensional Geometry and Its Applications*. New York: Springer; 2009.
4. Wang XZ, Medasani S, Marhoon F, Albazzaz H. Multidimensional visualization of principal component scores for process historical data analysis. Ind Eng Chem Res 2004;43(22):7036–7048.
5. Inselberg A. The plane with parallel coordinates. Vis Comput 1985;1(2):69–91.
6. Oksel C, Ma C, Wang X. Current situation on the availability of nanostructure–biological activity data. SAR QSAR Environ Res 2015;26:79–94.
7. Puzyn T et al. Nanomaterials–the next great challenge for Qsar modelers. In: Puzyn T, Leszczynski J, Cronin MTD, editors. *Recent Advances in QSAR Studies: Methods and Applications*. Springer Science + Business Media B.V.; 2010. p 383–409.
8. Puzyn T, Leszczynska D, Leszczynski J. Toward the development of "Nano-QSARs": Advances and challenges. Small 2009;5(22):2494–2509.
9. Poater A et al. Computational methods to predict the reactivity of nanoparticles through structure–property relationships. Expert Opin Drug Deliv 2010;7(3):295–305.
10. Burello E, Worth AP. QSAR modeling of nanomaterials. Wiley Interdiscip Rev Nanomed Nanobiotechnol 2011;3(3):298–306.
11. Burello E, Worth AP. A theoretical framework for predicting the oxidative stress potential of oxide nanoparticles. Nanotoxicology 2011;5(2):228–235.

12. Fourches D, Pu D, Tropsha A. Exploring quantitative nanostructure-activity relationships (QNAR) modeling as a tool for predicting biological effects of manufactured nanoparticles. Comb Chem High Throughput Screen 2011;14(3):217–225.
13. Gajewicz A et al. Advancing risk assessment of engineered nanomaterials: Application of computational approaches. Adv Drug Deliv Rev 2012;64(15):1663–1693.
14. Winkler DA et al. Applying quantitative structure-activity relationship approaches to nanotoxicology: current status and future potential. Toxicology 2012;313(1):15–23.
15. Fourches D et al. Quantitative nanostructure – Activity relationship modeling. ACS Nano 2010;4(10):5703–5712.
16. Sayes C, Ivanov I. Comparative study of predictive computational models for nanoparticle-induced cytotoxicity. Risk Anal 2010;30(11):1723–1734.
17. Puzyn T et al. Using nano-QSAR to predict the cytotoxicity of metal oxide nanoparticles. Nat Nanotechnol 2011;6(3):175–178.
18. Epa VC et al. Modeling biological activities of nanoparticles. Nano Lett 2012;12(11):5808–5812.
19. Liu R et al. Development of structure-activity relationship for metal oxide nanoparticles. Nanoscale 2013;5(12):5644–5653.
20. Zhang H et al. Use of metal oxide nanoparticle band gap to develop a predictive paradigm for oxidative stress and acute pulmonary inflammation. ACS Nano 2012;6(5):4349–4368.
21. Wang XZ et al. Principal component and causal analysis of structural and acute in vitro toxicity data for nanoparticles. Nanotoxicology 2014;8(5):465–476.
22. Rasulev B et al. Nano-QSAR: Advances and challenges. In: Leszczynski J, Puzyn T, editors. *Towards Efficient Designing of Safe Nanomaterials*. 2012. p 220–256.
23. Reinsch B et al. Sulfidation of silver nanoparticles decreases Escherichia coli growth inhibition. Environ Sci Technol 2012;46(13):6992–7000.
24. Suresh AK et al. Cytotoxicity induced by engineered silver nanocrystallites is dependent on surface coatings and cell types. Langmuir 2012;28(5):2727–2735.
25. Devillers J. *Neural Networks in QSAR and Drug Design*. Academic Press; 1996.
26. Baskin II, Palyulin VA, Zefirov NS. Neural networks in building QSAR models. In: Livingstone DJ, editor. *Artificial Neural Networks*. Springer; 2009. p 133–154.
27. Brooks R. Viewing process information multidimensional for improved process understanding, operation and control. in Aspen World Conference. Boston 1997.
28. Albazzaz H, Wang XZ, Marhoon F. Multidimensional visualisation for process historical data analysis: a comparative study with multivariate statistical process control. J Process Control 2005;15(3):285–294.
29. Brooks, R.W. and J.G. Wilson, *Multi-variable Processes. US6879325*, 2005.
30. Brooks, R.W. and J.G. Wilson, Multi-variable operations. US patent 7,443,395. 2008.
31. Brooks, R.W. and J.G. Wilson, Method and a system for operating a controllable multi-variable process. US 7,916,140. 2011.
32. Inselberg A, Reif M, Chomut T. Convexity algorithms in parallel coordinates. J Assoc Comput Mach 1987;34(4):765–801.
33. Inselberg A. Visual data mining with parallel coordinates. Comput Stat 1998;13(1).
34. Wegman EJ. Hyperdimensional data analysis using parallel coordinates. J Am Stat Assoc 1990;85(411):664–675.

35. Inselberg A. Visualization and data mining of high-dimensional data. Chemom Intell Lab Syst 2002;60(1):147–159.
36. Heinrich J, Weiskopf D. State of the art of parallel coordinates. In: Sbert M, Szirmay-Kalos L, editors. *Eurographics 2013-State of the Art Reports*. The Eurographics Association; 2012.
37. PPCL, P.P.C.L. 2011. Products and about PPCL. Available at http://www.ppcl.com. Accessed 2015 Oct 22.
38. Thiele, G., M. Poston, and R. Brown. 2013. A case study in sizing nanoparticles. Available at http://www.micromeritics.com/repository/files/a_case_study_in_sizing_nano_particles.pdf. Accessed 2013 Dec.
39. Jones T et al. The physicochemical characterisation of microscopic airborne particles in south Wales: A review of the locations and methodologies. Sci Total Environ 2006;360:43–59.
40. Price H et al. Airborne particles in Swansea, UK: Their collection and characterization. J Toxicol Environ Health A Curr Issues 2010;73:355–367.
41. Cho W-S et al. Differential pro-inflammatory effects of metal oxide nanoparticles and their soluble ions in vitro and in vivo; zinc and copper nanoparticles, but not their ions, recruit eosinophils to the lungs. Nanotoxicology 2012;6(1):22–35.
42. Nel AE et al. Understanding biophysicochemical interactions at the nano–bio interface. Nat Mater 2009;8(7):543–557.
43. Donaldson K et al. Asbestos, carbon nanotubes and the pleural mesothelium: a review of the hypothesis regarding the role of long fibre retention in the parietal pleura, inflammation and mesothelioma. Part Fibre Toxicol 2010;7(5):5.
44. Poland CA et al. Carbon nanotubes introduced into the abdominal cavity of mice show asbestos-like pathogenicity in a pilot study. Nat Nanotechnol 2008;3(7):423–428.
45. Poland CA et al. Length-dependent pathogenic effects of nickel nanowires in the lungs and the peritoneal cavity. Nanotoxicology 2012;6(8):899–911.
46. Verma A et al. Surface-structure-regulated cell-membrane penetration by monolayer-protected nanoparticles. Nat Mater 2008;7(7):588–595.
47. Hornung V et al. Silica crystals and aluminum salts activate the NALP3 inflammasome through phagosomal destabilization. Nat Immunol 2008;9(8):847–856.

INDEX

Nanomaterial Characterization: An Introduction, First Edition. Edited by Ratna Tantra.

Printed and bound by CPI Group (UK) Ltd, Croydon, CR0 4YY

16/04/2025

14658589-0001